ISBN 978-1-332-83283-5
PIBN 10157907

This book is a reproduction of an important historical work. Forgotten Books uses
state-of-the-art technology to digitally reconstruct the work, preserving the original format
whilst repairing imperfections present in the aged copy. In rare cases, an imperfection in
the original, such as a blemish or missing page, may be replicated in our edition. We do,
however, repair the vast majority of imperfections successfully; any imperfections that
remain are intentionally left to preserve the state of such historical works.

1 MONTH OF
FREE
READING

at

www.ForgottenBooks.com

By purchasing this book you are eligible for one month membership to ForgottenBooks.com, giving you unlimited access to our entire collection of over 700,000 titles via our web site and mobile apps.

To claim your free month visit: www.forgottenbooks.com/free157907

English
Français
Deutsche
Italiano
Español
Português

www.forgottenbooks.com

Mythology Photography **Fiction**
Fishing Christianity **Art** Cooking
Essays Buddhism Freemasonry
Medicine **Biology** Music **Ancient**
Egypt Evolution Carpentry Physics
Dance Geology **Mathematics** Fitness
Shakespeare **Folklore** Yoga Marketing
Confidence Immortality Biographies
Poetry **Psychology** Witchcraft
Electronics Chemistry History **Law**
Accounting **Philosophy** Anthropology
Alchemy Drama Quantum Mechanics
Atheism Sexual Health **Ancient History**
Entrepreneurship Languages Sport
Paleontology Needlework Islam
Metaphysics Investment Archaeology
Parenting Statistics Criminology
Motivational

Department of the Interior:

U. S. NATIONAL MUSEUM.

BULLETIN

OF THE

UNITED STATES NATIONAL MUSEUM.

No. 20.

BIBLIOGRAPHIES OF AMERICAN NATURALISTS.—I. THE PUBLISHED
WRITINGS OF SPENCER FULLERTON BAIRD, 1843–1882.

BY

GEORGE BROWN GOODE,

ASSISTANT DIRECTOR OF THE NATIONAL MUSEUM.

————◄►————

WASHINGTON:
GOVERNMENT PRINTING OFFICE.
1883.

Department of the Interior:

U. S. NATIONAL MUSEUM.

— 23 —

BULLETIN

OF THE

UNITED STATES NATIONAL MUSEUM.

No. 20.

PUBLISHED UNDER THE DIRECTION OF THE SMITHSONIAN INSTITUTION.

WASHINGTON:
GOVERNMENT PRINTING OFFICE.
1883.

ADVERTISEMENT.

This work is the twenty-third of a series of papers intended to illustrate the collections of natural history and ethnology belonging to the United States, and constituting the National Museum, of which the Smithsonian Institution was placed in charge by the act of Congress of August 10, 1846.

It has been prepared at the request of the Institution, and printed by authority of the honorable Secretary of the Interior.

SPENCER F. BAIRD,
Secretary of the Smithsonian Institution.

SMITHSONIAN INSTITUTION,
    Washington, February 1, 1883.

II

BIBLIOGRAPHIES OF AMERICAN NATURALISTS.

I.

THE PUBLISHED WRITINGS

OF

SPENCER FULLERTON BAIRD,

1843–1882.

BY

GEORGE BROWN GOODE,

ASSISTANT DIRECTOR OF THE NATIONAL MUSEUM.

WASHINGTON:
GOVERNMENT PRINTING OFFICE.
1883.

PREFATORY NOTE.

The writer, since 1874, has been collecting materials for a work to be entitled "An Index Bibliography of American Ichthyology," the object of which will be to render as easily accessible as possible to the investigator and the student the literature relating to American fishes. Its scope embraces not only anatomical and descriptive ichthyology, but the literature of the fisheries, angling, fishery legislation and diplomacy, fishery statistics, and the commerce of the fisheries. References will be made not only to separate works and papers in scientific periodicals and the transaction of societies, but to all accessible articles and paragraphs in narratives of voyages and books of travel in America, and to pertinent references in literary and sporting periodicals. Briefly, it is the intention to enumerate by title every writing ever published which refers to American fish or fisheries. The work will be called an "index-bibliography," because it is intended to form a comprehensive index to the works it describes. Each title will be followed by a brief descriptive or critical paragraph, which will supplement the title and indicate in a general way the scope of its author's effort. In the case of an important paper a synopsis of its contents may be given. Under each descriptive paper will be printed the names of the new species described, with the locality whence the types were obtained, and the page of the reference. In important works, containing critical remarks, similar references may be given for each species discussed. References to every engraving published will be made.

The pressure of other engagements has delayed the completion of this work, which it was at first intended to publish in 1876. It is now expected that it may be finished in 1884. Titles of works published before July, 1884, will be included. In the mean time it is proposed, from time to time, to publish special bibliographies of the writings of the most prominent naturalists, for convenience of current reference, and to invite criticism of methods, corrections of any kind, and the co-operation of those who are interested in the successful completion of the undertaking.

The first of these special bibliographies is now presented. No one will be likely to question the propriety of selecting for its subject the works of Professor BAIRD, since it is he who, more than any one else, has carried on the work of identifying and describing the material in the National Museum, and since he has probably been one of the most prolific of all contributors to the systematic zoology of the United States.

The writer has in preparation special bibliographies of CHARLES GIRARD and THEODORE GILL, but it is possible that before these can be published others, by other writers, will be assigned a place in the series. The one now most nearly ready for publication is that of ISAAC LEA, prepared by Mr. Newton P. Scudder.

WASHINGTON, *January* 1, 1882.

BIOGRAPHICAL SKETCH OF SPENCER FULLERTON BAIRD.

A N A L Y S I S .

I. Outline of his public career.

II. Honors and dignities.

III. Ancestry and development of character.

IV. Early friendships and their influence.

V. Analysis of his work and its results.

VI. Contributions to science and scientific literature.

VII. Educational and administrative works.

VIII. Work as Commissioner of Fisheries.

IX. Epilogue.

I.

Spencer Fullerton Baird was born in Reading, Pennsylvania, February 3, 1823. In 1834 he was sent to a Quaker boarding-school kept by Dr. McGraw, at Port Deposit, Maryland, and the year following to the Reading Grammar School. In 1836 he entered Dickinson College, and was graduated at the age of seventeen. After leaving college, his time for several years was devoted to studies in general natural history, to long pedestrian excursions for the purpose of observing animals and plants and collecting specimens, and to the organization of a private cabinet of natural history, which a few years later became the nucleus of the museum of the Smithsonian Institution. During this period he published a number of original papers on natural history. He also read medicine with Dr. Middleton Goldsmith, attending a winter course of lectures at the College of Physicians and Surgeons, in New York, in 1842. His medical course was never formally completed, although in 1848 he received the degree of M. D., *honoris causa*, from the Philadelphia Medical College. In 1845 he was chosen professor of natural history in Dickinson College, and in 184– his duties and emoluments were increased by election to the chair of natural history and chemistry in the same institution. In 1848 he declined a call to the professorship of natural science in the University of Vermont. In 1849 he undertook his first extensive literary work, translating and editing the text for the "Iconographic Encyclopedia," an English version of Heck's *Bilder Atlas*, published in connection with Brockhaus's *Conversations Lexikon*.

July 5, 1850, he accepted the position of Assistant Secretary of the Smithsonian Institution, and October 3, at the age of twenty-seven years, he entered upon his life work in connection with that foundation—"the increase and diffusion of useful knowledge among men." *

His work as an officer of the Institution will be discussed more fully below. It was constant and arduous, but did not prevent the publication of many original memoirs, among the most elaborate of which are the Catalogue of North American Serpents (1853); the "Birds of North America" (1858); the "Mammals of North America" (1859); the "Review of North American Birds" (1864-'66); the "Geographical Distribution of North American Birds (1865); the History of North American Birds, in connection with Thomas M. Brewer and Robert Ridgway (1874), and the preparation of numerous official reports. From 1870 to 1878 he was scientific editor of the periodicals published by Harper Brothers, of New York, and the author of their yearly cyclopedia of science, entitled "The Annual Record of Science and Industry." In 1871 he was appointed by President Grant to the position of United States Commissioner of Fish and Fisheries, an unsalaried office, to the duties of which he has for eleven years devoted a large portion of his time. In 1876 he served as one of the Government Board of Commissioners to the International Exhibition at Philadelphia, and was also a member of the international jury. In 1877 he was present, as advisory counsel, at the session of the Halifax Fishery Commission.

In May, 1878, after the death of Professor Henry, he was, by the unanimous vote of the Regents, elected Secretary of the Smithsonian Institution.

II.

Professor Baird, in 1856, received the degree of Doctor of Physical Science from Dickinson College, and in 1875 that of Doctor of Laws from Columbian University. He was, in 1878, awarded the silver medal of the Acclimatization Society of Melbourne; in 1879 the gold medal of the *Société d'Acclimatation* of France, and in 1880 the Erster Ehrenpreiz of the *Internationale Fischerei Ausstellung* at Berlin, the gift of the Emperor of Germany. In 1875 he received from the King of Norway and Sweden the decoration of "Knight of the Royal Norwegian Order of St. Olaf." He was one of the early members of the National Academy of Sciences, and ever since the organization has been a member of its council. In 1850 and 1851 he served as permanent secretary of the American Association for the Advancement of Science, and since 1878 has been one of the trustees of the Corcoran Gallery of Art in Washington. He has been president of the Cosmos Club, and for many years a trustee of Columbian University. Among his honorary relations to numerous scientific societies of the United States and other countries are included those of foreign membership in the Linnæan

---

* The motto of the Smithsonian Institution and of its founder, JAMES SMITHSON.

Society of London, and the Zoological Society of London, honorary membership in the Linnæan Society of New South Wales, and corresponding membership in the *K. K. Zoologisch-botanische Gesellschaft*, Vienna; the *Sociedad de Geographia*, Lisbon; the New Zealand Institute, the *Koninklijke Natuurkundige Vereeniging in Nederlandsch Indië*, Batavia; the *Magyar Tudományos Akadémia*, Buda-Pesth; the *Société Nationale des Sciences Naturelles*, Cherbourg; the *Academia Germanica Naturæ Curiosorum*, Jena; the *Naturforschende Gesellschaft*, Halle; the *Naturhistorische Gesellschaft*, Nuremburg; the Geographical Society of Quebec; the Historical Society of New York; the *Deutsche Fischerei Verein*, Berlin.

The nomenclature of zoology contains many memorials of his connection with its history. A partial enumeration shows that over twenty-five species and one genus of fishes bear his name.

A post-office in Shasta County, California, located near the McCloud River Salmon Hatching Station of the United States Fish Commission, was named "Baird" by the Postmaster-General in 1877.

III.

His ancestry upon the one side was English, upon the other Scotch and German. His paternal grandfather was Samuel Baird, of Pottstown, Pa., a surveyor by profession, whose wife was Rebecca Potts. The Bairds were from Scotland, while the Potts family removed from Germany to Pennsylvania at the close of the seventeenth century. His great grandfather on the mother's side was the Rev. Elihu Spencer, of Trenton, one of the war preachers of the Revolution, whose patriotic eloquence was so influential that a price was set on his head by the British Government; his daughter married William M. Biddle, a banker, of an English family for many generations established in Pennsylvania, and identified with the banking interests of Philadelphia. Samuel Baird, the father of the subject of this sketch, established himself as a lawyer at Reading, Pennsylvania, and died when his son was ten years old. He was a man of fine culture, a strong thinker, a close observer, and a lover of nature and of out-of-door pursuits. His traits were inherited by his children, but especially by his sons Spencer and William. The latter, who was the elder, was the first to begin collecting specimens, and as early as 1836 had in hand a collection of the game-birds of Cumberland County. His brother soon became his companion in this pursuit, and six years later they published conjointly a paper entitled "Descriptions of two species, supposed to be new, of the Genus Tyrranula Swainson, found in Cumberland County, Pennsylvania." *

There are still in the museum at Washington specimens of birds prepared by these boys forty-five years ago by a simple process of evisceration, followed by stuffing the body-cavities full of cotton and arsenical

---

*See list on a subsequent page.

soap. The brother, William M. Baird, diverged into other paths, and at the time of his death in 1872 was United States collector of internal revenue at Reading.

The inheritance of a love of nature and a taste for scientific classifica-tion, the companionship of a brother similarly gifted, tended to the development of the young naturalist, and a still more important element was the encouragement of a judicious mother by whom he was per-mitted to devote the five years immediately following his graduation to his own devices and plans instead of being pushed at once into a profession. In 1841, at the age of eighteen, we find him making an ornithological excursion through the mountains of Pennsylvania, walk-ing 400 miles in twenty-one days, the last day 60 miles between day-light and rest. The following year he walked more than 2,200 miles. His fine physique and consequent capacity for work are doubtless due in part to his outdoor life during these years.

IV.   •

An important stimulus to the efforts of this young naturalist was the friendship which he formed as early as 1838 with Audubon, with whom he was for many years in correspondence, and who, in 1842, gave to him the greater part of his collection of birds, including most of his types of new species. Young Baird contributed many facts and speci-mens for the History of North American Quadrupeds at that time in preparation, as well as to the Ornithological Biography, and was only prevented by ill health from accompanying Audubon as his secretary on his six months' expedition to the Yellowstone in 1840. In those days were formed many of the friendships and partnerships with scien-tific men which influenced his after life. Among his early correspond-ents were George N. Lawrence (1841), John Cassin (1843), John G. Morris (1843), Thomas M. Brewer (1845), and S. S. Haldeman (1845). In 1847 he met Agassiz, then just arrived from Switzerland in company with Desor and Girard. At this time or a year later was projected the work of Agassiz and Baird on "The Fresh-water Fishes of the United States," which was, however, never published, although a number of illustrations and some pages of text were elaborated. In 1843 he trans-lated Ehrenberg's "Corals of the Red Sea" for J. D. Dana, who was then preparing his reports for the United States exploring expedition. As early as 1846 we find him engaged in the preparation of a synonymy of North American birds, and visiting Boston to consult the libraries of Amos Binney and the Boston Society of Natural History for works not possessed by the Philadelphia Academy of Natural Sciences. This material was utilized twelve years later in the "Birds of North America."

As professor of natural history in Dickinson College he taught the seniors in physiology, the sophomores in geometry, and the freshmen in zoology. He found time, however, to carry on the works begun in previous years, and to make in summer extended collecting expeditions: To the Adirondacks in 1847 ; to Ohio in 1848, to collect, in company with Dr. Kirtland, from the original localities of the types, the fishes de-scribed by him in his work on the fishes of Ohio; to the mountains of Virginia in 1849; and to Lake Champlain and Lake Ontario in 1850.

When in 1850, upon the urgent recommendation of the late George P. Marsh, he was elected an officer of the Smithsonian Institution, he brought with him to Washington methods of work, developed in his personal experience, which became at once the methods of the establish-ment, and are still employed in many of its departments.

V.

There may be noted in the career of Professor Baird several distinct phases of activity, namely, (1) a period of twenty-six years, 1843–1869, occupied in laborious investigation and voluminous publication upon the vertebrate fauna of North America; (2) forty years of continuous con-tribution to scientific literature, of which at least ten were devoted to scientific editorship; (3) five years, 1845–1850, devoted to educational work; forty years, 1842–1883, devoted to the encouragement and promo-tion of scientific enterprises, and the development of new workers among the young men with whom he was brought into contact; (5) thirty-three years, 1850–1883, devoted to administrative work as an officer of the Smithsonian Institution, and in charge of the scientific collections of the government—twenty-eight as principal executive officer and five as Secretary and responsible head; (6) twelve years as head of the Fish Commission, a philanthropic labor for the increase of the food-supply of the world, and incidentally in promoting the interests of biological and physical investigation of the waters.

VI.

The extent of Professor Baird's contributions to science and scien-tific literature may be at least partially comprehended by an examina-tion of the succeeding pages of the present work. The list of his writings is complete to the end of the year 1882, and contains 1,063 titles. Of this number 775 are brief notices and critical reviews con-tributed to the "Annual Record of Science and Industry," while under his editorial charge, 31 are reports relating to the work of the Smith-sonian Institution, 7 are reports upon the American fisheries, 25 are schedules and circulars officially issued, and 25 are volumes or papers edited. Out of the remaining 200 the majority are formal contributions to scientific literature.

It seems scarcely necessary to remark that most of the official reports above referred to, as well as many of the brief articles in the Annual Record, contain important original matter.

Nineteen of the descriptive papers were published conjointly with Charles Girard, while the most elaborate work, "The Birds of North America," was prepared in its first edition with the aid of Messrs. Cassin and Lawrence, and in its second with that of Messrs. Brewer and Ridgway.

Of the total number of papers enumerated in the list 73 relate to mammals, 80 to birds, 43 to reptiles, 431 to fishes, 61 to invertebrates (these being chiefly reviews), 16 to plants, 88 to geographical distribution, 46 to geology, mineralogy, and paleontology, 45 to anthropology, 31 to industry and art, 109 to exploration and travel.

While the number of new species described does not necessarily afford any clew to the value of the work accomplished, it may not be uninterest-ing to refer to it as an indication of the pioneer work which it was ne-cessary to do even in so prominent a group as the vertebrates. I note among mammals 49, birds 70, reptiles 186, fishes 56. Forty-nine of 220, or nearly one-fourth, of the mammals discussed in the "Mammals of North America," were there described for the first time. In the cata-logue of serpents not more than 60 per cent. had been named, and in in preparation for studying the specimens, each was carefully ticketed with its locality, and then the 2,000 or more individuals were thrown indiscriminately into one great pile, and the work of sorting them out by resemblances was begun. Not the least valuable have been the nu-merous accurate figures of North American vertebrates, prepared under Professor Baird's supervision. These include representatives of 170 species of mammals and 160 species of reptiles, besides still many hun-dreds of birds.*

VII.

Passing to the consideration of the influence of Professor Baird on the encouragement of scientific enterprise, it seems scarcely necessary to call attention to the manner in which this influence has been exerted, since the relation of the Smithsonian Institution to scientific explora-tion, particularly in the lines of natural history and ethnology, is a part of the scientific history of the country, and since this department of the work of the Institution was always from its inception under the direction of the assistant secretary. The first grant made by the In-stitution for scientific exploration and field research was in 1848 to Spen-cer F. Baird, of Carlisle, for the exploration of the bone caves and the local natural history of Southeastern Pennsylvania.

From the start the Department of Explorations was under his charge; and in his reports to the Secretary, published year by year in the an-nual report of the Institution, may be found the only systematic record of government explorations which has ever been prepared. From 1850 to 1860 several extensive government expeditions were sent to

---

* In the bibliography below all these figures are thoroughly indexed.

the western territories, and it became the duty of Professor Baird to enlist the sympathies of the commanders of these expeditions in the objects of the Institution, to supply them with all the appliances for collecting, as well as with instructions for their use, and also in most cases to organize the natural history parties, nominate the collectors, employ and supervise the artists in preparing the plates, and in many instances to edit the zoological portions of the reports.

The fitting out of such expeditions was only a small part of the work; from the beginning until now there have been numerous private collectors, deriving their materials, their literature, and, to a considerable extent, their enthusiasm from the Smithsonian Institution, and consequently in correspondence with its officers. The Smithsonian "Instructions to Collectors," which has passed through several large editions, as well as numerous circulars written with a similar purpose, were prepared by Professor Baird in connection with this department of his work.

As a result of this extensive work of organization, a large number of young men have been trained as collectors and observers, and among them not a few have become eminent in various departments of science.

In addition to this special branch of his work, the assistant secretary had, from the start, the charge of certain departments of the routine work of the Institution; the system of international exchanges, for instance, which had ever been one of the leading objects of the Smithsonian Institution, was organized by him in its details. His first task, after entering upon his duties, was to distribute the second volume of the "Smithsonian Contributions to Knowledge." Already in connection with his private enterprises he had developed a somewhat extensive system of exchanges with European and American correspondents, and the methods thus established were expanded for the wider needs of the Institution. The main duty of the assistant secretary, however, was the development of the natural history collections. As has already been indicated, the private collection which he brought with him to Washington formed the nucleus of the Smithsonian museum. The only specimens in possession of the Institution at the time of his arrival were a few boxes of minerals and plants. The collections of the Wilkes Exploring Expedition, which constitute the legal foundation of the National Museum of the United States, were at that time under the charge of the National Institute; and, although by the act of incorporation the Smithsonian Institution was the legal custodian of the national cabinet of curiosities, it was not until 1857 that the Regents finally accepted the trust and the National Museum was definitely placed under the control of the Smithsonian Institution and transferred to its building. Until this time Congress had granted no funds for the support of the Smithsonian cabinets, and the collections had been acquired and cared for at the expense of the endowment fund. They had, however, become so large and important in 1857 that the so-called "National Collection" at that time acquired were small in comparison.

The National Museum then had a double origin. Its actual although not its legal nucleus was the collection gathered in the Smithsonian building prior to 1857. Its methods of administration, which were in fact the very same that had been developed by Professor Baird in Carlisle as early as 1845, are those which are still in use, and which have stood the test of thirty years without any necessity for their modification becoming apparent. In the bibliography below is reprinted from the fifth annual report of the Smithsonian Institution, now exceedingly rare, a report by the assistant secretary in charge of the natural history department for the year 1850, which enumerates the specimens belonging to the Museum on January 1, 1851, including a full account of his own deposit.

Having thus almost from the very outset been associated with Professor Henry in the organization of the Smithsonian Institution, his course since his accession to the secretaryship has been a consistent continuation of that which had for twenty-eight years been adopted.

VIII.

The work of the Fish Commission, in one of its aspects, may perhaps be regarded as the most prominent of the present efforts of the government in aid of aggressive biological research.

On the 9th of February, 1874, Congress passed a joint resolution which authorized the appointment of a Commissioner of Fish and Fisheries. The duties of the Commissioner were thus defined: "To prosecute investigations on the subject (of the diminution of valuable fishes) with the view of ascertaining whether any and what diminution in the number of the food-fishes of the coast and the lakes of the United States has taken place; and, if so, to what causes the same is due; and also whether any and what protective, prohibitory, or precautionary measures should be adopted in the premises, and to report upon the same to Congress."

The resolution establishing the office of Commissioner of Fisheries required that the person to be appointed should be a civil officer of the government, of proved scientific and practical acquaintance with the fishes of the coast, to serve without additional salary. The choice was thus practically limited to a single man. Professor Baird, at that time assistant secretary of the Smithsonian Institution, was appointed and at once entering upon his duties soon developed a systematic scheme of investigation.

The Fish Commission now fills a place tenfold more extensive and useful than at first. Its work is naturally divided into three sections:

1. The systematic investigation of the waters of the United States and the biological and physical problems which they present. The scientific studies of the Commission are based upon a liberal and phi-

losophical interpretation of the law. In making his original plans the Commissioner insisted that to study only the food-fishes would be of little importance, and that useful conclusions must needs rest upon a broad foundation of investigations purely scientific in character. The life history of species of economic value should be understood from beginning to end, but no less requisite is it to know the histories of the animals and plants upon which they feed or upon which their food is nourished; the histories of their enemies and friends, and the friends and foes of their enemies and friends, as well as the currents, temperatures, and other physical phenomena of the waters in relation to migration, reproduction, and growth. A necessary accompaniment to this division is the amassing of material for research to be stored in the national and other museums for future use.

2. The investigation of the methods of fisheries, past and present, and the statistics of production and commerce of fishery products. Man being one of the chief destroyers of fish, his influence upon their abundance must be studied. Fishery methods and apparatus must be examined and compared with those of other lands, that the use of those which threaten the destruction of useful fishes may be discouraged, and that those which are inefficient may be replaced by others more serviceable. Statistics of industry and trade must be secured for the use of Congress in making treaties or imposing tariffs, to show to producers the best markets, and to consumers where and with what their needs may be supplied.

3. The introduction and multiplication of useful food-fishes throughout the country, especially in waters under the jurisdiction of the general government, or those common to several States, none of which might feel willing to make expenditures for the benefit of the others. This work, which was not contemplated when the Commission was established, was first undertaken at the instance of the American Fish Cultural Association, whose representatives induced Congress to make a special appropriation for the purpose.

IX.

Comment upon the facts presented in this biographical sketch seems to be unnecessary. Future historians of American science will be better able than are we to estimate justly the value of the contributions to scientific literature which are enumerated in the bibliography; but no one not living in the present can form an accurate idea of the personal influence of a leader upon his associates, and upon the progress of thought in his special department, nor can such an influence as this well be set down in words. This influence is apparently due not only to extraordinary skill in organization, to great power of application and concentration of thought constantly applied, and to a philosophical and comprehensive mind, but to an entire and self-sacrificing devotion to the interests of his own work and that of others.

A LIST OF GENERA AND SPECIES NAMED IN HONOR OF PROFESSOR BAIRD.

Bairdiella, GILL. Proc. Acad. Nat. Sci. Phila., xiii, 1861, p. 83. (Type, *Bodianus argyroleucus,* Mitchill.)

> A genus of the family *Sciænidæ* was represented by one species on the east coast of the United States.*

Acanthidops Bairdi, RIDGWAY. Proc. U. S. National Museum, iv, . 1882, p. 336.

> A bird of the family Dendrocolaptidæ, inhabiting Costa Rica.

Actodromus Bairdii, COUES. Proc. Acad. Nat. Sci. Phila., 1861, p. 494. (*Atodromus*) Sclater, Proc. Zool. Soc. 1867, p. 332.

> A bird of the family *Scolopacidæ,* inhabiting North America, chiefly in the interior.

Alepocephalus Bairdii, GOODE & BEAN. Proc. U. S. National Museum, ii, p. 55, July 1, 1879.

> A fish of the family *Alepocephalidæ,* inhabiting the deep waters of the northwestern Atlantic.

Anchitherium Bairdii, LEIDY. Owen's Rep. Geol. Surv. Wisc., &c., 1852, p. 572. Ext. Vert. Fauna, Wash. Terr., 1873, p. 322, pl. vii, fig. 15.

> A fossil mammal of the order *Perissodactyli,* family *Anchitheriidæ,* found in the Mauvaises Terres of White River, Dakota, and the tertiary formations on John Day's River, Oregon.

Archaster Bairdii, VERRILL. Amer. Journ. Sci., xxiii, p. 139, February, 1882.

> A star-fish of the family *Astropectinidæ,* inhabiting the deeper waters off the New England coast.

Buteo Bairdi, HOY. Proc. Acad. Nat. Sci. Phila., vi, 1853, 451.

> A bird of the family *Falconidæ,* inhabiting
> = *Buteo Swainsoni,* Bonap.

Calliostoma Bairdii, VERRILL & SMITH. Amer. Journ. Sci., xx, p. 396, November, 1880.

> A gastropod mollusk of the family *Trochidæ,* inhabiting the deep waters off the New England coast.

Campephilus Bairdi, CASSIN. Proc. Acad. Nat. Sci. Philadelphia, 1863, p. 322.

> A bird of the family *Picidæ,* inhabiting

Certhiola Bairdi, CABANIS. Journ. Ornitho., 1865, p. 412.

> A bird of the family *Sylvicolidæ,* a member of the West Indian fauna.
> =*Certhiola bahamensis,* Reich.

Coccygus Bairdi, SCLATER. Proc. Zool. Soc., March, 1864, p. 120.

> A bird of the family *Cuculidæ,* described from Jamaica.

Coluber Bairdi, YARROW. Bull. U. S. Nat. Museum, No. 17, 1880, p. 41.

> A serpent of the family *Colubridæ,* inhabiting Texas.

---

* The name BAIRDIA was dedicated to Dr. Baird of the British Museum.

Cottus Bairdii, GIRARD. Proc. Amer. Assoc. Adv. Sci. ii, 1850, p. 410. Proc. Acad. Nat. Sci. Phila., iii, 1850, p. 189. Smithsonian Contributions, iii, 1852. Mon. Cott., p. 44, pl. i, figs. 5, 6.

A fish of the family *Cottidæ*, inhabiting the streams of Ohio and Cayuga Lake, N. Y.

Delphinus Bairdii, DALL. Proc. Cal. Acad. Sci., v, Jan., 1873. Scammon, Marine Mammals of the Northwest Coast, 1874, p. 283 (and 99), pl. xix, fig. 1.

A cetacean of the family *Delphinidæ*, found in the Pacific waters of the United States.

Dolium Bairdii, VERRILL & SMITH. Amer. Journ. Sci., xxii, p. 296, Oct., 1881.

A gastropod mollusk of the family *Buccinidæ*, inhabiting the deep waters off the New England coast.

Elasmognathus Bairdii, GILL. Proc. Acad. Nat. Sci. Phila., 1865, p. 183.

A mammal of the family *Tapiridæ*, inhabiting Central America.

Emberiza Bairdii, AUDUBON. Birds of North Am., vii, 1843, p. 359, pl. 500.

Baird, Brewer & Ridgway. Birds of North America, i, 1874, p. 531, pl. xxv, fig. 3.

A bird of the family *Fringillidæ*, inhabiting the central plains of North America. = *Centronyx Bairdii*, (Audubon).

Empidonax Bairdii, SCLATER. Proc. Zool. Soc. Lond., 1858, p. 301.

A bird of the family *Tyrannidæ*, inhabiting the mountains of Central America.

Graculus Bairdi, "GRUBER, MSS.", COOPER. Proc. Phil. Acad. 1865, p. 5.

A bird of the family *Graculidæ*, inhabiting the Farallone Islands, California.

= *Phalacrocorax Bairdi*, (Gruber).

Junco Bairdi, BELDING, MSS. Proc. U. S. Nat. Mus., 1883.

A bird of the family *Fringillidæ*, inhabiting Lower California

Lepus Bairdi, HAYDEN. Amer. Nat., iii, 1869, 115. Bull. Essex Inst., vi, 1874, pp. 61–66.

A mammal of the family *Leporidæ*, inhabiting the Rocky Mountains. = *Lepus americanus*, var. *Bairdi*, ALLEN.

Macrurus Bairdii, GOODE & BEAN. Amer. Journ. Sci. and Arts, xiv, p. 471, Dec., 1877.

A fish of the family *Macruridæ*, inhabiting the deep waters of the northwestern Atlantic.

Melanerpes formicivorus Bairdi, RIDGW. Bull. U. S. Nat. Mus., No. 21, 1881, 34.

A bird of the family *Picidæ*, inhabiting California.

Mus Bairdii, HOY & KENNICOTT. Agricultural Report, U. S. Patent Office for 1856 (1857), p. 92, pl. xi.

A mammal of the family *Muridæ*, inhabiting the Mississippi valley. = *Hesperornys michiganensis*, Wagner (A. & B.).

Octopus Bairdii, VERRILL.   Amer. Journ. Sci. and Arts, 1873, p. 5.
>A cephalopod mollusk of the family *Octopodidæ*, inhabiting the deep waters off the New England coast.

Palæotherium Bairdii, LEIDY.   Proc. Acad. Nat. Sci. Phila., v, p. 122, and 6th Ann. Rep. Smithsonian Institution, 1852, p. 64.
>A fossil mammal of the family *Palæotheriidæ*, found in the territory of the Mauvaises Terres of Dakota.

Papilio Bairdii, EDWARDS.   Proc. Ent. Soc. Phila., 1866.
>A butterfly of the family *Papilionidæ*, inhabiting Arizona and New Mexico.

Picus Bairdi, SCLATER (*MS*).   Malherbe, Mon. Pic., 1, p. 188, vol. xxvii, figs. 7, 8.—*Picus scalaris*, Wagler.
>A bird of the family *Picidæ*.

Pomacentrus Bairdii, GILL.   Proc. Acad. Nat. Sci. Phila., 1862, 148.
>A fish of the family *Pomacentridæ*, inhabiting the waters of Lower California.

Pyrula Bairdi, MEEK & HAYDEN.   Proc. Acad. Nat. Sci. Phila., vii, 1856, p. 66, fig. —, in Meek's Invertebrate Paleontology, U. S. Geol. Surv. of the Territories, ix, pl. xxxi, fig. 10 *a. b.*
>A fossil mollusk of the family *Pyrulidæ* from the Fox Hills group of the Upper Missouri cretaceous series.
>=*Pyropsis Bairdi*, (M. & H) MEEK.

Salmo Bairdii, SUCKLEY.   Am. Lyc. Nat. Hist. N. Y., vii, 1869, p. 309 (*Salmo*).
>A fish of the family *Salmonidæ*, inhabiting the streams of the Pacific coast of North America.

Saurophagus Bairdii, GAMBEL, Journ. Phila. Acad., i, second ser., 1847, p. 40.

Syngnathus Bairdianus, DUMERIL.   Hist. Nat. Poiss., ii, 1870, p. 574.
>A fish of the family *Syngnathidæ*, inhabiting the Pacific coast of Mexico.

BIBLIOGRAPHY OF THE PUBLICATIONS

OF

PROFESSOR SPENCER FULLERTON BAIRD, LL. D.

I. CHRONOLOGICAL CATALOGUE.

1.

1843. BAIRD, SPENCER F., *and* WILLIAM M. BAIRD.* Descriptions of two Species, supposed to be new, of the Genus Tyrannula Swainson, found in Cumberland County, Pennsylvania. By William M. & Spencer F. Baird, of Carlisle, Pa. < *Proc. Acad. Nat. Sci. Phila.*, i, pp. 283–285, 1843. Presented for publication July 11; ordered printed July 25.

> *Tyrannula flaviventris*, Baird, n. s... 283
> Carlisle, Pa., 1840.
> *Tyrannula minima*, Baird, n. s... 284
> Near Carlisle, Pa.. May, 1839.

2.

1844. BAIRD, SPENCER F., *and* WILLIAM M. BAIRD. List of Birds found in the vicinity of Carlisle, Cumberland County, Penn., about Lat. 40° 12′ W., Lon. 77° 11′ W. By William M. & Spencer F. Baird. < *Amer. Journ. Sci. and Arts*, xlvi, 1844, No. 2, Jan.-Mar., art. vi, pp. 261–273.

> 201 species are enumerated, the times of appearance and relative abundance of each being mentioned. Breeders marked. The following summation is made:—
> "Species spending the summer 112
> Species resident all the year 38
> Winter visitors .... 14." ·
> 4 other species are indicated, as well as a number supposed to have been observed by B. S. Barton and others.

3.

1844. BAIRD, SPENCER F., *and* WILLIAM M. BAIRD. Descriptions of two species, supposed to be new, of the genus Tyrannula (Swainson), found in Cumberland Co., Penn. By Wm. M. & Spencer F. Baird, of Carlisle, Pa. < *Amer. Journ. Sci. and Arts*, xlvi, 1846, No. 2, Jan.-Mar., pp. 273–276.

> Same as No. 1.
> *Tyrannula flaviventris* 274
> *Tyrannula minima* 275

4.

1844. BAIRD, SPENCER F. On the application of bi-chromate of potassa to photographic purposes. < *Literary Record and Journal of the Linnæan Association of Pennsylvania College*, i, No. 2, Dec., 1844,, pp. 17–19.

> "By Spencer F. Baird of Carlisle, Pa." Describes the process of copying other flat objects on paper sensitized by bi-chromate of potassa.

---

* WILLIAM M. BAIRD, brother of Prof. S. F. Baird, born in Reading, Pa., Aug. 4, 1817, died in Reading Oct. 19, 1872. Entered Lafayette College 1834. Graduated at Dickinson College 1837. Admitted to Berks County bar, Reading, April 12, 1844. Mayor of Reading 1855–56. Collector of Internal Revenue, 8th District of Pennsylvania, 1869–1872.

1844. BAIRD, SPENCER F.—Continued.

"To the purpose of copying a coarse print, a piece of music, an embroidering pattern or a leaf this process is admirably adapted. It is for the latter object that the art has been mostly used by the writer, who last summer copied leaves of nearly all the trees and shrubs of Cumberland County (Pennsylvania), amounting to nearly two hundred species. These photographs are as valuable for scientific purposes as good engravings of the same would be, perhaps more so, as not only is the outline perfectly given, but in most cases the fine and delicate nervation, whose arrangement frequently forms a specific character, is distinctly preserved."

This collection of leaf photographs, still preserved in the National Museum, has been one of the standard resources of American paleophytologists, and has been used in the preparation of many of the works on the Fossil Botany of the United States.

5.

1845. BAIRD, SPENCER F. Contributions towards a catalogue of the trees and shrubs of Cumberland County, Pa. < *Lit. Rec. and Journ. Linnœan Assoc. Pennsylvania College*, i, No. 4, Feb., 1845, pp. 57–63.

After introductory remarks on the geology and topography of Cumberland County and the relations of peculiarities of vegetation to the soil, a list of the trees and shrubs is given, with common names, stations, and notes regarding abundance. The whole number of species enumerated is 150, 20 of which had been found in Chester County.

The collection of woods, cut and polished, prepared in connection with this paper is preserved in the National Museum.

6.

1845. BAIRD, SPENCER F. Catalogue of birds found in the neighborhood of Carlisle, Cumberland Co., Pa. < *Lit. Rec. and Journ. Linnœan Assoc. Pennsylvania College*, i, No. 12, Oct., 1845, pp. 249–257.

A revision of No. 2.

A list of the species of birds collected by the writer between 1840 and 1845 (with a very few exceptions) within a few miles of Carlisle. "None are admitted without having been actually killed and preserved; in no case have any been admitted on the authority of others. A residence nearer the Susquehanna would no doubt have enabled us to increase this number considerably, as we have heard of several not in this list which have been killed about Harrisburg. Some of these will be found in a catalogue of additional species procured at Marietta, Pa. (25 miles south of Cumberland Co., by Mr. J. Libhart), most if not all of which touch on our eastern border. The nomenclature employed is based on that of Prince Bonaparte, . . . with the authority for each species, and various alterations called for by a strict regard to the law of priority. The name immediately succeeding each species is that of its first describer, and if in parenthesis, under a different genus. The second name is his who first placed that particular specific appellation under its present genus. A (†) prefixed shows that this bird breeds here."

The list of species is supplemented by references to Audubon's names, by the common names, and notes on relative abundance and season of appearance. The paper is summed up as follows (p. 257):

"Total of species in Cumberland Co 202
of which:—Breeding.. 104
        Summer visitors 66
        Resident... 38
        Transitory in spring 90
        Seen in autumn only .... 8."

The Marietta list includes 9 species, and 6 others are mentioned "of whose existence in our vicinity we have had strong proof (but) are not included for want of specimens." None of these are included in the summation.

The following combinations appear in this paper for the first time:—

Myiodioctes pusillus (Wils.) Baird. R................................. 252
Carpodacus purpureus (Gm.) Baird. R................................... 254
Pluvialis virginiacus (Bork) Baird. R.
Nycticorax discors (Nutt.) Baird. R 255
Porzana carolina (Linn.) Baird. R.
Porzana noveboracensis (Gm.) Baird. R.
Porzana jamaicensis (Briss.) Baird. R................................. 257

7.

1846. B[AIRD], S[PENCER] F. The sea-serpent in Norway. < *Lit. Rec. and Journ. Linnœan Assoc. Pennsylvania College*, ii, No. 5, Mar., 1846, pp. 106–107.

Notice of information obtained by Dr. H. Boie, while on a tour through Norway in 1817.

8.

1846. BAIRD, SPENCER F. Hints | for Preserving | Objects of Natural History | prepared | by Prof. S. F. Baird, | for | Dickinson College, | Carlisle, Pa. | Carlisle: | Printed by Gitt & Hinckley, | 1846. 8vo. pp. 12.

Reptiles and Fish, pp. 7, 8.

A supplementary schedule without title-page was published in 1848.

9.

1847. BAIRD, SPENCER F. Dr. Leidy read a letter from Prof. Spencer F. Baird, of Carlisle, Pa., describing a Hybrid between the Canvass back Duck and the Common Duck. < *Proc. Acad. Nat. Soc. Phila.*, iii, 1846 and 1847, p. 209 (May 4, 1847).

Nothing additional to the above ever published.

10.

1849. BAIRD, SPENCER F. Revision of the North American Tailed-Batrachia, with descriptions of new genera and Species. < *Journ. Acad. Nat. Sci. Phila.*, 2d ser., i, pp. 281–294, Oct., 1849.

Issued also as excerpt, without title or repagination.

"The following notes, introductory to a more detailed memoir on the same subject, will, it is hoped, throw some light upon this obscure portion of American zoology. To this we would refer for the descriptions of species, with their anatomical and physiological characters, *giving only in this place brief outlines of the genera and the synonomy of the species.*"

The various genera are defined, and a synonymic list of species is given, with notes upon habitats.

Desmognathus (Baird), n. g. (on *S. niger* and *S. fuscus*, auct.)........................ 282

The following new combinations are made:—

Ambystoma punctata (Linn.) Baird. R... 283
Ambystoma opaca (Gravenhast) Baird. R.
Ambystoma jeffersoniana (Green) Baird. R.
Ambystoma macrodactyla, Baird, n. s. (name given only, species described below).
Ambystoma tigrina (Green) Baird. R 284
Ambystoma lurida (Sager) Baird. R.
Ambystoma mavortia, Baird, n. s. (name only, described below).
Ambystoma episcopus, Baird, n. s. (name only, described below).
Notophthalmus viridescens (Rafin) Baird. R.
Notophthalmus torosus (Eschscholtz) Baird. R.
Plethodus erythronota (Green) Baird. R.. 285
Desmognathus niger (Green) Baird. R.
Desmognathus fuscus (Rafin.) Baird. R.
Desmognathus auriculatus (Holbrook) Baird. R.
Pseudotriton montanus, Baird, n. s. (name only, described below).................... 287
Pseudotriton salmoneus (Stover) Baird. R.
Spelerpes longicauda (Green) Baird. R.
Spelerpes guttolineata (Holbrook) Baird. R.
Spelerpes bilineata (Green) Baird. R.
Spelerpes cirrigera (Green) Baird. R.
Batrachoseps quadridigitata (Holbrook) Baird. R.
Necturus lateralis (Say) Baird. R.. 290
Necturus maculatus (Barnes) Baird. R.
Siredon maculatus (Owen) Baird. R....................................... 292

11.

1849. BAIRD, SPENCER F. Descriptions of four new species of North American Salamanders, and one new species of Scink. < *Journ. Acad. Nat. Sci. Phila.*, 2d ser., i, Oct., 1849, pp. 292–294.

Ambystoma macrodactyla, Baird. n. s.. 292

Astoria, Oreg. J. K. Townsend, M. D.

1849. BAIRD, SPENCER F.—Continued.

> *Ambystoma mavortia*, Baird, n. s.
>> New Mexico. Dr. Wizlizenus.
>
> *Ambystoma episcopus*, Baird, n. s 293
>> Kemper County, Miss. Clinton Lloyd.
>
> *Pseudotriton montanus*, Baird, n. s.
>> South Mountain, near Carlisle, Pa. S. F. Baird.
>
> *Plestiodon anthracinus*, Baird, n. s 294
>> South Mountain, near Carlisle, Pa. S. F. Baird.

12.

1850. BAIRD, SPENCER F. Descriptions of four new species of North American Salamanders, and one new species of Scink. $<$*Amer. Journ. Sci. and Arts*, ix, 2d ser., Jan., 1850, pp. 137–139.

> The memoir in the Journal of the Academy of Natural Sciences, (2) i, pp. 292-294, is given in full, prefixed by a commendatory notice by the editor of the Journal, of this paper and the one on the Tailed Batrachians.

13.

1850. BAIRD, SPENCER F. On the Bone Caves of Pennsylvania. $<$*Proc. Amer. Assoc. Adv. Sci.*, ii, 1850, pp. 352–355. (Cambridge Meeting, Aug., 1849.) Read Aug. 20, 1849.

> Explorations of a cave near Carlisle, Pa., and two others in Pennsylvania.

14.

1850. BAIRD, SPENCER F. On the Urodelian Batrachians. $<$*Proc. Amer. Assoc. Adv. Sci.*, ii, 1850, p. 402.

> A verbal abstract of this paper was presented to the Association and only the title was published.

14a.

1850. [BAIRD, SPENCER F.] Registry of Periodical Phenomena.

> (Printed on a half-sheet of thin, blue letter-paper.) Published by the Smithsonian Institution.
>
> "The Smithsonian Institution, being desirous of obtaining information with regard to the periodical phenomena of animal and vegetable life in North America, respectfully invites all persons who may have it in their power to record their observations and to transmit them to the Institution. The points to which particular attention should be directed are the first appearance of leaves and of flowers in plants; the dates of appearance and disappearance of migratory or hybernating animals—Mammalia, Birds, Reptiles, Fishes, Insects, &c.; the time of nesting of Birds; of moulting of and littering of Mammalia; of utterance of characteristic cries among Reptiles and Insects, and anything else which may be deemed noteworthy.
>
> "A list of plants is appended to which particular reference should be had in making observations. It has been prepared from materials furnished by Dr. John Torrey and others, and will be found to contain many species distributed throughout the United States, together with an indigenous number to, or cultivated in Europe. For the present, attention may be paid alone to the time of flowering of these species, this period in all cases being indicated by the first appearance of the anther in the expanding flower.
>
> "The Smithsonian Institution is also desirous of obtaining detailed list of all the animals and plants of any locality throughout this continent. These, when practicable, should consist of the scientific names as well as those in common use; but when the former are unknown, the latter alone may be employed. It is in contemplation to use the information thus gathered in construction of a series of species, showing the geographical distribution of the animal and vegetable kingdoms of North America."

15.

1850. BAIRD, SPENCER F. General Directions for Collecting and Preserving Objects of Natural History.

> Published by the Smithsonian Institution. Printed on a half-sheet of blue letter-paper. On the back a list of "special desiderata."
>
> A letter from the Quartermaster-General, dated March 31, 1848, granting facilities for transportation in accordance with request of Professor Baird.

16.

1851. BAIRD, SPENCER F. Report of the Assistant Secretary in charge of the natural history department [of the Smithsonian Institution] for the year 1850. <*Fifth Annual Report of the Secretary of the Smithsonian Institution for the years* 1850, 1851, pp. 41-50.

Contains a list of the principal accessions to the Museum of the Smithsonian Institution made prior to January 1, 1851, p. 41, and summary of Specimens belonging to S. F. Baird and deposited by him in the cabinet of the Smithsonian Institution.*

To JOSEPH HENRY, LL. D.,
Secretary of the Smithsonian Institution:

SIR: I beg leave to present to you a report of operations up to January 1, 1851, in the Department of Natural History, assigned to my charge.

I commence with a list of the most important specimens of natural history received at the Smithsonian Institution prior to January 1, 1851. The dates of reception have not been given, owing to the fact that most had arrived before July 1, 1850, the period when my official connection with the Institution commenced. More detailed accounts of these objects will hereafter be furnished, as well as of those which may in future be received.

LIST OF THE PRINCIPAL ACCESSIONS TO THE MUSEUM OF THE SMITHSONIAN INSTITUTION MADE PRIOR TO JANUARY 1, 1851.

Lieutenant Lynch, U. S. N. Sealed bottles containing water from the Dead Sea; cones of the cedar of Lebanon.

Miss D. L. Dix. Box of minerals from North Carolina.

Dr. F. B. Hough. Box of minerals and fossils from St. Lawrence County, New York.

Mr. Guest. Box of minerals from same locality.

Mr. Polkinhorn Box of Tertiary fossils from North Carolina.

Dr. James Eights, Albany, N. Y. Box of sands, clays, and concretions from the vicinity of the city of Albany, N. Y.

Dr. William B. Smith, Indiana. Silurian fossils from Indiana.

William Phillips, Esq., Augusta, Ga. Box of minerals from Georgia.

Oscar Freeman and Gilbert Taylor, civil engineers. Very large specimens of crystallized calcareous spar coated with quartz, found in tunnelling St. Anthony's Nose, Peekskill, N. Y.

Faxon D. Atherton, Esq. Specimens of native silver from Chili.

Maj. B. Alvord, Fort Gratiot, Mich. Keg containing fishes from Lake Huron, caught in the vicinity of Fort Gratiot.

Col. J. J. Abert. Box of minerals from Arkansas.

R. J. Pollard, Washington City. Skeleton of gazelle (*Antilope saiga* Pall.) from Turkey.

Thomas Whelpley, Brest, Mich. Cask of Unionidæ and other shells from Lake Erie.

John G. Pendergast, Sackett's Harbor. Box of minerals.

Dr. Jared P. Kirtland, Cleveland, Ohio. Jar of rare salamanders.

Maj. J. H. Carleton, Fort Leavenworth. Skull of bighorn, (*Ovis montana*); horns of bighorn; antlers of black-tailed deer (*Cervus macrotis*); skull of Antelope (*Antilope americana*), from Black Hills, Fort Laramie.

W. Pidgeon, Iowa. Crania and other relics from various aboriginal mounds; paintings in oil on cloth, of various mounds in the Northwest, by a native Sioux Indian.

Robert Howell, Nichols, Tioga County, N. Y. Box of minerals and fossils from Tioga County, N. Y.

THE FOLLOWING SPECIMENS HAVE BEEN RECEIVED FROM COLLECTORS WHO WERE ASSISTED IN THEIR EXPLORATIONS BY THE SMITHSONIAN INSTITUTION:

Augustus Fendler. Collections of plants made in the vicinity of Santa Fé, N. Mex., in 1846 and 1847.

Charles Wright. Plants collected in an expedition from Texas to El Paso in 1849.

Thaddeus Culbertson. Skins, skulls, and skeletons of mammalia from the Upper Missouri fossil vertebrate animals from White River.

Many specimens brought back by Mr. Culbertson were presented to the Institution through him by members of the American Fur Company. Among them may be mentioned Messrs. Alexander Culbertson, Ferdinand Culbertson, Edward T. Denig, Schlegel and Gilbert. Messrs. Denig and F. Culbertson, at the request of Mr. Alexander Culbertson, prepared skins of the grizzly bear and other large mammalia.

---

* This list and summary, not having been included with the Secretary's report in the volume of reprints issued in 1857, is here reproduced, being of much interest as a record of the collections which formed the nucleus of the National Museum.

1851. BAIRD, SPENCER F.—Continued.

THE FOLLOWING SPECIMENS HAVE BEEN DEPOSITED IN CONFORMITY WITH THE PROVISIONS OF THE ACT ESTABLISHING THE SMITHSONIAN INSTITUTION:

General Land Office. Minerals illustrating the geological survey of the mineral region of Lake Superior, by Dr. Charles T. Jackson, contained in nineteen boxes.

Possessing myself large collections in different branches of zoology, I have deposited them with the Institution. The following list contains a brief enumeration of the most important of these. It will thus be evident to the naturalist that the objects already secured by the Smithsonian Institution, if not as numerous as in other collections, are yet valuable as being more than usually complete in certain neglected branches of natural history.

SUMMARY OF SPECIMENS BELONGING TO S. F. BAIRD, AND DEPOSITED BY HIM IN THE CABINET OF THE SMITHSONIAN INSTITUTION.

MAMMALIA.

Skins of the principal mammalia of the Northern and Middle States and of Eastern Europe, with numerous specimens of the smaller species preserved in alcohol.

BIRDS.

A collection of about five hundred species of North American birds in skins, consisting of about twenty-five hundred specimens in the various stages of age, sex, and season.

About two hundred and fifty species of European birds in one thousand specimens.

Eggs of about one hundred and fifty species of North American birds. Duplicates of many of them, in some cases amounting to over a hundred of a single species. The nests accompany the eggs of many of these species.

Nests and eggs of about seventy-five species of European birds, likewise in duplicate.

REPTILES AND FISHES.

A collection of the reptiles and fishes of the United States, at present contained in more than five hundred glass jars, and in numerous barrels, kegs, and tin vessels. Most of these species are represented by numerous specimens, amounting in certain cases to hundreds and even thousands of a single species. No approximation can, at present, be formed as to the number, either of the specimens or of the species. Very many, especially of the fishes, are still undescribed. Most of them have been personally collected in special expeditions to various parts of the country, such as Eastern and Western Pennsylvania, the great lakes, Northern and Eastern Ohio, Southern and Western Virginia, &c.; others have been furnished by contributors in Georgia, Florida, Mississippi, Arkansas, and other States. This collection is especially rich in batrachian reptiles, which are preserved in large numbers in all their peculiar conditions of transformation. There is, in addition to these, a good collection of the fresh-water fishes and reptiles of Central and Eastern Europe.

EMBRYOLOGY.

Embryos of many birds, mammals, and batrachian reptiles.

OSTEOLOGY.

Skulls and skeletons of many North American vertebrata, amounting to some six hundred specimens. A considerable number also belonging to European species.

Also, microscopical sections of teeth and bone of various species of North American vertebrata.

FOSSIL REMAINS.

A large collection of fossil bones from various caves in Pennsylvania and Virginia. This includes nearly all the species of mammalia now living in the United States, with quite a number of those which are now entirely extinct. Chelonian remains likewise in large number.

*   *   *   *   *   *   6   *   *   *

Some general suggestions in regard to future operations of the Smithsonian Institution in the Department of Natural History, as follows:

"I beg leave, in conclusion, to present some general suggestions in regard to future operations of the Smithsonian Institution, in the Department of Natural History. It is a fundamental principle in its organization, as presented in the programme and your annual reports, not to

1851. BAIRD, SPENCER F.—Continued.

attempt complete collections of all natural objects, but rather to gather up such materials for investigation as have been comparatively neglected by others. It may, indeed, be desirable, for purposes of general examination, to have extensive series of specimens from the three kingdoms of nature—animal, vegetable, and mineral—so far as they can be procured and exhibited without undue expense of time, money, and space. For the present, however, attention should be directed mainly to such branches as hitherto may not have had their due share of attention.

"A prominent object in making collections should be to furnish to travelers the means of determining the character of objects collected in various parts of North America.

"Hitherto, officers of the Army returning to Washington have generally been obliged to send or carry these obejects out of the city for the purpose of identification or verification, thus involving a considerable loss of time and credit. These specimens becoming widely scattered rarely return hither, and, when another occasion arises, the whole labor has to be repeated. By retaining them here and combining with them such series of specimens from North America, and other parts of the world, as may be specially procured for the purpose, very little delay in making up reports need hereafter arise. It will, of course, be necessary to call in the aid of the library in procuring all the general and special works which may be required in these investigations. Towards such help the rich collection of transactions of learned societies, already in the library of the Institution, and augmenting daily, will greatly tend.

"Collections illustrating the general natural history of North America become, then, an object of primary importance. Much valuable material of this kind is now on hand and much, it is hoped, will be procured in the various ways hereafter specified. An exceedingly important aid to this is furnished by the act of Congress establishing the Smithsonian Institution, which specifies that all objects of natural history belonging, or hereafter to belong, to the United States, in whosoever's custody the same may be, shall be delivered to such persons as are authorized by the Board of Regents to receive them. This entrusts to the Institution the custody of all collections publicly and officially made; but there are many valuable specimens procured in a private way whose acquisition must depend on the co-operation and assistance of officers of the various expeditions and of heads of departments. Officers stationed at the various military posts have it in their power to do much by procuring the objects of natural history in their vicinity and forwarding them to Washington. It is earnestly hoped that this co-operation may be obtained generally.

"Next in importance to North American objects of nature are those of Europe. The ties uniting the two continents are not merely those of moral, civil, and political relationship, for the connection existing between the natural history of the two is almost as intimate. A large proportion of the genera found in the one occur in the other—often the same species—or those that are very closely allied. This is true of all orders of animals and of most families of plants.

"Next to Europe comes Japan, a region which, in some respects, is more closely allied to our country than even Europe. This is especially the case with respect to reptiles, some of which, as species of Plestiodon and others, have been considered by eminent herpetologists absolutely identical with North American. Unfortunately there are at the present time almost insuperable difficulties in the way of procuring Japanese specimens; the Dutch naturalists being the only ones who have succeeded in exploring even the shores of this country. Little can be done, therefore, except by exchange with the museums of Holland.

"With regard to collections from other countries than those specified, the best rule will be to seek for those series which the other museums of the country do not possess. What these are I do not, at present, feel prepared to state; but hope to have it in my power in a future report to illustrate more fully this subject in a general account of the different collections in North America.

"It may, perhaps, be well to indicate briefly the branches of North American natural history which have received most attention. Mammalia have been ably investigated by Godman, Harlan, Audubon, Bachman, and others; the present state of our knowledge of the subject being exhibited in the works of the two last-named gentlemen. There is, however, no good collection of these animals; that of the Academy of Natural Sciences of Philadelphia being much the best in the country. The private collection of Mr. Audubon is more complete than any other. It is a mortifying fact that this gentleman was obliged to have recourse to foreign museums for the purpose of figuring and describing certain North American species which should have been accessible in one collection, at least, in this country.

"The ornithological collections of the country, both public and private, are very numerous. Among the former that of the Academy of Natural Sciences is by far the best. The New York Lyceum and the Boston Natural History Society have pretty good collections. Of private collections, among the best are those of Messrs. Bell, Giraud, and Lawrence, of New York. The ornithology of North America, east of the Mississippi, has been pretty well worked up, but much remains to be done west of this boundary.

1851. BAIRD, SPENCER F.—Continued.

"General collections of North American reptiles are very rare in this country; that of the Philadelphia Academy, as usual, being the best among public museums.

"Fishes have been preserved in several museums throughout the country. The Boston Natural History Society has the best series of North American marine species. The New York Lyceum comes next. Neither possesses many fresh-water species, being vastly exceeded in this respect by the collections of Professor Agassiz and my own. There is more difficulty in preserving alcoholic specimens, as collections of reptiles and fishes must for the most part necessarily be, than those that are dried; it is to this fact that the scanty representation of these classes of vertebrata is owing.

"Among insects, Coleoptera have been almost exclusively studied. The private collections of Messrs. Leconte, Haldeman, Morris, Harris, Melsheimer, and many others are rich in species. The Messrs. Leconte, father and son, have the largest of these, embracing many hundreds and indeed thousands of undescribed species. The public collection of the Academy of Natural Sciences at Philadelphia and others are of less value. Lepidoptera, or butterflies and moths, come next. The best collection, perhaps, of these is that of Mr. Titian Peale, of Washington. Messrs. Harris, Morris, and Haldeman, and the Academy of Natural Sciences of Philadelphia, have also good collections.

"Comparatively little is known of the other orders of insects. The Neuroptera and Orthoptera of New England have been collected by Dr. Harris; Diptera, Hemiptera, and Hymenoptera have been almost entirely neglected. Say is almost the only American naturalist who has occupied the whole field of entomology.

"Spiders have been ably investigated and abundantly collected by Hentz, who is still continuing his labors in this department. Much, however, remains to be done.

"The Podophthalmian Crustacea are preserved in various cabinets, although many species yet await discovery. Messrs. Say, Dana, and Gibbes are the principal workers in this field. The remaining orders, as Amphipoda, Entomostraca, Isopoda, &c., &c., have been almost wholly neglected.

"The North American worms have never been collected to any extent.

"Of all invertebrata the hard parts of mollusca or shells have received most attention in this country. There are numerous valuable cabinets, public and private, including both domestic and foreign species. The best public collection of American species is probably that of the Academy of Natural Sciences. Among private ones, may be named those of Dr. John C. Jay, John S. Philips, Isaac Lea, Major John Leconte, J. G. Anthony, Professor Haldeman, and others. Most of these gentlemen have had especial reference to Unionidæ in their collections. Nothing, however, has been done towards preserving a series of the animals of shells.

"Very little is known of the Radiata of North America. A few species are preserved in public museums, but by far the most extensive collection is that belonging to Professor Agassiz.

"Phanerogamic plants have received much attention, and the private collections of Doctors Torrey, Gray, and others, with numerous public ones of greater or less extent, leave comparatively little to be desired in this respect. Great additions are continually being received from the country west of the Mississippi, in collections made by officers of the Army and private individuals. Among these should be mentioned Colonel Fremont, Colonel Emory, Captain Stansbury, Major Rich, Messrs. Lindheimer, Wright, Fendler, Gregg, Wislizenus, Drummond, and others.

"Cryptogamic botany has been considerably neglected, until within a few years past. The best collections are in the hands of private individuals, as Messrs. Sullivan, Tuckerman, Curtis, Bailey, Lesquereux, and others. A great deal remains still to be done in this branch of botany. The work of Doctor Harvey on North American Algæ, in preparation for the Smithsonian Institution, will tend greatly to stimulate collectors to pay attention to this order.

"Collections in palaeontology are quite numerous, though principally local. The best general collection is that of the Academy of Natural Sciences. Their museum is incomparably richer, than any other in this country, in collection of fossil vertebrata. The only collection of any extent, of the fossil bones found in the caves of the United States, is in the cabinet of this Institution. Of the interesting Eocene species of the Upper Missouri, Doctor Evans, of Washington, has made an exceedingly valuable collection under direction of the Land Office. Next to this comes a similar one made by Mr. Culbertson for the Smithsonian Institution. An excellent collection of Tertiary fossils is in possession of Prof. F. S. Holmes, of Charleston, S. C. The Tertiary and Cretaceous fossil shells in the Philadelphia Academy are very numerous in species.

"Many of the mineralogical collections of this country are very complete, both as respects domestic and foreign species. Such are the cabinets of Yale College, of the Academy of Natural Sciences, of Dartmouth College, of Bowdoin College, of Messrs. Markoe, Vaux, Clay, Ashmead, Alger, Bouve, and others. The general interest in the subject of mineralogy is such as scarcely to require any additional stimulus, except so far as relates to geology.

1851. BAIRD, SPENCER F.—Continued.

"There are various ways in which collections may be made by the Smithsonian Institution the principal of which are as follows:

"Deposits by government,
" by individuals,
Exchange,
Purchase,
Employment of collectors,
Donations.

"To the first of these, I have already briefly referred. Up to the present time nothing has been received, save the series of specimens illustrating Dr. Jackson's report on the mineral lands of Lake Superior.

"In some collections, specimens deposited by individuals form a conspicuous feature. These, when of considerable extent and completeness, or when illustrating some special researches or publications, are often very important, particularly as they are, in most cases, ultimately presented. Single specimens, unless of much value, are not generally desirable as deposits. Free choice must, of course, be left the Institution, to say what shall be received and what rejected.

"To the individual collector, exchange with other individuals or with societies forms the principal mode of forming his cabinet, beyond what may be personally procurable. This of course implies that the specimens be gathered in larger quantities than would be necessary for a single collection. By a judicious system of exchange, based upon a large stock of duplicates, it seems possible to procure almost any species, domestic or foreign, at little expense beyond that of transportation. To this end it is desirable to secure large numbers of such objects as may be specified hereafter.

"Purchase is an excellent method of increasing a collection in a short time. It not unfrequently happens, however, that acquisitions thus made are of comparatively little value, as is found to be the case in regard to most of the miscellaneous museums, public and private, which are offered for sale. It is of course different with respect to collections made for a specific purpose by practised naturalists, particularly when they contain undescribed species or serve as the types of standard works. Considerable operations of this kind require large sums of money, as will be seen by reference to the annual statement of expenditures made by the British Government in behalf of the National Museum;* and, with numerous

---

* *Expenditures by the British Government for the specimens of natural history in the British Museum.*

FROM 1753 TO 1846, INCLUSIVE.

Natural history in general	£10,405	3	8		
Minerals and fossils	17,238	12	1		
Zoological specimens	12,751	4	11		
Botanical	1,204	11	7		
				£41,599	12	3

1847.

Minerals and fossils	672	2	9		
Zoological specimens	1,295	17	8		
Botanical specimens	31	15	0		
Preparation of specimens	1,317	7	5		
				3,297	2	10

1848.

Minerals and fossils	1,111	16	9		
Zoological specimens	1,085	5	10		
Botanical specimens	40	1	3		
Preparation of specimens	1,259	11	6		
				3,496	15	4

1849.

Minerals and fossils	701	12	0		
Zoological specimens	1,080	6	1		
Botanical specimens	40	8	3		
Preparation of specimens	945	14	7		
				2,768	00	11

Total	51,161	11	2

TOTAL EXPENDITURES OF ALL KINDS, NATURAL HISTORY, SPECIMENS, BOOKS, FINE ARTS, &C.

From 1753 to 1846, inclusive	£816,063	11	00
1847	49,854	7	10
1848	49,845	2	11
1849	47,791	3	4
	963,555	5	1

1851. BAIRD, SPENCER F.—Continued.

drafts on its income, it is not deemed expedient for the Smithsonian Institution ever to do much more for its cabinet by direct purchase. It is confidently believed, too, that the museum will increase almost as rapidly as accommodations can be furnished, by donations of individuals who may have it in their power to make collections, as well as by the special efforts of its officers. This hope is strengthened by the actual experience of other institutions.

"The employment or assistance of collectors in visiting particular portions of country is productive of very important results at very little expense. In illustration of this, I would refer to the acquisitions made by the Institution through Messrs. Lendler, Lindheimer, Wright, Culbertson, and others. In this I am also borne out by my own experience. For several years past I have been in the habit of visiting different portions of the United States, mainly in search of vertebrate animals. Accompanied on such occasions by zealous volunteers, I have succeeded in accumulating very extensive collections, including very many rare and even undescribed species, besides obtaining much valuable information in regard to the general history of animals and plants.

"It is mainly to the employment of collectors that the great European museums owe their richness. In most of these a regular corps is employed continually in traveling through various portions of the world and gathering large numbers of duplicates, which are ultimately distributed in exchange to other institutions.

"In cases where memoirs containing descriptions of animals or plants are presented to the Institution for publication, it should, as far as possible, be made a condition of their acceptance that a series of the objects described be deposited for the purpose of being placed on record and as authenticating the species. These should be labeled by the author, and the names thus attached be ever afterwards retained, even though they may have been incorrect or may have been modified by subsequent discoveries. Individuals, too, should be requested to present similar specimens, to be kept in the same manner, illustrating descriptions published elsewhere than by the Smithsonian Institution.

"At some future period, when the number of duplicates is sufficiently large, it may be possible to furnish lyceums, schools, and other institutions with series of specimens, properly labeled and arranged, of various branches of natural history. Individuals, too, engaged in special investigations may hereafter find it practicable to procure objects in such quantities or of such character as to render material if not indispensable aid. This feature will, however, require the cordial co-operation of naturalists and collectors to render it practicable.

"I may remark that, for the assistance of those who may be unskilled in the collecting, preservation, and packing of specimens, a pamphlet containing the directions is now in preparation and will shortly be issued by the Institution. This will be of considerable size, and, in addition to the merely taxidermical portions, will contain notices of special desiderata in particular portions of the world, a brief indication of the principal divisions of natural history, and notices of the most accessible sources to which the beginner must apply for information respecting the different branches of the subject, the whole illustrated by figures.

"Respectfully submitted.

"DECEMBER 31, 1850."   "SPENCER F. BAIRD.

17.

1851. BAIRD, SPENCER F. [Note prefatory to catalogues of specimens of Natural History collected in the Mauvaises Terres and on the Upper Missouri, by T. A. Culbertson.] < *Fifth Annual Report Smithsonian Institution for the year* 1850, p. 133.

18.

1851. BAIRD, SPENCER F. (*editor*). Proceedings | of | the American Association | for the | Advancement of Science | Fourth Meeting, | held at New Haven, Conn. August 1850. | — | Washington City. | Published by S. F. Baird, | New York: G. P. Putnam. | — | 1851. [Edited by Spencer F. Baird, Permanent Secretary.] 8vo. pp. xxxiv, 415, folding map.

19.

1851. BAIRD, SPENCER F. (*editor*). Proceedings | of | the American Association | for the | Advancement of Science. | Fifth Meeting, | held at Cincinnati, Ohio, May 1851. | — | Published by the liberality of the Citizens of Cincinnati | — | Washington City : | Published by S. F. Baird. | Cincinnati: Ward & Gaylor.] — | 1851. [Edited by Spencer F. Baird, Permanent Secretary.] 8vo. pp. xxiv, 261.

20.

1851. BAIRD, SPENCER F. Iconographic | Encyclopædia | of | Science, Literature, and Art. | Systematically arranged by | J. G. Heck, | translated from the German with additions | and edited by | Spencer F. Baird, A. M., M. D., | Professor of Natural Sciences in Dickinson College, Carlisle, Pa. | Illustrated by five-hundred steel plates, | containing upwards of twelve thousand engravings. | In four vols. | New York: 1852. | Rudolph Garrigue, Publisher | 2 Barclay St., Astor House. In four vols. 8vo of text and two vols. oblong 4to of plates.

Vol. II contains the part relating to "Zoology," pp. 502, plates 74-118.

PREFACE.

"The text of the work which is now presented to the American public is based upon the well known "Bilder Atlas zum Conversations Lexicon," just published in Leipsic, by F. A. Brockhaus, and edited by Mr. John G. Heck. The engravings are impressions from the original steel plates.

"The object steadily kept in view in preparing the Iconographic Encyclopædia has been to furnish a book to which the general reader may apply for an explanation of the principal physical facts which come under his notice. To do this satisfactorily, pictorial representation is necessary, which it is hoped the five hundred quarto plates, with their 12,000 figures, will abundantly furnish.

"Much of the utility of an Encyclopædia depends on its arrangement. The method which the editor's experience of works of this kind has shown to be most convenient, is that of a systematic grouping of distinct treatises, according to their natural affinities. The work thus becomes, as it were, a series of text-books, capable of being used as such, and to which recourse may be had for all the general information required on a given subject.

"To enable the reader, however, to refer readily to any individual fact a copious alphabetical index, or series of indexes, is indispensable. By including numerous cross references, it will be possible to furnish all the facilities of a strictly alphabetical arrangement without any of its disadvantages.

"This, then, is the plan which has been adopted in the arrangement of the Iconographic Encyclopædia. Each article falling within its scope has been treated of independently, and, as far as it goes, is complete in itself. It will not be expected that in the extensive range of subjects involved, even with the exclusion of biography, speculative philosophy, and all abstract sciences in general, any one can be treated in its fullest extent. All that has been aimed at, and indeed all that could have been looked for, was to present a general view of each subject, essentially popular in character, and fitted, more particularly, for those who wish to have the principal facts of numerous works condensed in a single one. Nevertheless, it will be found, on examination, that many of the subdivisions of this Encyclopædia are much fuller in their details than most of the text-books or popular treatises of the day.

"Tables of contents and indexes have been prepared for each volume, and no pains have been spared to make these more than usually accurate. The indexes do not refer to words merely, but to facts and ideas, so that the text can be readily consulted upon any given topic. The lists of the figures on the plates will be found under the contents of the text which they are intended to elucidate, with references to the pages in the letter-press where explanations may be looked for. They furnish an immediate explanation of any figure that may arrest the eye. A glossary of German terms and phrases used in a few of the plates is also added to these lists. It would undoubtedly have been more convenient if the few plates which have caused the necessity of such translations had been re-engraved in English; but the expense of doing so would have more than doubled the price of the work, whose unparalleled cheapness could only be secured by a liberal contract for impressions from the excellent German plates.

"To Mr. Heck belongs exclusively the credit of the conception and execution of the original work; and whether we regard its magnitude, or the regularity and efficiency of its performance, it is one that has rarely, if ever, been excelled.

"In undertaking an English version of the Iconographic Encyclopædia it was soon found that a literal translation of the original would not satisfy the wants of the American public. Written in and for Germany, the different subjects were treated of much more fully in relation to that country than to the rest of the world. In some articles, too, owing to the lapse of time or other causes, certain omissions of data occurred, which did not allow of their being considered as representing the present state of science, or as suiting the wants of the United States. This, therefore, has rendered it necessary to make copious additions, alterations, and abridgements in the respective translations; while, in some instances, it has been thought proper to rewrite entire articles. Several of these original papers have been

1851. BAIRD SPENCER F.—Continued.

prepared by the editor, and the remainder kindly furnished by some of his friends. Some of these again have relieved him of the burden of translating, and have added much to the merit of their work by judicious alterations and additions; while others have revised his MSS. and enriched them with important suggestions. The authority and value of the assistance thus obtained will be sufficiently evident from the names of those who have so kindly rendered it. To all he here takes the opportunity of returning his warmest acknowledgements.

"The second volume, or the one containing botany, zoology, and anthropology, has been entirely rewritten. The articles in it not prepared by the editor are *Invertebrate Zoology*, by Prof. S. S. Haldeman; *Ornithology*, by John Cassin, Esq.; and *Mammalia*, by Charles Girard, Esq.

"The friends to whom he is indebted for careful revision of his MSS. are Prof. Wolcott Gibbs (*Chemistry*); Prof. J. D. Dana (*Mineralogy*); Prof. L. Agassiz (*Geognosy and Geology*); Dr. Asa Gray (*Botany*); Dr. T. G. Wormley (*Anatomy*); and Herman Ludewig, Esq. (*Geography*).

"Those who have assisted him by translating and editing entire articles are Wm. M. Baird, Esq. (*Ethnology of the Present Day*); Major C. H. Larned, U. S. Army (*Military and Naval Sciences*); F. A. Petersen, Esq. (*Architecture*); Prof. Chas. E. Blumenthal (*Mythology and Religious Rites*); Prof. Wm. Turner (*Fine Arts*); and Samuel Cooper, Esq. (*Technology*).

The editor is likewise under very great obligations to the publisher, not only for affording him every facility in the prosecution of his task, but for unwearied and invaluable assistance in the discharge of his editorial duties. He here also takes occasion to acknowledge his indebtedness to Mr. Wm. H. Smith for revision of the proof-sheets and preparation of the alphabetical indexes; and also to Mr. Robert Craighead for the care which he has displayed in the typographical execution.

"WASHINGTON CITY, D. C., *April*, 1851."   "S. F. BAIRD."
[Pp. iii—vi.]

21.

1851. BAIRD, SPENCER F. Outlines | of | General Zoology | Mammals by Charles Girard | Birds " John Cassin | Reptiles " Spencer F. Baird | Fishes " Spencer F. Baird | Invertebrates " S. S. Haldeman | — | Reprinted from the Iconographic Encyclopædia | of Science, Literature, and Art | — ʃ —New York: | Rudolph Garrigue, Publisher, | 2 Barclay St., Astor House | 1851. [8vo. pp. xxii, 502, xvi.]

"This extract constitutes the Zoological portion of the work entitled 'Iconographic Encyclopædia, xxx.' Much of the Encyclopædia, instead of being translated, has been entirely rewritten, with special reference to adaptation to this country. The part on Zoology, among others, has been compiled entirely anew by its authors, and will be found to contain much original matter never before published. The references to the plates are retained in this extract, though the plates themselves are not supplied." (Spencer F. Baird in preliminary "Notice", p. vii.)

The section relating to Fishes, pp. 197-247, contains many allusions to the fishes of North America, besides presenting a comprehensive view of the state of ichthyological science at the time of its preparation, and many important critical biological and economical notes.

The chapter on Reptilia, pp. 244-289, contains a full discussion of the classification of the group and many important biographical and critical observations, particularly so with reference to the Amphibians.

A new genus of salamanders from the Sandwich Islands is described.

Aneides, new genus...p. 256
Aneides lugubris (Hallowell), Baird. R.

22.

1851. BAIRD, SPENCER F. IV. American Ruminants.—On the ruminating animals of North America and their susceptibility of domestication. < *Report of Commissioner of Patents for* 1851, pp. 104-128, 8 plates.

"In the present paper we propose to present, in a few words, the principal characteristics of the ruminating animals of North America, with especial reference to the economical employment of several species, as beasts of burden or draught, as furnishing food of excellent quality, or as yielding valuable materials for the useful arts." ....p. 104.

Tarandus Arcticus, Rich ....105, 2d plate.
"   *Hastalis*, Agassiz ....108.
"   *Furcifer*...109.

1851. BAIRD, SPENCER F.—Continued.

Alces Americana..112, 8th plate.
Elaphus canadensis, Ray..116, 6th plate.
Cervus Macrotis, Say...118.
"   Lewisii, Peale ....118, 5th plate.
"   Leucurus, Douglas ....119.
Capra Americana, Blainville..120, 4th plate.
Antilocapra Americana, Ord ....121, 1st plate.
Ovibos Moschatus, Blainville ....121, 7th plate, 1st fig.
Ovis Montana, Desm..123, 3d plate.
Bison Americanus, Gm ....124, 7th plate, 2d fig.

23.

1852. BAIRD, SPENCER F. Report of Assistant Secretary in charge of the Museum, &c. < *Fourth Annual Report of the Secretary of the Smithsonian Institution, for the year* 1851, pp. 40–52 (Appendices, pp. 52–65).

Name *Siredon lichenigerus*, Baird, n. s., proposed, but no description......................51

24.

1852. BAIRD, SPENCER F. An account of natural history explorations in the United States during 1851. < *Sixth Annual Report Smithsonian Institution, for the year* 1851, 1852, pp. 52–56. Appendix A to Report of Assistant Secretary.

25.

1852. BAIRD, SPENCER F. Directions | for | Collecting, Preserving, and Transporting | Specimens of Natural History | prepared for the use of the Smithsonian Institution. | [Seal.] | Smithsonian Institution: | Washington. | January, 1852. | 8vo. pp. 23.

Included in Smithsonian Catalogue of Publications as No. 34. This number was probably shared by the edition under consideration and a later edition published in 1857.

26.

1852. BAIRD, SPENCER F. (*editor*). Proceedings of the American Association for the Advancement of Science. Sixth meeting, held at Albany (N. Y.), August, 1851.—Published by the liberality of the citizens of Albany.—Washington City: | Published by S. F. Baird. New York: G. P. Putnam. | — | 1852. [Edited by S. F. Baird, Permanent Secretary.] 8vo. pp. lx, 412.

27.

1852. BAIRD, SPENCER F., *and* CHARLES GIRARD. Characteristics of some New Reptiles in the Museum of the Smithsonian Institution. [First Part.] < *Proc. Acad. Nat. Sci. Phila.*, vi, 1852–3 (1854), pp. 68–70. Presented for publication April 20; read April 27, 1852.

"Full descriptions and figures of these species will shortly appear in Capt. Stansbury's Report to Congress on the Great Salt Lake (Utah)." (Preliminary note.)

Siredon lichenoides, Baird, n. s 68
        Lake at the head of Santa Fé Creek, N. Mex. R. H. Rem.
Cnemidophorus tigris, B. and G., n. s 69
        Valley of Great Salt Lake. Capt. Stansbury.
Crotaphytus Wizlizenii, B. and G., n. s.
        Santa Fé. Dr. Wizlizenus.—Between San Antonio and El Paso del Norte. Ool.
        J. D. Graham.
Uta, B. and G., n. g.
Uta Stansburiana, B. and G., n. s.
        Valley of Great Salt Lake. Capt. Stansbury.
Sceloporus graciosus, B. and G., n. s.
        Valley of Great Salt Lake.
Elgaria scincicauda (Skilton), B. & G. R.. 69
Plestiodon Skiltonianum, B. and G., n. s.
        Oregon, with preceding. Rev. George Geary.

1852. BAIRD, SPENCER F., *and* CHARLES GIRARD—Continued.

Phrynosoma platyrhinos, G., n. s.
> Great Salt Lake, Stansbury's party.

Phrynosoma modestum, G., n. s.
> Valley of Rio Grande. Gen. Churchill.—Between, San Antonio, and El Paso del Norte. Col. Graham.

Churchillia, B. and G., n. g 70
Churchillia bellona, B and G., n. s.
> Rio Grande, 1846. Gen. Churchill.

Coluber mormon, B. and G., n. s.
> Valley of Great Salt Lake. Capt. Stansbury.

Heterodus nasicum, B. and G., n. s.
> Texas. Gen. Churchill.

28.

1852. BAIRD, SPENCER F. (*editor*). Zoology [of the Valley of the Great Salt Lake of Utah]. < *Stansbury's Exploration and Survey of the Valley of the Great Salt Lake of Utah, etc.*, Philadelphia, 1852, App. C, pp. 305–378, pll. i–x (Zoological plates numbered separately). For full title, see under STANSBURY, HOWARD.*

This part of the Stansbury Report was issued as a separate in June, 1852, with the following title:

Zoology | of the | Valley of the Great Salt Lake | of | Utah. | Mammals by Prof. S. F. Baird. | Birds by Prof. S. F. Baird. | Reptiles by Prof. Baird and C. Girard. | Insects by Prof. S. S. Haldeman | — | Extracts from Captain H. Stansbury's Report to the U. S. Senate, March 10, 1852. | — | Philadelphia: Lippincott, Grambo & Co. | June, 1852. 8vo. pp. 305–379. No title except that printed on cover.

Though the names of his collaborators appear on the title-page of the Appendix and of the separate, this report seems to have been submitted under the name of Professor Baird, who acted as editor. The introductory chapter, in which the zoological results of the survey are epitomized, pp. 307-8, is signed by him.

---

* 1852. STANSBURY, HOWARD, Captain Corps Topographical Engineers, U. S. Army. Special Session. March, 1851. } Senate. { Executive. No. 3. | — | Exploration and Survey | of the | Valley | or the | Great Salt Lake of Utah, | including | a reconnoissance of a new route through | the Rocky Mountains. | By Howard Stansbury, | Captain Corps Topographical Engineers, | U. S. Army. | Printed by order of the Senate of the United States. | Philadelphia: | Lippincott, Grambo & Co. | 1852. 8vo, *pp.* 487, 34 *plates of scenery, etc., some of them colored; 10 plates geology, numbered i–x; 9 plates botany, numbered i–x; 4 plates botany, numbered i–iv; map at p. 154.—All lithographic.* Report.—(Narrative). pp. 7–269. APP. A. Table of Distances Measured along the route, etc., pp. 270–294. APP. B. Latitudes and longitudes of principal triangulation stations, etc., pp. 295–304. APP. C. Zoology, pp. 305–379. Mammals, by Prof. Spencer F. Baird, pp. 309–313; Birds, by Prof. Spencer F. Baird, pp. 314–335; Reptiles, by Prof. Baird and Charles Girard, pp. 336–365; Insects, by Prof. Haldeman, pp. 366–378; On certain Insect Larvæ, by Titian R. Peale, p. 379. APP. D. Botany, by Prof. John Torrey, pp. 381–398. APP. E. Geology and Paleontology, by Prof. Jas. Hall, pp. 399–414. APP. F. Chemical Analysis, etc., by Dr. L. D. Gale, pp. 414–421. APP. G. Meteorological Observations, pp. 423–478. INDEX, pp. 479–487. *A supplementary volume, without titles, containing two maps.*

Published early in 1852, probably late March or early April.

"An edition of the same work was subsequently printed by the Public Printer for the House of Representatives in 1853. This is much inferior in typography and illustrations."—BAIRD.

1852. STANSBURY, HOWARD. An | expedition | to the |*valley of the Great Salt Lake | of Utah: | including | a description of its geography, natural history, and | minerals, and an analysis of its waters: | with an authentic account of the Mormon settlement. | Illustrated by numerous beautiful plates, | from drawings taken on the spot. | Also a reconnoissance of a new route through the Rocky Mountains and two large and accurate maps of that region. | — | By Howard Stansbury, | Captain Corps Topographical Engineers, United States Army. | — | Philadelphia: | Lippincott, Grambo & Co. | 1852.

A separate edition, issued from the same stereotype and lithographic plates as the official government edition; the text and illustrations are the same.

29.

1852. BAIRD, SPENCER F. Mammals [of the Valley of the Great Salt Lake]. <*Stansbury's Exploration and Survey of the Valley of the Great Salt Lake of Utah, etc.*, Philadelphia, 1852 [App. C.] pp. 309–313.

    1. *Vulpes macrourus*, Baird, n. s... 309
                 Salt Lake Valley. Capt. Stansbury.
    2. *Pistorius vison*, Linn 311
    3. *Pistorius erminea*, Linn.
    4. *Meles labradoria*, Sabine.
    5. *Gulo luscus*, Linn.
    6. *Fiber zibethicus*, Linn.. 312
    7. *Spermophilus 13 lineatus*, Mitchell.
    8. *Ovis montana*, Desm.

30.

1852. BAIRD, SPENCER F. (Mammals) Collected by Lieutenant Abert (in New Mexico). < *Stansbury's Exploration and Survey of the Valley of the Great Salt Lake of Utah, etc.* Philadelphia, 1852. [App. C.] p. 313.

    1. *Pseudostoma castanops*, Baird, n. s. 313
             Prairie road to Bent's Fort, N. Mex. Lieut. Abert.
    Included in the report on the mammals of the Valley of the Great Salt Lake.

31.

1852. BAIRD, SPENCER F. Birds [of the Valley of the Great Salt Lake]. < *Stansbury's Exploration and Survey of the Valley of the Great Salt Lake of Utah, etc.* Philadelphia, 1852. [App. C.] pp. 314–325 [+ 325–335 extraneous bird matter].

    1. *Buteo borealis*, Bp 314
    2. *Accipiter fuscus*. Bp.
    3. *Athene hypugœa*, Cassin.
    4. *Sialia macroptera*, Baird, n. s.
         Salt Lake City, March 18, 1850. Capt. H. Stansbury.
    5. *Parus septentrionalis*, Harris.. 316
    6. *Sturnella neglecta*, Aud.
    7. *Niphœa oregona*, Aud.
    8. *Pencœa lincolnii*. Aud.. 317
    9. *Leucosticte tephrocotis*, Bp.
    10. *Otocoris occidentalis*, McCall 318
    11. *Picus torquatus*, Wils 319
    12. *Tetrao urophasianus*, Bp.
    13. *Charadrius vociferus*, Linn.
    14. *Grus canadensis*, Temm.
    15. *Botaurus lentiginosus*, Montagu... 320
    16. *Numenius longirostris*, Wils.
    17. *Symphemia semipalmata*, Hart.
    18. *Recurvirostra americana*, Gm.
    19. *Cygnus americanus*, Sharpless 321
    20. *Anser erythropus*, L.
    21. *Anser canadensis*, Vieill.
    22. *Anas boschas*, Linn 322
    23. *Mareca americana*, Steph.
    24. *Querquedula carolinensis*, Bp.
    25. *Pterocyanea rafflesii*, King.
    26. *Dafila acuta*, Bp... 323
    27. *Fuligula affinis*, Eyton 324
    28. *Clangula albeola*, Bp.
    29. *Pelecanus trachyrrhynchus*, Lath.
    30. *Phalacrocorax dilophus*, Sw.
    31. *Colymbus glacialis*, L.

31.

1852. BAIRD, SPENCER F.   Birds collected in New Mexico by Lieut. Abert. *< Stansbury's Exploration and Survey of the Valley of the Great Salt Lake of Utah, etc.*
Philadelphia, 1852. [App. C.] pp. 325, 326.

*1. *Falco sparverius*, L 325
 2. *Pipilo aberti*, Baird, n. s.
            New Mexico. Lieut. J. W. Abert.
*3. *Agelaius xanthocephalus*, L...826
*4. *Picus varius*, L.
*5. *Columba leucoptera*, L.
*6. *Callipepla squamata*, Vig.
 7. *Callipepla gambeli*, Nutt.
*8. *Actiturus bartramius*. Wils.
 9. *Recurvirostra occidentalis*, Vig.

32.

1852. BAIRD, SPENCER F.   List of birds inhabiting America, west of the Mississippi,
not described in Audubon's Ornithology. *< Stansbury's Exploration and Survey of the Valley of the Great Salt Lake, etc.* Philadelphia, 1852. [App. C.]
pp. 327–335.

"The list includes a few specimens recently described from the regions east of the Mississippi."

153 species are enumerated by name, with citation of description and note on habitat.

"This list of 153 spp. contains a large proportion of synonyms or species not since satisfactorily determined to inhabit North America north of Mexico. The list makes no apparent
claim to critical precision, ostensibly showing which species have been ascribed to the region
in question, but not necessarily vouching for their occurrence there. 'California' long remained a vague term with ornithologists." COUES, 1878.

BIRDS INHABITING AMERICA, WEST OF THE MISSISSIPPI, NOT DESCRIBED
IN AUDUBON'S ORNITHOLOGY.

Archibuteo ferrugineus, Licht ....p. 327
Rosthramus sociabilis, Vieill.
Strix frontalis, Licht.
Acanthylis vauxii, Towns.
Chordeiles brasilianus, (Gm.)
Antrostomus nuttalli, Aud.
Ceryle americana, Boie.
Ornismya costæ, Bourcier.
Conirostrum ornatum, Lawr.
Picolaptes brunneicapillus, Laf.
Troglodytes albifrons, Giraud.
Vireo huttoni, Cassin ....p. 328
Vireo belli, Aud.
Vireo atricapilla, Woodhouse.
Vireosylva philadelphica, Cassin.
Vireosylva altiloqua, Vieill.
Sialia macroptera, Baird.
Lanius elegans, Sw.
Lanius excubitroides, Sw.
Hypocolius ampelinus, Bp.
Icteria velasquezii, Bp.
Oulicivora atricapilla, Sw.
Sylvicola olivacea, Giraud.
Vermivora brevipennis, Giraud.
Turdus rufopalliatus, Lafresn.
Merula olivacea, Brewer.
Mimus leucopterus, Vig.
Mimus longirostris, Lafresn.
Toxostoma rediviva, Gambel.
Toxostoma curvirostris, Swainson ....p. 329
Toxostoma lecontei, Lawr.

* Name only.

1852. BAIRD, SPENCER F.—Continued.

Motacilla leucoptera, Vig. Zool. of Blossom.
Agrodoma spraguei, Aud.
Saxicola œnanthoides, Vig.
Saurophagus sulphuratus, Swainson.
Saurophagus bairdii, Gambel .... p. 329
Tyrannus cassinii, Lawrence.
Tyrannula cayanensis. Gm.
Tyrannula lawrenceii, Giraud.
Tyrannula cinerascens, Lawrence.
Tyrannula flaviventris, Baird.
Tyrannula minima, Baird.
Pyrocephalus rubineus, Bodd.
Setophaga vulnerata, Wagler.
Setophaga belli, Giraud.
Setophaga rubra, Swainson.
Setophaga picta, Swainson.
Setophaga rubrifrons, Giraud.
Embernagra rufivirgata, Lawrence .... p. 330
Embernagra blandingiana, Gambel.
Saltator rufiventris, Vig. Zool.
Euphonia elegantissima, Bp.
Spermophila albogularis, Swainson.
Rhamphopis flammigerus, Jard.
Chrysopoga typica, Bp.
Fringilla meruloides, Vig.
Zonotrichia querula, Nutt.
Zonotrichia gambeli, Nutt.
Zonotrichia cassinii, Woodhouse.
Chrysomitris lawrenceii, Cassin.
Pipilo fusca, Sw.
Pipilo oregona, Bell.
Pipilo aberti, Baird.
Emberiza lecontei, Aud.
Emberiza bairdii, Aud.
Emberiza bilineata, Cassin.
Emberiza belli, Cassin. .... p. 331
Carpodacus obscurus, McCall.
Carpodacus familiaris, McCall.
Coccothraustes ferreo-rostris, Vig.
Cardinalis sinuatus, Bp.—Lawrence.
Pyrhula inornata, Vig.
Leucosticte greiseinucha, Brandt.
Plectrophanes maccownii, Lawrence.
Passerella unalaschensis, Bp.
Passerella rufina, Brandt.
Euspiza arctica, Bp.
Alauda rufa, Lath.
Otocoris occidentalis, McCall.
Sturnella neglecta, Aud.
Quiscalus macrourus, Sw.
Scolecophagus mexicanus, Sw.
Pendulinus californianus, Less.
Psarocolius auricollis, De Wied.
Xanthornus mexicanus, Briss.
Xanthornus affinis, Lawrence.
Icterus cucullatus, Sw.
Icterus melanocephalus, Wagler.
Icterus vulgaris, Daud.
Icterus frenatus, Licht.
Chamea fasciata, Gambel.
Lophophanes septentrionalis, Harris.

2 BD

1852. BAIRD, SPENCER F.—Continued.

Lophophanes inornatus, Gambel.
Lophophanes wollweberi, Bp.
Lophophanes atricristatus, Cassin.
Parus montanus, Gambel.
Gymnokitta cyanocephalla, De Wied.
Cyanocorax coronatus, Sw.
Cyanocorax luxuosus, Lesson.
Cyanocorax cassinii, McCall.
Garrulus californicus, Vig .... **p. 333**
Pica beecheyii, Vig.
Crotophaga ——— ?
Piaya cayanensis, Gambel.
Geococcyx affinis, Hartlaub.
Geococcyx viaticus, Wagler.
Melanerpes albolarvatus, Cassin.
Melanerpes formicivorus, Swainson.
Centurus santacruzii, Bp.
Centurus flaviventris, Swainson.
Centurus elegans, Sw.
Colaptes mexicanoides, Lafres.
Colaptes ayresii, Aud.
Colaptes collaris, Vig.
Picus scapularis, Vig.
Picus nuttallii, Gambel.
Picus scalaris, Wagler.
Picus lecontei, Jones.
Columba solitaria, McCall.
Columba flavirostris, Wagler. .... **p. 334**
Penelope policephala, Wagler.
Ortalida vetula, Wagler.
Cyrtonix massena, Gould.
Callipepla gambeli, Nutt.
Callipepla picta, Dougl.
Callipepla elegans, Less.
Callipepla douglassii, Vig.
Callipepla squamata, Vig.
Strepsilas melanocephalus, Vig.
Numenius rufiventris, Vig.
Macrorhamphus scolopaceus, **Lawrence.**
Recurvirostra occidentalis, Vig.
Anser nigricans, Lawr.
Anas europhasianus, Vig.
Dendrocygna arborea? Penn.
Dendrocygna autumnalis, Eyton.
Cyanopterus rafflesii, King.
Oidemia velvetina, Cassin .... **p. 335**
Larus brachyrhynchus, Gould.
Larus belcheri, Vig.
Sterna elegans, Gambel.
Sterna caspia, L.—**Lawrence.**
Procellaria meridionalis, Lawrence.
Thalassidroma furcata, Lath.
Thalassidroma fregetta, Kuhl.
Phalacrocorax perspicillatus, Pall.
Phalacrocorax penicillatus, Brandt.
Uria brevirostris, Vig.
Mergulus cirrocephalus, Vig.
Mergulus cassinii, Gambel.
Ptychorhamphus aleuticus, Brandt.
Brachyrhamphus wrangelli, Brandt.
Brachyrhamphus brachypterus, Brandt.

33.

1852. BAIRD, SPENCER F., *and* CHARLES GIRARD. Reptiles [of the Valley of the Great Salt Lake]. < *Stansbury's Exploration and Survey of the Valley of the Great Salt Lake.* Philadelphia, 1852. [App. C.] pp. 336–353 [+ 354–365 by GIRARD. A Monographic Essay on the genus Phrynosoma.] plates in following order: i, ii, iii, vi, v, iv [+ viii, vii in Girard's paper].

These species were described in a previous paper, No. 27.

Siredon lichenoides, Baird ....p. 336, pl. i.
Cnemidophorus tigris, B. and G.......................................p. 338, pl. ii.
Crotaphytus, Holbrook ....p. 339.
Crotaphytus Wislizenii, B. and G....................................p. 340, pl. iii.
Hoolbrookia, Girard ....p. 341.
Holbrookia maculata, Girard...p. 342, pl. vi, fig. 1–3.
Uta, Baird and Girard ....p. 344.
Uta Stansburiana, B. and G..p. 345, pl. v, fig. 4–6.
Sceloporus graciosus, B. and G......................................p. 346, pl. v, fig. 1–3.
Elgaria scincicauda, B. and G.......................................p. 348, pl. iv, fig. 1–3.
Plestiodon Skiltonianum, B. and G...................................p. 349, pl. iv, fig. 4–6.
Churchillia, B. and G...p. 350.
Churchillia bellona, B. and G.
Coluber mormon, B. and G..p. 351.
Heterodon nasicus, B. and G...p. 352.

34.

1852. BAIRD, SPENCER F. [Note in reference to Vulpes Utah, Aud. and Bach.] < *Proc. Acad. Nat. Sci. Phila.*, vi, 1852–3 (1854), p. 124. Read Aug. 3, 1852.

Abstract of communication claiming priority for name *Vulpes macrourus*, Baird, over *Vulpes utah*, Aud. and Bach.

35.

1852. BAIRD, SPENCER F., *and* CHARLES GIRARD. Characteristics of some New Reptiles in the Museum of the Smithsonian Institution. By Spencer F. Baird and Charles Girard. Second Part. Containing the species of the Saurian order, collected by John H. Clark, under Col. J. D. Graham, head of the Scientific Corps, U. S. and Mexican Boundary Commission, and a few others from the same adjoining territories, obtained from other sources, and mentioned under their special headings. < *Proc. Acad. Nat. Sci. Phila.*, vi, pp. 125–9. Presented for publication July 6, ordered printed Aug. 31, 1852.

Holbrookia texana (Trosch), B. and G. R............................... 125
Holbrookia affinis, B. and G., n. s.
    Along the Rio San Pedro.
Holbrookia propinqua, B. and G., n. s 126
    Texas, between Indianola and San Antonio.
Crotaphytus Gambelii, B. and G., n. s.
    California. Wm. Gambel.
Crotaphytus dorsalis, B. and G., n. s.
    Desert of Colorado, Cal. J. L. Leconte.
Uta ornata, B. and G., n. s.
    Rio San Pedro, Tex.
Sceloporus Poinsettii, B. and G., n. s 127
    Rio San Pedro, and the Province of Sonora.
Sceloporus Clarkii, B. and G., n. s.
Sceloporus Thayerii, B. and G., n. s.
    Indianola, Tex.
Sceloporus dispar, B. and G., n. s.
    Vera Cruz. Dr. Burroughs.
Cnemidophorus marmoratus, B. and G., n. s........................... 128
    Between San Antonio, Texas, and El Paso del Norte.
Cnemidophorus Grahamii, B. and G., n. s.
    Found with preceding species.

1852. BAIRD, SPENCER F., *and* CHARLES GIRARD—Continued.

Cnemidophorus gularis, B. and G., n. s.
Indianola, Tex., and the Valley of the Rio San Pedro.
Cnemidophorus perplexus, B. and G.
Valley of the Rio San Pedro.
Cnemidophorus gracilis, B. and G.
Desert of Colorado. J. L. Leconte.
Cnemidophorus præsignis, B. and G., n. s.
Chagres. C. B. Adams.
Plestiodon obsoletum, B. and G., n. s.
Valley of the Rio San Pedro.
Elgaria nobilis, B. and G., n. s.
Fort Webster, New Mexico.

36.

1852. BAIRD, SPENCER F., *and* CHARLES GIRARD. Characteristics of some New Reptiles in the Museum of the Smithsonian Institution. By Spencer F. Baird and Charles Girard. Third Part. Containing the Batrachians in the collection made by J. H. Clark, Esq., under Col. J. D. Graham, on the United States and Mexican Boundary. < *Proc. Acad. Nat. Sci. Phila.*, vi, p. 173. Presented for publication Oct. 21, ordered printed Oct. 26, 1852.

Amblystoma proserpine, B. and G., n. s... 178
On the route from Montgomery, Mexico.
Rana areolata, B. and G., n. s.
Rio San Pedro of the Gila.
Bufo punctatus, B. and G., n. s.
Rio San Pedro of the Rio Grande del Norte.
Bufo granulosus, B. and G., n. s.
Between Indianola and San Antonio, Texas.

37.

1852. BAIRD, SPENCER F., *and* CHARLES GIRARD. Descriptions of New Species of Reptiles, collected by the U. S. Exploring Expedition under the command of Capt. Charles Wilkes, U. S. N. First Part.—Including the species from the Western Coast of America. < *Proc. Acad. Nat. Sci. Phila.*, vi, pp. 174–177. Presented for publication Oct. 6, ordered printed Oct. 26, 1852.

Amblystoma tenebrosum, B. and G., n. s 174
Oregon.
Rana aurora, B. and G., n. s.
Puget Sound.
Rana draytonii, B. and G., n. s.
San Francisco, Cal., and on Columbia River.
Hyla regilla, B. and G., n. s.
Sacramento River, in Oregon and Puget Sound.
Bufo boreas, B. and G., n. s.
Columbia River and Puget Sound.
Sceloporus gracilis, B. and G., n. s... 175
Oregon.
Sceloporus occidentalis, B. and G., n. s.
California, probably Oregon.
Sceloporus frontalis, B. and G., n. s 175
Puget Sound.
Elgaria principis, B. and G., n. s.
Oregon and Puget Sound.
Elgaria formosa, B. and G., n. s.
California.
Elgaria grandis, B. and G., n. s.. 176
Tropidonotus ordinoides, B. and G., n. s.
Puget Sound.
Wenona, B. and G., n. g.
Wenona isabella, B. and G., n. s.
Puget Sound.

1852. BAIRD, SPENCER F., *and* CHARLES GIRARD—Continued.
　Calamaria tenuis, B. and G., n. s.
　　Puget Sound.
　Crotalus lucifer, B. and G., n. s.
　　Oregon and California.
　Emys marmorata, B. and G., n. s.
　　Puget Sound.

38.

1853. BAIRD, SPENCER F., *and* CHARLES GIRARD. List of Reptiles collected in California by Dr. John L. Leconte, with description of New Species. < *Proc. Acad. Nat. Sci. Phila.*, vi, pp. 300–302. Presented for publication Feb. 15, ordered printed Feb. 22, 1853

OPHIDIANS.

1. *Crotalus lucifer ? ?*, B. and G 300
　San Diego, Cal.
2. *Eutainia ordinoides*, B. and G.
　San Francisco.
3. *Bascanion vetustus*, B. and G., n. s. (name only).
　San José.
4. *Pituophis annectens*, B. and G., n. s.
　San Diego.
5. *Rhinocheilus Lecontei*, B. and G., n. s.
　San Diego.
6. *Contea mitis*, B. and G., n. s.
　San José.
7. *Diadophis amabilis*, B. and G., n. s.
　San José.
8. *Rena humilis*, B. and G., n. s.
　Vallecitas, Cal.

SAURIANS.

1. *Crotaphytus dorsalis*, B. and G 301
　Desert of Colorado.
2. *Sceloporus occidentalis*, B. and G.
　San Francisco.
3. *Uta Stansburiana*, B. and G.
　Valley of Great Salt Lake.
4. *Uta ornata*, B. and G.
　San Diego and San Francisco.
5. *Phrynosoma coronatum*, Blainv.
　San Diego.
6. *Cnemidophorous gracilis*, B. and G.
　Desert of Colorado.
7. *Elgaria scincicauda*, B. and G.
　California.
8. *Plestiodon Skiltonianum*, B. and G.
　San Diego.
9. *Anniella pulchra*, Gray.
　San Diego.

BATRACHIANS.

1. *Bufo halophila*, B. and G., n. s.. 301
　Benicia.
2. *Hyla regilla*, B. and G.
　San Francisca or San Francisco.
3. *Litoria occidentalis*, B. and G., n. s.
　San Francisco.
4. *Rana Lecontei*, B. & G., n. s.
　San Francisco.
5. *Aneides lugubris*, Baird.
　San Francisco.
7. *Taricha laevus*, B. and G., n. s.
　San Francisco.

39.

1853. BAIRD, SPENCER F., *and* CHARLES GIRARD. Catalogue | of | North American Reptiles | in the Museum of the | Smithsonian Institution. | — | Part I.—Serpents. | By | S. F. Baird and C. Girard. | — | [Seal of the Smithsonian Institution.] | Washington: | Smithsonian Institution. | January, 1853. | January 1, 1853. [Accepted for publication November, 1852.] 8vo, pp. xvi, 172.

CONTENTS.

Although published under the names of the two authors as a joint production, this work is a series of monographs for which the individual authors only claimed to be responsible. The record of authorship is lost. *In the following list of species discussed, the parts written by Professor Baird, so far as he can recall the matter to memory, are printed in italic type, those by Girard in heavy-face type.*

"In the present catalogue it is proposed to present a systematic account of the collection of North American Serpents in the museum of the Smithsonian Institution. In the Appendix will be found such species not in possession of the Institution, as could be borrowed for description, as well as notes on more or less authentic species of which no specimens could be found.

"A complete synonymy of all the species has not been attempted, as tending to swell the bulk of a catalogue too much. All those, however, necessary to a proper understanding of the history or character of the species, have been introduced, and all the synonyms quoted have been actually verified by original reference.

"Owing to the want of osteological preparations, it has been a difficult task to arrange the genera in a natural succession. In many cases forms are now combined which will hereafter necessarily be widely separated. The almost entire deficiency of modern general works upon the *Colubridæ*, has also been a serious obstacle to any correct idea of a natural system. The forthcoming work of M. M. Duméril will undoubtedly clear up much of the obscurity which now exists. But when systematic writers all carefully avoid the subject of the Ophidians, each waiting for the others to make the first step, the attempt to combine genera by well-marked, though perhaps artificial points of relation, will it is hoped be looked upon with indulgence, even after more comprehensive and extended investigations shall render it necessary to break up the combinations here adopted.

"The collections upon which the original descriptions of the present catalogue have been based are as follows :—

Spencer F. Baird.   Species from Massachusetts, New York, Ohio, and Pennsylvania.
Charles Girard.   Maine, Massachusetts, and South Carolina.
Rev. Charles Fox.   Species from Eastern Michigan.
Dr. P. R. Hoy.   Species from Eastern Wisconsin.
Prof. L. Agassiz.   Lake Superior, Lake Huron, and Florida.
Dr. J. P. Kirtland.   Northern Ohio.
G. W. Fahnestock.   Western Pennsylvania.
Miss Valeria Blaney.   Eastern Shore of Maryland.
Dr. C. B. R. Kennerly.   Northern Virginia.
John H. Clark.   Maryland, Texas, New Mexico, and Sonora.

1853. BAIRD, SPENCER F., *and* CHARLES GIRARD—Continued.

John Varden. District of Columbia and Louisiana.
Dr. J. B. Barratt. Western South Carolina.
Miss Charlotte Paine and Mrs. M. E. Daniel. Western S. Carolina.
Dr. S. B. Barker. Charleston, S. C.
Prof. F. S. Holmes and Dr. W. J. Burnett. South Carolina.
R. R. Cuyler and Dr. W. L. Jones. Georgia.
D. C. Lloyd. Eastern Mississippi.
Dr. B. F. Shumard and Col. B. L. C. Wailes. Mississippi.
James Fairie. Mexico and Western Louisiana.
Capts. R. B. Marcy and G. B. McClellan, U. S. A. Red River, Ark.
Ferdinand Lindheimer. Central Texas.
Col. J. D. Graham, U. S. A. The specimens collected while on the U. S. and Mex. Bound-
ary Survey, by Mr. J. H. Clark, viz., in Texas, New Mexico, and Sonora.
Maj. W. H. Emory. Specimens collected on the U. S. and Mexican Boundary Survey, by
Arthur Schott, at Eagle Pass, Tex.. and by J. H. Clark, in Texas and New Mexico.
Gen. S. Churchill, U. S. A. Valley of the Rio Grande.
Dr. L. Edwards, U. S. A. Northern Mexico.
Dr. Wm. Gambel. New Mexico and California.
Dr. John L. Le Conte. Littoral California.
Dr. C. C. Boyle and J. S. Bowman. Central California.
Dr. A. J. Skilton. Species collected in California by Henry Moores, Esq.
U. S. Exploring Expedition. Littoral California and Oregon.
Academy of Natural Sciences of Philad. Various unique specimens described by Dr. Hol-
brook.
Boston Society of Natural History. California.
"SMITHSONIAN INSTITUTION, *January* 5, 1853."

SYSTEMATIC INDEX OF WELL-ASCERTAINED SPECIES OF NORTH AMERI-CAN SERPENTS.

CROTALUS, Linn.

CROTALOPHORUS, GRAY.

AGKISTRODON, Beauv.

TOXICOPHIS, Troost.

ELAPS, FITZ.

EUTAINIA, B. & G.

OPHIBOLUS, B. & G.

1. *Ophibolus Boylii*, B. & G., n. s.............California........................ 82
2. " *splendidus*, B. & G., n. s...........Sonora 83
3. " . *Sayi*, B. & G.....................La., Miss., Ark., Texas............. 84
4. " *getulus*, B. & G ....S. C., Miss 85
5. " *rhombomaculatus*, B. & G..........Ga., S. C 86
6. " *eximius*, B. & G...................Mass., N. Y., Penna............ 87
7. " *clericus*, B. & G., n. s...............Va., Miss 88
8. " *doliatus*, B. & G ....Mississippi 89
9. " *gentilis*, B. & G., n. s...............Ark., La 90

GEORGIA, B. & G.

1. *Georgia Couperi*, B. & G ....Georgia................... 92
2. " *obsoleta*, B. & G..................Texas 158

BASCANION, B. & G.

1. *Bascanion constrictor*, B. & G ....Penna., Md., Miss., S. C., La........... 93
2. " *Fremontii*, B. & G., n. s...........California................... 95
3. " *Foxii*, B. & G., n. s.................Mich., Penna 96
4. " *flaviventris*, B. & G ....Texas, Cal 96
5. " *vetustus*, B. & G., n. s.............Cal., Oregon 97

MASTICOPHIS, B. & G.

1. *Masticophis flagelliformis*, B. & G ....South Carolina...............98, 149
2. " *flavigularis*, B. & G ....Texas, Ark 99
3. " *mormon*, B. & G....................Utah...................... 101
4. " *ornatus*, B. & G., n. s.............Texas 102
5. " *tæniatus*, B. & G.................California.................. 103
6. " *Schottii*, B. & G., n. s.............Texas 160

SALVADORA, B. & G.

1. *Salvadora Grahamiæ*, B. & G., n. s.........Sonora................... 104

LEPTOPHIS, BELL.

1. *Leptophis æstivus*, BELL ....Md., Va., S. C., Miss............ 106
2. " *majalis*, B. & G., n. s..............Texas, Ark 107

CHLOROSÓMA, WAGL.

1. *Chlorosoma vernalis*, B. & G...............Me., Mass., N. Y., Penna., Mich., Wisc.,
Miss.................................. 108

CONTIA, B. & G.

1. *Contia mitis*, B. & G., n. s..................Cal., Oregon 110

DIADOPHIS, B. & G.

1. *Diadophis punctatus*, B. & G................N. Y., Penna., Ga., S. C., Miss 112
2. " *amabilis*, B. & G., n. s...........California................... 113
3. " *docilis*, B. & G., n. s..............Texas 114
4. " *pulchellus*, B. & G., n. s...........California.................. 115
5. " *regalis*, B. & G., n. s.............Sonora.................... 115

LODIA, B. & G.

1. *Lodia tenuis*, B. & G......................Oregon 116

SONORA, B. & G.

1. *Sonora semiannulata*, B. & G., n. s.........Sonora.................... 117

RHINOSTOMA, FITZ.

1. *Rhinostoma coccinea*, HOLBR..............S. C., Ga., Miss., La 118

RHINOCHEILUS, B. & G.

1. *Rhinocheilus Lecontii*, B. & G., n. s.........California.................... 120

1853. BAIRD, SPENCER F., *and* CHARLES GIRARD—Continued.

HALDEA, B. & G.

1. Haldea striatula, B. & G..................:.......Va., S. C., Miss.   122

FARANICA, GRAY.

1. Farancia abacurus, B. & G ....:........ S. C., La   123

ABASTOR, GRAY.

1. Abastor erythrogrammus, GRAY.......... Ga.......................................   125

VIRGINIA, B. & G.

1. Virginia Valeriæ, B. & G., n. s ....Md., Va., S. C   127

OELUTA, B. & G.

1. Celuta amœna, B. & G......................Penna., Md., Va., S. C., Miss............   129

TANTILLA, B. & G.

1. Tantilla coronata, B. & G., n. s...........Mississippi   131
2. "   gracilis, B. & G., n. s.................Texas   132

OSOEOLA, B. & G.

1. Osceola elapsoidea, B. & G ....S. C., Miss   133

STORERIA, B. & G.

1. Storeria Dekayi, B. & G.....................Wisc., Mich., Ohio, Mass., N. Y., Pa.,
    Md., S. C., Ga., La., Tex   135
2. "   occipito-maculata, B. & G ....Me., N. Y., Lake Sup., Wisc., Pa., S. C.,
    Ga....................................   137

WENONA, B. & G.

1. Wenona plumbea, B. & G.................Oregon   139
2. "   isabella, B. & G ....Oregon   140

RENA, B. & G.

1. Rena dulcis, B. & G., n. s...................Texas   142
2. "   humilis, B. & G., n. s.................California...............................   143

SUMMARY.

GENERA. Old   13
    New   22
                                                                 ——
             Total   35
SPECIES. Old   65
    New ....:.   54
                                                                 ——
             Total...   119

40.

1852. BAIRD, SPENCER F. Description of a new Species of Sylvicola. < *Ann. Lyc. Nat. Hist. N. Y.*, v, 1852, pp. 217, 218, pl. vi.
    Sylvicola Kirtlandii, Baird, n. s.. p. 217, pl. vi.
        Shot near Cleveland, Ohio, by Mr. Charles Pease, May 13, 1851.

41.

1853. BAIRD, SPENCER F. Report of the Assistant Secretary, in charge of the Museum, &c. < *Seventh Annual Report Smithsonian Institution for the year* 1852, 1853, pp. 45–58. Appendices, pp. 58–73.

42.

1853. BAIRD, SPENCER F. Account of scientific explorations and reports on explorations, made in America during the year 1852. *< Seventh Annual Report Smithsonian Institution for the year* 1852, 1853, pp. 58–65. Appendix A of Assistant Secretary's Report.

43.

1853. BAIRD, SPENCER F., and CHARLES GIRARD. Descriptions of some new fishes from the River Zuni. *< Proc. Acad. Nat. Sci. Phila.*, vi, 1853, pp. 368, 369.

Gila, B. & G., n. g.. 368
   1. *Gila robusta*, B. and G., n s... 369
      River Zuni.
   2. *Gila elegans*, B. and G., n. s.. 369
      River Zuni.
   3. *Gila gracilis*, B. and G., n. s... 369
      River Zuni.

44.

1853. BAIRD, SPENCER F., and CHARLES GIRARD. "A communication * * * upon a species of frog, and another of toad * * * recently described from specimens in the Herpetological Collections of the U. S. Exploring Expedition." *< Proc. Acad. Nat. Sci. Phila.*, vi, 1853, pp. 378–379. Read Aug. 9, 1853.

Rana pretiosa, B. and G., n. s.
   Puget Sound.
Bufo columbianus, B. and G., n. s.
   Columbia River.

45.

1853. BAIRD, SPENCER F., and CHARLES GIRARD. Descriptions of New Species of Fishes collected by Mr. John H. Clark, on the U. S. and Mexican Boundary Survey, under Lt. Col. Jas. D. Graham. *< Proc. Acad. Nat. Sci. Phila.*, vi, 1853, pp. 387–390. Read Aug. 30, 1853.

   1. *Pileoma carbonaria*, B. and G., n. s... 387
      Rio Salado, Texas.
   2. *Boleosoma lepida*, B. and G., n. s... 388
      Upper tributaries of the Rio Nueces, Texas.
   3. *Pomotis aquilensis*, B. and G., n. s.
      Eagle Pass, Texas.
   4. *Catostomus latipinnis*, B. and G., n. s.
      Rio San Pedro, of the Rio Gila.
   5. *Gila Emoryi*, B. and G., n. s.
      Near mouth of the Gila.
   6. *Gila Grahamii*, B. and G., n. s.. 389
      Rio San Pedro, of the Gila.
   7. *Fundulus grandis*, B. and G., n. s... 389
      Brackish waters in the vicinity of Indianola, Tex.
   8. *Fundulus tenellus*, B. and G., n. s.. 389
      Prairie Mer Rouge, La., and Russellville, Ky.
   9. *Hydrargyra similis*, B. and G., n. s... 389
      Brackish waters in the vicinity of Indianola.
   10. *Cyprinodon elegans*, B. and G., n. s.. 389
      Rio Grande del Norte.
   11. *Cyprinodon macularius*, B. and G., n. s... 389
      Rio Gila.
   12. *Cyprinodon bovinus*, B. and G., n. s.. 389
      Leon's Springs, Rio Grande del Norte.
   13. *Cyprinodon gibbosus*, B. and G., n. s... 390
      Brackish waters of Indianola.
   14. *Heterandria affinis*, B. and G., n. s.. 390
      Rio Medina and Rio Salado.

1853. BAIRD, SPENCER F., *and* CHARLES GIRARD—Continued.

46.

1853. BAIRD, SPENCER F., *and* CHARLES GIRARD. Description of New Species of Fishes collected by Captains R. B. Marcy and Geo. B. McClellan in Arkansas. < *Proc. Acad. Nat. Sci. Phila.*, vi, 1853, pp. 390–392. Read Aug. 30, 1853.

47.

1853. BAIRD, SPENCER F., *and* CHARLES GIRARD. Fishes [of the Zuñi River]. < *Sitgreaves' Report of an Expedition down the Zuñi and Colorado Rivers*. Washington, 1853. pp. 148–152.

48.

1853. BAIRD, SPENCER F. [Directions for making collections in Natural History, prepared for the use of the parties engaged in the Exploration of a route for the Pacific Railroad along the 49th parallel.] 4to, about 10 pp. Printed on thin blue paper.

    I have not been able to find a copy of this paper. The above title is supplied from the memory of Professor Baird.—G. B. G.

49.

1853. BAIRD, SPENCER F., *and* CHARLES GIRARD. Reptiles [of the Red River Region]. < *Marcy and McClellan's Exploration of the Red River of Louisiana in the year* 1852. Washington, 1853. (Appendix F.) 8vo. pp. 217–244.

    18 (10 + 6 + 2) species are described, and of these 11 are figured. Those designated by a * are here figured for the first time. Generic and specific diagnoses and descriptions are given and a partial synonymy.

1853. BAIRD, SPENCER F., *and* CHARLES GIRARD—Continued.

LIZARDS.

BATRACHIANS.

50.

1853. BAIRD, SPENCER F., *and* CHARLES GIRARD. Fishes [of the Red River Region]. < *Marcy and McClellan's Exploration of the Red River of Louisiana in the year* 1852. Washington, 1853. (Appendix F.) pp. 245–252.

5 species are described and all are figured.

51.

1854. BAIRD, SPENCER F., *and* CHARLES GIRARD. Descriptions of new species of Fishes collected in Texas, New Mexico and Sonora by Mr. John H. Clark, on the U. S. and Mexican Boundary Survey, and in Texas by Capt. Stewart Van Vliet, U. S. A. Second Part. < *Proc. Acad. Nat. Sci. Phila.*, vii, 1854, pp. 24–29. Presented for publication March 7, ordered printed March 28, 1854.

1854. BAIRD, SPENCER F., *and* CHARLES GIRARD—Continued.

52.

1854. BAIRD, SPENCER F., *and* CHARLES GIRARD. Notice of a new genus of Cyprinidæ. *< Proc. Acad. Nat. Sci. Phila.*, vii, 1854, p. 158.

53.

1854. BAIRD, SPENCER F. Report of the Assistant Secretary (of the Smithsonian Institution) in charge of publications, exchanges and natural history. *< Eighth Annual Report Smithsonian Institution* (1853), 1854. pp. 34–37.

54.

1854. BAIRD, SPENCER F., *and* CHARLES GIRARD. Descriptions of new species of Fishes collected in Texas, New Mexico and Sonora, by Mr. John H. Clark, on the U. S. and Mexican Boundary Survey, and in Texas by Capt. Stewart Van Vleit, U. S. A. Second Part. *< Proc. Acad. Nat. Sci. Phila.*, vii, 1854–55, pp. 24–29. (Read March 28, 1854.)

55.

1854. BAIRD, SPENCER F. Descriptions of New Genera and Species of North-American Frogs. *< Proc. Acad. Nat. Sci. Phila.*, vii, pp. 59–62. Presented for publication April 4, ordered printed April 25, 1854.

    Seventeen new species and one new genus are characterized.

1854. BAIRD, SPENCER F.—Continued.
    Helocœtes, Baird, n. s.
        3. *Helocœtes feriarum*, Baird, n. s.
           Carlisle, Pa.
        4. *Helocœtes triseriatus* (Max. von Wied.), Baird, n. s.
           Michigan, Illinois, Wisconsin, and the Upper Missouri.
        5. *Holocœtes Clarkii*, Baird, n. s.
           Galveston and Indianola, Tex.
        6. *Hyla Richardii*, Baird, n. s.
           Cambridge, Mass.
        7. *Hyla Andersonii*, Baird, n. s.
           Anderson, South Carolina.
        8. *Hyla eximia*, Baird, n. s...p. 61
           City of Mexico.
        9. *Hyla Vanvlietti*, Baird, n. s.
           Brownsville, Tex.
      10. *Hyla affinis*, Baird, n. s.
           Northern Sonora.
    Ranidœ.
      11. *Rana montezumœ*, Baird, n. s.
           City of Mexico.
      12. *Rana septentrionalis*, Baird, n. s.
           Northern Minnesota.
      13. *Rana sinuata*, Baird, n. s.
           Sackett's Harbor, New York.
      14. *Rana pretiosa*, B. and G. 62
           Washington Territory.
      15. *Rana cantabrigensis*, Baird, n. s.
           Cambridge, Mass.
      16. *Rana Boylii*, Baird, n. s. 62
           California, interior.
      17. *Scaphiopis Couchii*, Baird, n. s.
           Coahuila and Tamaulipas.

56.

1854. BAIRD, SPENCER F. Descriptions of New Birds collected between Albuquerque, N. M., and San Francisco, California, during the winter of 1853–54, by Dr. C. B. R. Kennerly and H. B. Möllhausen, Naturalists attached to the Survey of the Pacific R. R. Route, under Lt. A. W. Whipple. < *Proc. Acad. Nat. Sci. Phila.*, vii, pp. 118–20. Presented for publication June 20, ordered printed June 27, 1854.

    Eight new species are characterized.

    Cypselus melanoleucus, Baird, n. s...p. 118
        Camp 123, West of San Francisco Mountains.
    Culicivora Plumbea, Baird, n. s... 118
        Bill-Williams' Fork.
    Psaltria Plumbea, Baird, n. s.
        Little Colorado, N. M.
    Cyanocitta macrolopha, Baird, n. s.
        100 miles west of Albuquerque, N. M.
    Carpodacus Cassinii, Baird, n. s... 119
        Pueblo Creek, N. M.
    Zonotrichia fallax, Baird, n. s.
        Pueblo Creek, N. M.
    Pipilo mesoleucus, Baird, n. s.
        Copper Mines, n. s.
    Centurus uropygialis, Baird, n. s... 121
        Bill-Williams' Fork of Colorado, N. M.

57.

1854. BAIRD, SPENCER F., *and* CHARLES GIRARD. [Cyprinidæ of Heerman's Collection.] < *Girard's Descriptions of New Fishes collected by Dr. A. S. Heerman (Proc. Acad. Nat. Sci. Phila.*, 1854, pp. 129-156) = (pp. 135-138).

Six new species are described.

17. *Gila conocephala*, B. and G., n. s..p. 135
    Rio San Joaquin, Cal.
18. *Pogonichthys maequilobus*, B. and G., n. s...136
    San Joaquin River, Cal.
19. *Pogonichthys symmetricus*, B. and G., n. s.
    Fort Miller, Cal.
20. *Lavinia exilicauda*, B. and G., n. s...137
    Sacramento River, Cal.
21. *Lavinia crassicauda*, B. and G., n. s.
    San Francisco, Rio San Joaquin, &c.
22. *Leucosomus occidentalis*, B. and G., n. s.
    Posa and Grove Creeks, Cal.

58.

1854. BAIRD, SPENCER F., *and* CHARLES GIRARD. Notice of a new genus of Cyprinidæ. < *Proc. Acad. Nat. Sci. Phila.*, vii, 1854, p. 158.

Cochlognathus, B. and G., n. g...156
Cochlognathus ornatus, B. and G., n. s.
    Brownsville, Tex.

59.

1854. BAIRD, SPENCER F. Characteristics of some New Species of Mammalia, collected by the U. S. and Mexican Boundary Survey, Major W. H. Emory, U. S. A., Commissioner. Part I. < *Proc. Acad. Nat. Sci. Phila.*, vii, pp. 331-333. Read and ordered printed April 24, 1854.

The following eleven species are described:

Sciurus limitis, Baird, n. s ....p. 384
    Devil's River, Texas.
Sciurus castanotus, Baird, n. s.
    Mimbres.
Tamias dorsalis, Baird, n. s.
    On the Mimbres.
Spermophilus spilosoma, Bennet.
    El Paso.
Spermophilus Couchii, Baird, n. s.
    Santa Caterina, Mex.
Perognatus flavus, Baird, n. s.
    El Paso.
Geomys Clarkii, Baird, n. s.
    Presidio del Norte, on the Rio Grande.
Thomomys umbrinus (Rich.), Baird.
    El Paso.
Sigmodon Berlanderi, Baird, n. s...333
    Between San Antonio and El Paso.
Neotoma mexicana, Baird, n. s.
    Chihuahua.
Neotoma micropus, Baird, n. s.
    Charco Escondido and Santa Rosalio, Mex.

60.

1854. BAIRD, SPENCER F. Characteristics of some New Species of North American Mammalia, collected chiefly in connection with the U. S. Surveys of a Railroad Route to the Pacific. Part I. < *Proc. Acad. Nat. Sci. Phila.*, vii, pp. 333-6. Read and ordered printed April 24, 1854.

Fifteen species are described, twelve of which are new.

Lepus washingtonii, Baird, n. s 333
    Puget Sound and Shoalwater Bay.

1854. Baird, Spencer F.—Continued.

Lepus Trowbridgii, Baird, n. s.
　　Coast of California.
Sciurus Suckleyi, Baird, n. s.
　　Puget Sound.
Tamias Oooperi, Baird, n. s..p. 334
　　Cascade Mountains, Wash.
Spermophilus Gunnisoni, Baird, n. s.
　　Cochitope Pass, Rocky Mountains.
Spermophilus grammurus, Say.
　　Western Texas.
Spermophilus Beecheyi, Richardson.
　　California.
Dipodomys montanus, Baird, n. s.
　　Fort Massachusetts.
Dipodomys agilis, Gambel 334–35
　　San Diego, or Monterey.
Geomys breviceps, Baird, n. s 335
　　Morehouse Parish, La.
Thomomys bottæ (Egd. and Gerv.), Baird. R.
　　Monterey.
Thomomys rufescens, Max.
　　Fort Pierre.
Thomomys laticeps, Baird, n. s.
　　Humboldt Bay.
Neotoma occidentalis, Cooper (mss.), n. s.
　　Shoalwater Bay.
Reithrodon montanus, Baird, n. s.
　　Rocky Mountains.
Hesperomys Boylii, Baird, n. s.
　　American River, Cal.
Hesperomys austerus, Baird, n. s 336
　　Fort Steilacoom, Puget Sound.

61.

1855. Baird, Spencer F. Report of the Assistant Secretary (of the Smithsonian Institution) for the year 1854. < *Ninth Annual Report Smithsonian Institution* (1854), 1855, pp. 31–46.

62.

1855. Baird, Spencer F. Report on American Explorations in the years 1853 and 1854. < *Ninth Annual Report Smithsonian Institution* (1854), 1855, pp. 79–97.
This series of reports was not separately made after this year, but the same material was incorporated in the regular reports of the Assistant Secretary.

63.

1855. Baird, Spencer F. Report on the fishes observed on the coasts of New Jersey and Long Island during the summer of 1854, by Spencer F. Baird, Assistant Secretary of the Smithsonian Institution. <*Ninth Annual Report of the Smithsonian Institution* [*for* 1854], 1855, pp. 317–325 + *337·
Reprinted as a pamphlet, with an index, and the following title:
Report | to | the Secretary of the Smithsonian Institution, | on the | Fishes of the New Jersey Coast, | as observed in the Summer of 1854, | by Spencer F. Baird, | Assistant Secretary S. I. | From the Ninth Annual Report of the Smithsonian Institution | for 1854. | — | Washington: | Beverly Tucker, Senate Printer, | June, 1855. | [8vo, 40 pp.]
68 species, of which 58 are marine, or brackish-water, and 10 fresh-water, were observed, and valuable notes on habits and color in a fresh state were recorded, during a period of six weeks spent on the coast of New Jersey, principally at Beesley's Point and Long Island, New York, and also on the Hudson River.
Notice > *Ninth Annual Report Smithsonian Institution* (1854), 1855, pp. 16–38.
"A period of six weeks spent on the coast of New Jersey, principally at Beesley's Point,

3 BD

1855. BAIRD, SPENCER F.—Continued.

and Long Island, New York, furnished an opportunity of studying the habits and distribution of the principal species of fishes that are found on that portion of our shores during the summer.

"Although many others, doubtless, are to be found in the same region, yet none have been introduced except those which were actually caught and carefully examined. A considerable number of the species whose habits and peculiarities are given at some length, have hitherto had nothing placed on record concerning them; and it is hoped that the present article may be found to contain some interesting information, given here for the first time, in addition to its character as a contribution to our knowledge of the geographical distribution of species.

"The difference of the names applied to the same species of fish at various points of our coast, even when these happen to be connected very closely, both commercially and geographically, must strike every one with astonishment.

"It is scarcely too much to say that no one species of fish bears the same vernacular appellation from Maine to Maryland, still less to Florida or the coast of Texas. This is probably owing to the fact that our shores have been originally settled by various nations from widely remote parts of Europe, each introducing its peculiar nomenclature, or deriving names from the equally isolated aboriginal tribes with their various languages. Thus the names of blue-fish, white fish, perch, blackfish, bass, king-fish, porgee, hake, tailor, whiting, horse-mackerel, shad, smelt, dog-fish, &c., may apply equally to two or more very different species. Among the synonyms of the species will be found the vernacular equivalents in the regions visited, together with some from other localities. It will be sufficiently evident, therefore, that before any species referred to under a trivial name can be identified, the origin of the fish or that of the writer must be ascertained.

"Although most of the facts recorded in the following paper have reference to Great Egg Harbor, New Jersey, during a period extending from the middle of July to the end of August, it has been thought not amiss to incorporate the results of a visit to Brooklyn, Riverhead, and Greenport, Long Island, as well as to some points on the Hudson River, in September. Some valuable information was thus obtained, tending to illustrate more fully the natural history and distribution of the species found on the New Jersey coast.

"And here I take occasion to render an acknowledgment for much kind assistance and important information derived from various gentlemen at the different points of operation. Among these I will particularly mention Messrs. Samuel and Charles Ashmead, at Beesley's Point, who devoted all their time to the furtherance of my objects in this exploration. I may also mention Messrs. John Stites, Willis Godfrey, Washington Blackman, John Johnson, in fact, most of the residents of Beesley's Point. Much benefit was derived at Greenport, Long Island, from the companionship of Mr. E. D. Willard, of the National Hotel, Washington; while to Mr. J. Carson Breevort, of Bedford, Long Island, well known as the first ichthyologist in New York, and surpassed by no one in his knowledge of our marine species, I am under the greatest obligations. Through the kindness of Mr. John G. Bell, of New York, and Smith Herring, of Piermont, I was enabled to make a complete collection of the fishes of the Upper Hackensack and Sparkill.

"It must be understood that the present article does not aim at giving a complete account of the species referred to. Such descriptions of color as have been given were in every case taken from the fresh and living fish, the object being to place on record features not usually preserved in alcoholic specimens. Of the species whose colors were known not to fade or alter in spirits, no notes of their peculiarities in this respect were taken, while the tints of others were so evanescent as to have escaped or altered before a description could be noted down.

"Very little respecting the habits or history of the species has been added from other authors, nor does the nomenclature profess to be at all final as to critical accuracy. To have accomplished this latter object, would have required more time than is at present at my disposal, involving, as it would, the entire revision of American ichthyology generally. The names given are principally those of De Kay in his history of the fishes of New York, and can thus be readily identified.

"As will be seen in the course of the article, several of the species collected appear new to science; to these I have been obliged to give names for the sake of proper reference, without at the same time furnishing a complete scientific description.   *   *   *

"The coast of New Jersey is well known to consist, for most of its extent, of a low beach, with sand-hills, separated from the mainland by a wide strip of low meadows filled with small ponds, and intersected by creeks and thoroughfares, which traverse it in every direction. There is no rock or stone of any description, and, consequently, there is a deficiency in the plants and animals which frequent rocky localities. At Beesley's Point there is scarcely a pebble of the smallest size to be seen.

1855. BAIRD, SPENCER F.—Continued.

"The meadows are densely coated with grass, and are covered with water only during unusually high tides.

"Beesley's Point is situated at the mouth of Egg Harbor river, where it empties into Great Egg Harbor bay. The water is, of course, salt at this point, though somewhat diluted by the volume of fresh water brought down by the river.

"The distance from the mouth of the river, or head of the bay, to the inlet on the beach, is about two or three miles; the extreme width about the same, although extending into thoroughfares, through which a boat may be taken to Absecom on the one side, and to Cape May on the other, without going outside of the beach. The mouth of the river is occupied by very extensive beds of oysters, which are celebrated for their excellent flavor. The bottom of the bay is in some part hard and shelly, in others sandy, or again, consists of a soft mud; the latter condition prevails near the shore, or wherever the current is of little strength.

"There are numerous mud-flats or sand-bars in the bay, some of them bare at low tide, or nearly so, and occupied by various species of water-fowl. These flats, continuing to increase in height, and at length acquire a growth of grass, which fixes still more the accumulating mud and sand, so that in time what was formerly a bar becomes an island elevated some feet above the water.

"This transition is, in fact, so rapid that many of the inhabitants now living have known islands several acres in extent to form within their own recollection.

"The greater part of the bottom of the bay and of the thoroughfares, generally, is a soft mud, rich in organic matter, and covered with a profuse growth of *Zostera marina* and algæ of various species. Mr. Samuel Ashmead, who has been engaged for some years in studying the sea-weeds of our coast, has found a much greater variety of species at Beesley's Point than Professor Harvey allots to the New Jersey coast. The water being generally shallow except in the channels, the submarine vegetation can be seen to great advantage, while sailing over the surface. The water becomes very warm during the summer, and supplies all the conditions necessary for the development of young fishes of many species. The young of all the large fish of the bay may thus be found in greater or less numbers along or near the shore.

"The ponds in the meadows, like the waters of the bay itself, are generally muddy at the bottom, sometimes bare of vegetation, and sometimes covered with a thick growth. The fishes found in these ponds consist almost entirely of cyprinodonts of various species, with occasional specimens of *Atherina*, small mullets, or sticklebacks. The creeks likewise contain cyprinodonts, generally of different species from those of the ponds, with young fish of various kinds. Crabs and eels are found everywhere.

"The line of beach is two or three miles from the mainland, and consists of a clear white sand raised into hills ten to thirty feet high, a few hundred feet from the water's edge. It is in the inlets at the ends of these beaches that the greatest variety of fish is to be found, particularly in the small indentations, protected from the roughness of the waves, and the bottom of which is covered with *Ceramium* or sea-cabbage.

"Corson's inlet, frequently mentioned in the following pages, is situated at the southern end of Peck's beach, which begins directly opposite Beesley's point at the entrance to the harbor, and extends to this inlet over a distance of about five miles.

"The only fresh water near Beesley's point is Cedar Swamp creek. This stream, rising in a cedar swamp, and flowing with a very sluggish current, (the water of a chocolate color), is cut off from the tide by a dam at Littleworth, three miles from the point. The bottom is very muddy. But little variety of fresh-water fish is to be found in this stream. Several species of *Esox*, two *Leuciscus*, one eel, three *Pomotis*, one each of *Aphredoderus*, *Labrax*, *Etheostoma*, and *Melanura*, and several cyprinodonts. The species are nearly all different from those found in the interior of Pennsylvania on the same latitude.

"Another Cedar Swamp Creek occurs on the opposite side of Egg Harbor River, in Atlantic County. In many respects it differs from that first mentioned in being of more rapid current, and the bottom, at some distance from the tide-water dam, consisting of sand or small pebbles. The water, too, in small quantity is clear, though where of considerable depth it appears almost black. Fewer species of fish were found here than in the other, the only additional one being the *Catastomus tuberculatus*.

"Ludley's Run is a small run crossing the road to Cape May, about eight miles from Beesley's point; fresh at low tide, but flooded at high water. The only fish found in it consisted of two cyprinodonts and the *Gasterosteus quadracus*."

The following species are discussed. The figures in parenthesis refer to the pages of the separate edition:

1. *Labrax lineatus*, Cuv. and Val .... (7)   321
2. *Labrax mucronatus*, Cuv. and Val .... (8)   322
3. *Centropristes nigricans*, Cuv. and Val .... (9)   323

1855. BAIRD, SPENCER F.—Continued.

    62. *Diodon maculato-striatus*, Mitch.
    63. *Diodon fuliginosus*, De Kay.
    64. *Tetraodon turgidus*, Mitch .... (38)   p. 352,
    65. *Carcharias cœruleus*, De Kay.
    66. *Mustelus canis*, De Kay... (39)   353
    67. *Zygœna tiburo*, Val.
    68. *Pastinaca hastata*, De Kay.
    Index .... (40)

64.

1855. BAIRD, S. F.   Mammals [of Chili].   <*Gillis, Naval Astronomical Expedition*, ii, pp. 153–162.*

    Descriptions and synonymy of the following species, specimens of which were obtained by the expedition:

    Felis concolor, L...p. 153
    Canis magellanicus, Gray ....p. 154
    Canis azarœ, Max.
    Galictis vittata, Bell...p. 155
    Dydelphys elegans, Waterhouse.
    Cavia australis, Geoff ....p. 156
    Lagidium Cuvieri, Wagn.
    Spalacopus Pœppigii, Wagl ....p. 157
    Myopotamus coypus, Geoff.
    Hesperomys.
    Chlamyphorus truncatus, Harl ....p. 158, pl. xi
    Auchenia llama, Desm ....p. 159

65.

1855. BAIRD, SPENCER F.   List of Mammalia found in Chili.   < *Gillis, Naval Astron. Exp.*, ii, pp. 163–171.*

    This paper gives a full synonymy of each species, with a statement of geographical range and list of common names.

    Cheiroptera.
      Insectivora.
        Stenoderma chilensis, Gay..p. 163
        Desmodus D'Orbignyi, Waterh.
        Dysopes nasutus, Temm.
        Nycticejus varius, Schinz.
        Nycticejus macrotis, Fisch.
        Vespertilio velatus, Fisch.
        Vespertilio chiloensis, Waterh.

      Rapacia.
        Carnivora.
          Felidœ.
            Felis concolor, L 164
            Felis pajeros, Desm
            Felis guigna, Mol.
            Felis colocolo, Mol.
        Canidœ.
          Canis fulvipes, Martin.
          Canis magellanicus, Gray.
          Canis azarœ, Max.
        Mustelidœ.
          A. *Martinœ.*
            Galictis vittata, Bell.. 165

---

* 23d Congress, } House of Representatives. { Ex. Doc. | — | The | U. S. Naval Astronom-
  1st Session. }   { No. 121.
ical Expedition | to | the Southern Hemisphere, | during | the years 1849-'50-'51-'52. | — |
Lieut. J. M. Gillis, Superintendent. | — | Volume II. | — | Washington: | A. O. P. Nicholson,
Printer. | MDCCCLV.

1855. BAIRD, SPENCER F.—Continued.
  B. *Melinæ.*
      Mephitis chilensis, ——.
      Mephitis patagonica, Licht.
      Mephitis? molinæ, Licht.
  C. *Lutrinæ.*
      Lutra felina (Molina), Gay.
      Litra huidobria, Gay.
  Pinnipedia.
  Phocidæ.
      Otaria porcina (Molina), Desm.. 166
      Otaria jubata (Schreb.), Desm.
      Otaria flavescens (Shaw), Desm.
      Otaria ursina (Linn.), Desm.
      Stenorhynchus leptonyx (Blainv.), F. Cuv.
      Macrorhinus leoninus (Linn.), F. Cuv.
  Marsupialia.
      Dydelphys elegans, Waterh.
  Rodentia.. 167
  Hystricidæ.
      Caviinæ.
          Cavia australis, Is. Geoff.
      Chinchillinæ.
          Lagidium Cuvieri (Benn.), Wagn.
          Lagidium criniger (Gay).   R.
          Lagidium pallipes (Bennet), Wagner.
          Chinchilla lanigera (Molina), Benn.
      Octodontinæ.
          Habrocoma Bennetti, Waterh.. 168
          Habrocoma Cuvieri, Waterh.
          Octodon degus (Molina), Waterh.
          Octodon Birdgesii, Waterh.
          Schizodon fuscus, Waterh.
          Spalacopus Pœppigii, Wagler.
          Ctenomys magellanicus, Benn.
      Echymyinæ.
          Myopotamus coypus (Molina), Commers... 169
  Muridæ.
  Murina.
      Oxymicterus scalops, Gay.
      Oxymicterus megalonyx (Waterh.), Gay.
      Hesperomys longipilis, Waterh.
      Hesperomys renggeri, Waterh.
      Hesperomys brachyotis, Waterh.
      Hesperomys? rupestris, ——.
      Hesperomys xanthorhinus, Waterh.
      Hesperomys Darwinii, Waterh.
      Hesperomys leutescens, Gay... 170
      Hesperomys longicaudatus (Bennett), Waterh.
      Reithrodon chinchilloides, Waterh.
  Edentata.
  Effodientia.
      Dasypus minutus, Desm.
      Chlamyphorus truncatus, Harl.
  Ruminantia.
  Camelidæ.
      Auchenia llama (Waterh.), Desm.
  Cervidæ.
      Cervus pudu, Gerv... 171
      Cervus chilensis, Gay and Gervais.
  Cetacea.
  Delphinidæ.
      Delphinus lunatus, Less.
      Delphinus albimanus, Peale.

1855. BAIRD, SPENCER F.—Continued.
    Physeteridæ.
      Physeter macrocephalus, L.
    Balænidæ.
      Balæna antarctica, Klein.

66.

1855. BAIRD, SPENCER·F. [Report to Capt. John Pope on the zoological collection made by his party of the Pacific R. R. Survey between El Paso and Fort Smith.] Dated Washington, Oct. 1, 1854. < *Pope's Report of Exploration of Route for P. R. R. near 32d Parallel.* [H. Doc. 129.] 1855(?), p. 129.

67.

1856. BAIRD, SPENCER F. List of Fishes inhabiting the State of New York: Sent to the New York State Cabinet of Natural History by the Smithsonian Institution in May, 1855 (by Professor S. F. Baird). < *Ninth Annual Report of the Regents of the University of the State of New York on the Condition of the State Cabinet of Natural History and the Historical and Antiquarian Collection annexed thereto.* * * * 1856, pp. 22–29.

A list of 70 species with partial synonymy, corresponding closely with the list given in the paper on the fishes of the New Jersey coast.

68.

1856. BAIRD, SPENCER F. [Report of the Assistant Secretary of the Smithsonian Institution for the year 1855.] < *Tenth Ann. Rep. Smithsonian Institution* (1855), 1856, pp. 36–61.
    Principal desiderata of this museum 54, 55
    Premiums for collections 55, 56

69.

1856. BAIRD, SPENCER F. Report of Professor S. F. Baird, late Permanent Secretary of the American Association for the Advancement of Science, on the Distribution and Disposal of the Volumes of Proceedings. < *Proc. Amer. Assoc. Adv. Science,* ix, 1855, pp. 287–292.

70.

1856. BAIRD, SPENCER F. [A description of the genus Ceratichthys.] < *Proc. Acad. Nat. Sci. Phila.,* viii, 1856, p. 212.

71.

1857. BAIRD, SPENCER F. [Report of the Assistant Secretary of the Smithsonian Institution for the year 1856.] < *Eleventh Ann. Rep. Smithsonian Institution for the year* 1856, 1857, pp. 47–68.

72.

1857. BAIRD, SPENCER F. [Name *Tamias pallasii* proposed instead of *Sciurus striatus,* Pallas, nec Linn.] < *Eleventh Ann. Rep. Smithsonian Institution,* 1857, p. 55.

73.

1857. BAIRD, SPENCER F. Directions for collecting, preserving, and transporting specimens of natural history. Prepared for the use of the Smithsonian Institution. < *Eleventh Ann. Rep. Smithsonian Institution* (1856), 1857, pp. 235–253.

74.

1857. BAIRD, SPENCER F.   American Oology. $<$ *Edinb. New Philos. Journal,* new ser., v, 1857, p. 374.

Extract from a letter relating to T. M. Brewer's work.

75.

1857. BAIRD, SPENCER F.   Catalogue | of | North American Mammals, | chiefly in the Museum of the | Smithsonian Institution. | By | Spencer F. Baird, | Assistant Secretary of the Smithsonian Institution. | — | Washington : | Smithsonian Institution, | July, 1857. | 4to.   pp. 21.

This catalogue is essentially the systematic list which forms the first twenty-one pages of the General Report on North American Mammals published the same year.

A large edition of this check-list catalogue was printed, and it remains to the present time the only check-list and the principal standard of authority for labeling collections.

76.

1857. BAIRD, SPENCER F.   Explorations and surveys for a railroad route from the Mississippi River to the Pacific Ocean. | War Department. | — | Mammals : | By Spencer F. Baird, | Assistant Secretary of the Smithsonian Institution. |

| — | Washington, D. C., 1857. $<$ 33d Congress, 2d Session. $\}$ House of Representatives. $\{$ Ex. Doc. No. 91. | — | Reports | of | explorations and surveys, | to | ascertain the most practicable and economical route for a railroad | from the | Mississippi River to the Pacific Ocean. | Made under the direction of the Secretary of War, in | 1853–6, | according to acts of Congress of March 3, 1853, May 31, 1854, and August 5, 1854. | — | Volume VIII. | — | Washington : | *A. O. P. Nicholson, Printer. | 1857.   4to.   pp. xlviii, 757, pl. xviii–lx. †

The special title-page quoted above is on page xxi.

CONTENTS.

Dated Washington July 20, 1857.

"It is a systematic account of the mammals (exclusive of *Cetacea, Chiroptera, Sirenia,* and *Pinnipedia*) of North America, about 220 in number, and is by far the most important and most authoritative treatise which has ever appeared on the subject." (GILL and COUES.)

Review.—*Annals and Magazine of Natural History,* May, ser. 3, vol. i, 1858, pp. 369–373, reprinted in *American Journal Science and Arts,* xxvi, 1858, pp. 141–146.

The following description of the work is taken from the preface:

"The present report is intended to embrace a systematic account of all the species of North American mammals collected or observed by the different parties organized under the direction of the War Department for ascertaining the best route for a railroad from the Mississippi River to the Pacific Ocean. It was originally proposed to furnish a separate report in detail, on the collections of each party, but a consideration of the fact, that, with scarcely an exception, almost every species was found on two or more lines of survey, and thus not peculiar to any one expedition, led to an abandonment of the first intention. It was considered to be worse than useless to repeat the same descriptions and details over and over again, while, at the same time, under the circumstances, it would have been difficult to say in what report any particular article could be best placed. As, too, the interest of North American zoology depends not merely on the character of the species, but also on their generic and family affinities, as well as on their relationships to latitude and longitude, climate, soil, elevation, &c., it would have been impossible to do justice to the subject by cutting up the

---

* Other copies have the imprint of Beverly Tucker.

† In Gill and Coues's "Material for a Bibliography of North American Mammals," pp. 992, this edition is cited as containing plates i-lx, which is erroneous, the first 17 plates being given in other volumes of this series of reports.

1857. BAIRD, SPENCER F.—Continued.

report into several isolated portions without any special connection as parts of a systematic whole.

"At the same time, however, as it was desirable to present a picture of the zoological character of the several routes, as well as to show what each party accomplished, and, as many very important notes of habits and local peculiarities were made by the naturalists of the different lines, it would have been clearly an act of injustice to these gentlemen as well as to their chief officers to merge all their results into one common report. For these and other reasons it was finally determined that there should be prepared one general report on the entire collections of the railroad surveys, to consist solely of the technical description of the families, genera, and species, and of such remarks as might be necessary to show their place in the systems, each species to be preceded by its synonymy, and followed by an enumeration of all the specimens collected, so arranged in tables as to show their geographical distribution.

"In addition to this general report, however, special reports by the naturalists of each line were also to be prepared and published, to embrace the systematic and vernacular names of their species, with a list of the specimens collected. To these special reports, were to be confined all the biographies of the animals seen, all notices of their habits, peculiarities, and distribution, as observed and recorded during the route. In order that there might be no misconception of the species referred to, it was concluded to give a short diagnosis of each, with a reference to the page of the general report where the purely zoological details might be found more at length.

"The present report, therefore, is the first of the series of general reports referred to, to be followed, as soon as practicable, by the remainder of the Vertebrata. The special reports on the zoölogy of each line of survey will be found in connection with the other reports belonging to their respective parties, in their full notices of the life of our western animals, possess a general and popular interest far greater than can attach to the present account of mere zoölogical and technical details.

"The large size of this report on the mammals collected by the railroad parties is owing to several causes. In the first place, the amount of new or little known material obtained was extraordinarily great. The summary of the species, at the beginning of the systematic list hereafter presented, will show that very many entirely undescribed animals were procured, and that, of a large number of others, previously little known, the specimens were sufficient to furnish many new and interesting details of characters, both external and internal.

"As, too, the object in calling for complete reports from the several parties was not merely to show the actual results of the several expeditions, but likewise to ascertain the general character of the Western Territories, I have not hesitated to include in this work all such materials derived from officers stationed at military posts, and other persons elsewhere in the West, as fell under my notice.

"In view of the large amount of new or little known species at hand, in the preparation of the present report, sometimes embracing entire genera and even families, it soon became evident that none of the published descriptions of the old and standard species were sufficiently minute and detailed to furnish the necessary means of comparison. With the discovery of forms very closely allied to or intermediate between those already known, the descriptions of the latter on record did not show sufficiently in what the differences consisted. It became necessary, therefore, to redescribe, as far as they could be procured, all such species, which, in fact, proved finally to be nearly all previously known. The present monograph of American mammals has, in the end, grown out of the necessities referred to.

"It will be sufficiently evident that, without the extraordinarily rich and full collection of North American mammals belonging to the Smithsonian Institution, the monographs and comparisons of species, in the present report, could not have been prepared. Independently of the specimens brought in by the Pacific Railroad surveying parties, the series in its Museum, from other sources, was found to embrace nearly all the previously known species, and many entirely new ones.

"I have also made free use of the collections and library of the Philadelphia Academy of Natural Sciences, for which every facility has been furnished in its hall. The examination of the specimens collected by Townsend, and described by Dr. Bachman, has contributed to settle some quite doubtful points, while in some rare or very costly works of its unequaled natural history library I have been enabled to verify many references which would otherwise have remained uncertain.

"I regret not to have been able to examine any of the types of the new species of Audubon and Bachman, as presented by the latter gentleman to the Charleston Museum. The rules of that establishment do not permit specimens to leave its hall, and it was not possible to visit it during the preparation of this report.

"I have endeavored to make all acknowledgments of aid from systematic writers in the body of the report, although it may be well to mention here that much use has been made

1857. BAIRD, SPENCER F.—Continued.

throughout of the works and articles of Wagner, Waterhouse, Gray, Brandt, Burmeister, Keyserling and Blasius, Giebel, Richardson, Agassiz, Englemann, and others, as enumerated in the synonymy and list of authorities. To the labors of Messrs. Audubon and Bachman, however, either singly or collectively, are acknowledgments especially due for whatever facilities may have previously existed for the preparation of a report on American mammals. The necessity or propriety of such a report is only to be found in the fact that, when the crowning work of these gentlemen, 'The Viviparous Quadrupeds of North America,' was prepared, the materials at their command were far less extensive than have been at mine, and that many species, which they could only examine in the museums of London, Paris, Berlin, and Leyden, are now to be found in the Smithsonian collection in a profusion of specimens of the most satisfactory and perfect character.

"An apology is necessary for the delay which has taken place in the completion of the general reports on the zoology of the Pacific Railroad surveying parties. This has arisen from the fact that, from the first organization of these expeditions, in the spring of 1853, nearly to the present time, one or more has been in the field, and engaged in fresh examinations; so that until all the specimens expected were received, the general systematic account of zoological results could not conveniently be prepared. The examination of the materials was actually commenced early in 1855, and many of the articles written in that year in 1856. With the continuous accession of additional specimens, it became finally necessary to rewrite, alter, or extend all that had been prepared prior to the present year (1857). It is to this that the frequent want of uniformity is due, the time allowed not being sufficient in many cases to permit the reworking of the whole matter. The measurements of the specimens were at first made in inches and lines, but that of hundredths was finally adopted; it is to this fact that the presence of the two different divisions of the inch is attributable, it not being convenient or possible to make the measurements conform throughout, as would have been desirable.

"It is, perhaps, unnecessary to state that the matter of the present report is entirely original throughout, the few cases in which extracts from other authors are made being so indicated. With very few exceptions, all the references in the synonymy have also been personally made and verified. Where this was not possible, the synonym is inclosed between quotation marks.

"In explanation of the too frequent occurrence of typographical errata in the body of this report, it is proper to state that, owing to various circumstances, the work was necessarily passed through the press with a rapidity probably unexampled in the history of natural-history printing, allowing very little opportunity for that critical and leisurely examination so necessary in correcting a work of the kind. For most of the time the proof has been furnished and read at the rate of twenty-four to thirty-two pages per day, nearly four hundred pages having been set up, read, and printed during the first half of July alone. The same cause has also interfered with the preservation of perfect uniformity of arrangement and detail throughout. In some cases, accidents to the form while on press have caused the loss or transposition of letters, words, or paragraphs; as, among others, the exchange of characters of Orders VIII and IX, on page 1, referred also to on page 625. For excuse of errors in the use of technical terms, in the formation and inflection of scientific names, and for all other shortcomings, the writer can only throw himself upon the kind indulgence of his readers, partly in consideration of the fact that, owing to the urgent necessity for a speedy completion of the volume, no time was allowed for any revision of the manuscript as a complete work, nor, indeed, of its separate portions, and that for much of the time the preparation of much of the manuscript was only a few hours in advance of its delivery to the compositor.

"A few words in explanation of the plan adopted in preparing the articles of the present report may not be out of place. I have usually made the entire detailed description of the species from one particular specimen, often indicating it by number, mentioning afterwards the variations presented from this type by the others before me. The specific diagnoses alone contain a combination or selection from the characters of several specimens. The numbers attached to the specimens, as enumerated, are those which they bear in the Smithsonian Museum Catalogues. Each class of animals has its separate catalogue and succession of numbers, from 1 upwards, in this series, the same number being never used twice for different objects in its class, and thus constituting an essential part of the specimen. There is also a special catalogue of the osteological collections. Thus, the skin of a mammal will have one number, and its skull, if separated, another; each specimen having both numbers attached, its own as numerator of a fraction. Thus, when a skull is labeled or entered $\frac{421}{1149}$, it is to be understood that 421 is its number as entered in the catalogue of skins (or entire specimens in alcohol), while 1149 is the number of the skull as entered in the osteological catalogue. The skull itself would in this case be marked $\frac{1149}{421}$.

1857. BAIRD, SPENCER F.—Continued.

"The column of 'original numbers' embraces those attached to specimens in the field by collectors. These are always retained as being referred to in the field-notes of the different parties.

"The measurements have, in all cases, been made in English inches,* divided either into lines or twelfths or into one-hundredths. All the skulls, and in most cases the smaller skins, have been measured with dividers or callipers. The measurements of the body have been made to the insertion of the tail into the rump, or nearly to the very base of the caudal vertebræ; the animal usually with the head, body, and tail extended into the same straight line, avoiding, as far as possible, all curvature.

"Where measurements are recorded as made before skinning, they are, in most cases, to be understood as having been furnished by the collector.

"For the sake of illustrating more fully the character of the species described in the present report, I have prepared the three lists as follows:

"The first list is that of the higher groups characterized in the following pages.

"The second list contains all the species of mammals, found in North America north of Mexico, that I have had an opportunity of examining while preparing the present work, together with a few that belong to the northern provinces of the last-mentioned State. These are inserted as very probably existing within the limits of the United States, even though not yet detected. The indications of geographical distribution are chiefly those furnished by the specimens before me, although I have occasionally given statements in this respect from Audubon, Bachman, and Richardson. I have not pretended to define with critical accuracy the complete range of the species, the facts on record not being sufficient for the purpose.

"The third list embraces the species which have not fallen under my notice. Some of these have little claim to a place in the fauna of North America north of Mexico, while others are, in all probability, the same as those mentioned in the first list. A few are unquestionably additional and good species, especially such as *Sorex fimbripes* and *palustris*, *Putorius nigripes*, *Arctomys pruinosus*, *Thomomys talpoides?* *Arvicola borealis*, *drummondii*, *richardsonii*, and *xanthognathus*, and some others.

"The following table will serve to indicate the additions in the list to the known species of North American mammals, as compared with the latest general work on the subject:

Species described here as new for the first time 35

Species described in 1855, from the same collection................. 17

Total of new species in the Smithsonian collection not mentioned by Audubon and
    Bachman... 52

Recognized species previously described, not mentioned by Audubon and Bachman...... 18

Total of North American species additional to the list of Audubon and Bachman.. 70

"The entire number of species mentioned by Audubon and Bachman in the Quadrupeds of North America, exclusive of varieties, is 197, of which about 160 were figured, the remainder consisting either of species previously described by the authors, but not procurable for purposes of illustration, or else copied from others, to render their work complete,

"The total number of species of North American mammals represented in the Smithsonian collection is very nearly 220. Washington, D. C., July 20, 1857." < *Extracts from* (*Preface*) *Exploration and Survey for the Pacific R. R.*, vol. viii, 1853-6, pp. xxv, xxvi, xxvii, xxviii, and xxix.

"GENERAL SKETCH OF LINES EXPLORED.

"Shortly before the close of the session of Congress in March, 1853, an appropriation of $150,000 was made to defray the expenses of the survey of the various routes along which it was supposed that a railroad might be constructed from the Mississippi River to the Pacific Ocean. For this purpose six parties were organized by the War Department for the survey of four main routes, and in a short space of time they were in the field. All the parties were fitted out in the most complete manner; the natural-history apparatus and material prepared under the direction of the Smithsonian Institution, which also furnished the necessary instructions as to the objects most important to be collected. In its efforts to secure the assignment to these parties of persons capable of making collections and observations it was met by the hearty co-operation of the War Department, through the heads of the different expeditions and Captain Humphreys, in charge of the Pacific Railroad office.

*   *   *   *   *   *   *

"A more detailed account of the collections and expeditions referred to above will be found in the Reports of the Smithsonian Institution—(sixth to the eleventh, 1851-1856). A notice

---

* The English inch used is about equal to 11.26 French lines, .9383 French inches, or to 25.40 millimeters. On the other hand, the French inch is equal to 1.0657 English inches, the French line to .0888 English inches, and the millimeter to .03937 English inches. The French meter is equivalent to 39.37 English inches, or to 3.28 feet.

1857. BAIRD, SPENCER F.—Continued.

of the collections from the eastern portions of the United States, used for purposes of comparison, will also be found in the same series.

"The very rich collections made by the United States and Mexican Boundary Survey in Texas, New Mexico, and California are described in detail in its report, which also embraces notices of the results of the explorations in Texas and Northern Mexico by Dr. Berlandier, Mr. John Potts, Major Rich, and Lieut. Couch. Incidental mention of these is also made occasionally in the present volume, wherever necessary to complete the indications of geographical distribution.

"The collections of all the government parties just mentioned were transmitted, from time to time, to the Smithsonian Institution, and were there properly cared for until the return of the several expeditions. They were then placed in the hands of the naturalists selected to elaborate them, and the necessary drawings prepared, within its walls, under the direction of the Institution, to which, also, was committed the general supervision of the engraving and printing of the plates. Every facility has been furnished by the War Department, through Captain Humphreys, in charge of the Pacific Railroad office, and the heads of the different expeditions, for bringing the results properly before the world." < *Extracts from Introduction, Exploration and Surveys for the Pacific Railroad*, vol. viii, 1853-6, parts of pp. xiii, xvi, and xvii.

The definitions of the higher groups are inserted in their proper places, and it is not thought necessary to index them.

A.—*List of examined and identified species of North American Mammals.*

1. NEOSOREX, BAIRD.

2. SOREX, L.

3. BLARINA, GRAY.

4. SCALOPS, CUV.

a. SCALOPS.

b. SCAPANUS, Pomel.

5. CONDYLURA, ILL.

6. UROTRICHUS, TEMM.

1857. BAIRD, SPENCER F.—Continued.

3 BD

1857. Baird, Spencer F.—Continued.

B.—*List of described North American species not seen and identified.*

 Note.—Of the above *Arvicolae*, those most likely to prove additional species are Nos. 25, 27, 31, 32, and 34.

1857. BAIRD, SPENCER F.—Continued.

45. LEPUS, LIN.

77.

1857. BAIRD, SPENCER F. Catalogues of Fishes, copied from a "Report on the Fishes observed on the Coasts of New Jersey and Long Island during the summer of 1854. By Spencer F. Baird, Assistant Secretary of the Smithsonian Institution." From the Ninth Annual Report of the Smithsonian Institution for 1854. < *Catalogue of Zoological and Botanical Productions of the County of Cape May, in Geology of the County of Cape May, State of New Jersey,* 1857. pp. 146–148.

A name catalogue only, the scope of which is explained by its title.

78.

1858. BAIRD, SPENCER F. Explorations and Surveys for a Railroad Route from the Mississippi River to the Pacific Ocean. | War Department. | — | Birds : | By Spencer F. Baird, | Assistant Secretary Smithsonian Institution, | with the co-operation of | John Cassin and George N. Lawrence. | — | Washington, D. C. | 1858. pp. i–lvi, 1–1005. (No illustrations.) Dated Washington, Oct. 20, 1853. < 33d Congress, 2d Session. } House of Representatives. { Ex. Doc. No. 91. | — | Reports | of | Explorations and Surveys, | to | ascertain the most practicable and economical route for a railroad | from the | Mississippi River to the Pacific Ocean. | Made under the direction of the Secretary of War, in | 1853–6, | according to acts of Congress of March 3, 1853, May 31, 1854, and August 5, 1854. | — | Volume IX. | — | Washington: | A. O. P. Nicholson, Printer. | 1858. 4to, pp. lvi, 1005.

"This report is complete in itself, and entirely independent of the various special articles by different naturalists of the several Surveys; an elaborate formal treatise on all the birds of North America north of Mexico. It represents the most important single step ever taken in the progress of American ornithology in all that relates to the technicalities. The nomenclature is entirely remodeled from that of the immediately preceding Audubonian period, and for the first time brought abreast of the then existing aspect of the case. It was adopted by the Smithsonian Institution, and thousands of separately printed (4to and 8vo) copies of the 'List of Species' were distributed during succeeding years to institutions and individuals; the names came at once into almost universal employ, and so continued, with scarcely appreciably diminished force, until about 1872. The synonymy of the work is more extensive and elaborate and reliable than any before presented. The compilation was almost entirely original, very few citations having been made at second-hand, and these being indicated by quotation-marks. The general text consists of diagnoses or descriptions of each species, with extended and elaborate criticisms, comparisons, and commentary. Of the general character of the specific determinations, it may be said that the author's tendency was to push specific discriminations beyond a point now usual; so that, though the work contains notably few purely nominal species, it has many that have proven to be simply geographical races. Tabular lists of the specimens examined, with localities where procured, collector, date of collection, and many measurements, are given. The work includes no biographical matter, nor is it illustrated.

"The appearance of so great a work, from the hands of a most methodical, learned, and sagacious naturalist, aided by two of the leading ornithologists of America, exerted an influence perhaps stronger and more widely felt than that of any of its predecessors, Audubon's and Wilson's not excepted, and marked an epoch in the history of American ornithology.

1858. BAIRD, SPENCER F.—Continued.

The synonymy and specific characters, original in this work, have been used again and again by subsequent writers, with various modifications and abridgment, and are in fact a large basis of the technical portion of the subsequent *History of North American Birds* by Baird, Brewer, and Ridgway. Such a monument of original research is like to remain for an indefinite period a source of inspiration to lesser writers, while its authority as a work of reference will always endure."—ELLOT COUES.

The following extract from the preface explains the scope of this work:

"The present report is a continuation of a systematic account of the vertebrate animals of North America, collected or observed by the different parties organized under the direction of the War Department for ascertaining the best route for a railroad from the Mississippi River to the Pacific Ocean.

"The collections of these expeditions having been deposited with the Smithsonian Institution by the War Department, in compliance with an act of Congress, the undersigned was charged by the Secretary of the Institution with the duty of furnishing the series of general reports upon them, as called for by the department. The account of the Mammals having been published in 1857, that of the Birds is herewith furnished, prepared according to the plan announced in the preface to that volume."

"As in the volume on the Mammals, by the insertion of the comparatively few species not noticed by the expeditions, this report becomes an exposition of the present state of our knowledge of the Birds of North America, north of Mexico. This addition, while rendering the work more valuable to the reader, was absolutely necessary for the proper understanding of the Western Fauna, the species of which are generally so closely allied to the Eastern forms as to require in most cases more minute and detailed descriptions of the latter than have been published."

"Certain portions of the report have been prepared by Mr. John Cassin, of Philadelphia, and Mr. George N. Lawrence, of New York, well known as the leading ornithologists of the United States. Mr. Cassin has furnished the entire account of the *Raptores*, from p. 4 to 64, of the *Grallæ*, from p. 689 to 753, and of the *Alcidæ*, from p. 900 to 918, in all about 135 pages. Mr. Lawrence has written the article on the *Longipennes*, *Totipalmes*, and *Colymbidæ*, from p. 820 to 900, making 80 pages."

"To Mr. P. L. Sclater, of London, acknowledgments are due for the examination of certain specimens in European museums, and for other valuable aid in determining points of synonymy; some of his notes received too late for insertion in their proper places will be found in Appendix A. Much assistance has also been rendered in various ways by Dr. J. G. Cooper."

"In the introduction to the general report upon the Mammals will be found a detailed account of the different railroad surveying parties from which Zoological collections were received, with their respective routes."   *   *   *

"A collection of about 150 species received from Mr. John Gould, of London, contains many rare birds from the Northwest and Arctic regions (some of them types of the 'Fauna Boreali-Americana'), as well as others from Mexico and Guatamala. The latter have proved of great service for comparison with closely allied species of the United States, as have also specimens from Mr. P. L. Sclater, of London, Mr. J. P. Verreaux, of Paris, and Messrs. J. H. Gurney and Alfred Newton, of Norwich, England."

"The types of Eastern birds have been furnished by the collection of the author deposited in the Smithsonian Institution. This consists of a full collection of birds of Central Pennsylvania, with sex, date, and measurements before skinning. It also embraces a large number of Mr. Audubon's typical specimens used in the preparation of his 'Birds of America,' including many of those from the Columbia River and Rocky Mountains, furnished him by Mr. J. K. Townsend."

"In addition to the collections just mentioned, with others not enumerated, all in charge of the Smithsonian Institution, and amounting to over 12,000 specimens, types have been supplied for the occasion by Mr. Cassin, Mr. Lawrence, Mr. John G. Bell, Dr. Michener, and others. The ornithological gallery of the Philadelphia Academy of Natural Sciences, believed to be the richest in the world, has also furnished the means of making many essential comparisons."

"The measurements of the specimens have usually been made in hundredths of the English inch,* mostly with the dividers. All the measurements in the list of specimens are as made before the bird was skinned, each collector being responsible for the accuracy of his work. The comparative tables of measurements show, in many cases, the change of dimensions produced in the dried skin.   "S. F. B.

"WASHINGTON, *October* 20, 1853."

< Preface, *Expl. and Surv. for the Pacific Railroad*, vol. ix, 1853-'56, pp. xiii and xv, xvi.

1858. BAIRD, SPENCER F.—Continued.

NOTE.—When authorities are inclosed in parentheses it shows that the species was first described under a different genus. A second authority (or a single one uninclosed) is that of the name as adopted. Extra limital species have their current number similarly inclosed.

1858. BAIRD, SPENCER F.—Continued.

1858. BAIRD, SPENCER F.—Continued.

1858. BAIRD, SPENCER F.—Continued.

The following birds are enumerated in the preceding list which are not legitimately entitled to a place in the fauna of North America (exclusive of Mexico). Some of them have been described in the report for the purpose of comparison with closely-allied species of the United States; others are mentioned because introduced by previous writers, though probably on erroneous data. Future investigations will doubtless result in the removal of others from the list now retained there:

1858. BAIRD, SPENCER F.—Continued.

244. *Vireo virescens, Vieill.[1]
292. Parus meridionalis, Sclater.
297. Psaltriparus melanotus, Bon.
298. Carpodacus hæmorrhous, Sclater.
311. *Chrysomitris Stanleyi, Bonap.
312. *Chrysomitris Yarrelli, Bonap.
350. Junco cinereus, Sclater.
405. Trupialis militaris, Bon.
408. *Icterus vulgaris, Daud.[1]
410. Icterus melanocephalus, Gray.
494. Butorides Brunnescens, Baird.
498. *Ibis rubra, L.[1]
514. *Haematopus ater, Vieill.

Total of species, 23; of which one is not mentioned in the list, leaving 22. Of the 23 species, nine marked with an asterisk are given by Mr. Audubon.

The following species, claiming to be actually inhabitants of North America, have not been described from the specimens, none having been procurable for the purpose. Of several of them no specimens are known in any collection:

Haliaelitus Washingtonii.
Regulus Cuvierii.
Dendroica montana.
    carbonata.
Myiodioctes minutus.
    Bonapartii.
Ægiothus canescens.
Leucosticte griseinucha.
    arctous.
Lagopus americanus.
Chloephaga canagica.
Polysticta Stelleri.[1]
Oidemia binaculata.
Somateria v-nigra.
Graculus perspicillatus.
    cincinnatus.

Thalassidroma Hornbyi.
    melania.
Larus chalcopterus.
Rissa brevirostris.
    nivea.
Pagophila brachytarsi.
Rhodostethia rosea.
Creagrus furcatus.
Xema Sabinni.
Chroicocephalus minutus.[1]
Podiceps auritus.[1]
Sagmatorhina labradoria.
Brachyrhamphus Kittlitzii.
    Wrangelii.
    brachypterus.
Total, 31 species.

The following species are probably accidental visitors only, and are not yet entitled to a permanent place in our fauna:

Milvulus tyrannus.
Saxicola œnanthe.
Chrysomitris magellanicus.
Philomachus pugnax.
Crex pratensis.

Heliornis surinamensis.[2]
Mareca penelope.
Nettion crecca.
Erismatura dominica.
Mergellus albellus.—Total, 10 species.

SUMMARY.

Species enumerated in the list 738
Of these, extralimital 22

Total of North American species[3] 716

North American birds given by Wilson in 1814 283
North American birds given by Bonaparte, 1838 471
North American birds given by Audubon in 1844 506

79.

1858. BAIRD, SPENCER F. Birds found at Fort Bridger, Utah. < *Pacific Railroad Report;* ix, 1858, App. B, pp. 926, 927.

Merely a list of 104 spp., collected by C. Drexler.

---

[1] No North American specimens seen.   [2] Not enumerated in the list.

[3] Of these no specimens at all, of 28 species, were procured in this country for examination, and extralimital ones only of 3 others. Many supposed species are referred to in different parts of the report, some of which may prove genuine.

5 BD

80.

1858. BAIRD, SPENCER F.   Catalogue | of | North American Birds, | chiefly in the Museum of the | Smithsonian Institution. | By | Spencer F. Baird, | Assistant Secretary of the Smithsonian Institution. | — | Washington : | Smithsonian Institution. | October, 1858. | 4to, pp. xv–lvi.

A large edition of this catalogue, a reprint of pp. xvii–lvi of the General Report on Birds, was printed in Oct., 1858, and distributed as No. 106 of the publications of the Smithsonian Institution.

"Separate reissue, *V. L. P.*, with new title-page, of pp. xvii–lvi of the author's main work. Besides the list of 738 spp., with habitats, these sheets contain a table of the higher groups, list of extralimital species (23) which are included in the work, and of those (31) claiming to be North American, but not so identified, and a summary of the total number as variously given by Wilson, Bonaparte, and Audubon. The species being all numbered, the brochure was much used for several years for practically convenient reference to the.species by number."—(COUES.)

81.

1858. BAIRD, SPENCER F.   [Report for 1857 of the Assistant Secretary of the Smithsonian Institution.]   < *Ann. Rep. Smithsonian Institution for the year* 1857, 1858, pp. 38–54.

Details of the system of exchanges... 44–46

82.

1858. BAIRD, SPENCER F.   Description of a Phyllostome Bat from California, in the Museum of the Smithsonian Institution.   < *Proc. Acad. Nat. Sci. Phila.*, x, 1858, pp. 116, 117.   Presented May 4, ordered published May 25, 1858.

Macrotus californicus, Baird, n. s... 116
  Fort Yuma, Cal. Maj. G. H. Thomas.

83.

1858. BAIRD, SPENCER F.   "Description of a new Sparrow collected by Mr. Samuels in California."   < *Proc. Bost. Soc. Nat. Hist.*, vi, pp. 379, 380, Aug., 1858. Read June 2.

Ammodromus Samuelis, Baird, n. s... 379
  Petaluma, Cal. E. Samuels.

84.

1858. [BAIRD, SPENCER F.]   Registry of Periodical Phenomena.   < *Directions for Metorological Observations and the Registry of Periodical Phenomena, Smiths. Misc. Coll.* (148), pp. 63–68.

Instructions for registering periodical phenomena of animal and vegetable life in North America.

85.

1858. [BAIRD, SPENCER F.]   United States | Exploring Expedition. | During the years | 1838, 1839, 1840, 1841, 1842. | Under the command of Charles Wilkes, U. S. N. Vol. XX. | — | Herpetology. | Prepared under the superintendency of | S. F. Baird. | With a folio Atlas. | — | Philadelphia: | Printed by C. Sherman & Son. | 1858. - 4to. pp. (4), v–ix, 492.

This book was not written by Professor Baird, who assures me that he did not touch pen to it. The work was done entirely by Dr. Charles Girard, but through some technicality his name was not allowed to appear on the title-page by the Naval authorities having the matter in charge, who insisted in publishing the book under the name of Professor Baird, to whom the original contract was given out. The matter is explained fully in the introduction, which is quoted entire:—

"INTRODUCTION.—The Joint Committee of the Library of Congress entered into an engagement with the undersigned, in 1857, to prepare the Report upon the Herpetological collections made by the United States Exploring Expedition. Finding that other duties would

1858. BAIRD, SPENCER F.—Continued.

interfere with the proper performance of the work, he was permitted to associate Dr. Girard with him in its execution, by whom the determinations and descriptions have been made, the drawings overlooked, and the work carried through the press. S. F. BAIRD. Washington, May, 1858. (p. vii.)"

As is well known only 100 copies of these reports were published by government, while the authors of the separate volumes were allowed to have 150 more printed at their own individual expense. In accordance with this ruling a special edition of this report was published by Girard with the following title:

1858. CHARLES GIRARD. United States | Exploring Expedition. | During the years | 1838, 1839, 1840, 1841, 1842. | Under the command of | Charles Wilkes, U. S. N. | Vol. XX. | — | Herpetology. | By | Charles Girard, | Doctor in Medicine and Surgery; Corresponding Member of the Boston | Society of Natural History; the Academy of Natural Sciences of Philadelphia; | the Lyceum of Natural History of New York; the Elliot Society of | Natural History of Charleston, S. C.; the California Academy | of Natural Sciences, San Francisco; the "Societe Helvetique | des Sciences Naturelles;" the "Naturforschende Gessell- | schaft in Zurich;" and the "Societe des Sciences | Naturelles de Neuchatel (Switzerland)," etc. | With a folio Atlas | — | Philadelphia | J. B. Lippincott & Co. | 1858. 4to. pp. i–xvii, 1–496, pll, i–xxxii.

Several additional plates were included in the Atlas to this edition.

86.

1859. BAIRD, SPENCER F. United States and Mexican | Boundary Survey, | under the order of | Lieut. Col. W. H. Emory, | Major First Cavalry, and United States Commissioner. | — | Mammals | of the Boundary, | by | Spencer F. Baird, | Assistant Secretary of the Smithsonian Institution. | With notes by the Naturalists of the Survey. | = | 4to, pp. (2) 3–62, pll. xxvii. $<$ 34th Congress, 1st Session. $\}$

House of Representatives. $\{$ Ex. Doc., No. 135. $|$ — | Report | on the | United States and Mexican Boundary Survey, | made under | the direction of the Secretary of the Interior, | by | William H. Emory. | Major first Cavalry and United States Commissioner. | — | Volume II. | Washington: | Cornelius Wendell, Printer, | 1859. | 4to. 1st part Botany: pp. (8) 9–270 + pll. 1–67 + pp. 1–78 (Cactaceae) + pll. 1–75. 2d part Zoology: pp. (2) + 3–62, pll. xxvii (Mammals); pp. (2) 3–33 (1), pll. xxv (Birds); pp. (2) 3–35 (1), pll. xli (Reptiles); + pp. 85 + (3) + (1–11), + pll. xli (Fishes).

Order CHEIROPTERA.

1859. BAIRD, SPENCER F.—Continued.

17. *Mephitis mesoleuca*, Licht.
18. *Mephitis varians*, Gray.
19. *Mephitis bicolor*, Gray...pl. xvii, fig. p. 321
20. *Taxidea Berlandieri*, Baird.
21. *Procyon Hernandezii*, Wagler ....pl. xviii   22
22. *Procyon Hernandezii* var. *mexicana*, St. Hilaire.
23. *Ursus horribilis* var. *horriaeus*, Baird., new variety ....pl. xx   24
24. *Ursus cinnamoneus* ....pl. xix   29
25. *Didelphys virginiana*, Shaw   31
26. *Didelphys californica*, Bennett...pl. iii   32

Order RODENTIA.

27. *Sciurus limitis*, Baird ....pll. iv, xxi, fig. 1   34
28. *Sciurus ludovicianus*, Custis...   35
29. *Sciurus carolinensis??*
30. *Sciurus castanonotus*, Baird...pll. v, xxi, fig. 2   35
31. *Tamias dorsalis*, Baird...pl. vi, fig. 1   37
32. *Spermophilus grammurus*, Bach ....pl. vii, fig. 1; pl. xxii, fig. 1   38
33. *Spermophilus Couchii*, Baird...pl. xxi, fig. 3   38
34. *Spermophilus tereticauda*, Baird ....pl. vii, fig. 2; pl. xxi, fig. 4   38
35. *Spermophilus mexicanus*, Wagner ....pl. xxii, fig. 2   39
36. *Spermophilus spilosoma*, Bennett.................pl. vii, fig. 3; pl. xxii, fig. 3   39
37. *Cynomys ludovicianus*, Baird.
38. *Castor canadensis*, Kuhl...   40
39. *Geomys Clarkii*, Baird ....pl. ix, fig. 1; pl. xxiii, fig. 1   41
40. *Thomomys umbrinus*, Baird.................pll. viii, x, fig. 1; pl. xxxiii, fig. 5   41
41. *Thomomys fulvus*, Baird.
42. *Dipodomys Ordii*, Woodhouse.................pl. ix, fig. 3; pl. xxiii, fig. 3   42
    Dipodomys Ordii var. *montanus* ....pl. xxiii, fig. 4
43. *Dipodomys agilis*, Gambel...pl. xxiii, fig. 2   42
44. *Perognathus penicillatus*, Woodhouse.
45. *Perognathus hispidus*, Baird ....pl. ix, fig. 2; pl. xxiii, fig. 6   42
46. *Perognathus flavus*, Baird ....pl. x, figg. 4, 5   42
47. *Mus tectorum*, Say.
48. *Reithrodon megalotis*, Baird ....pl. vii, fig. 4; pl. x, fig. 6; pl. xxiv, fig. 4   43
49. *Hesperomys texanus*, Woodhouse.
50. *Hesperomys sonoriensis*, Leconte.
51. *Hesperomys eremicus*, Baird...
52. *Neotoma mexicana*, Baird ....pl. x, fig. 3; pl. xxiv, fig. 1
53. *Neotoma micropus*, Baird...pl. xxiv, fig. 2
54. *Sigmodon berlandieri*, Baird ....pl. vi, fig. 2; pl. x, fig. 2   44
55. *Fiber zibethicus*, Cuv...pl. xxiv, fig. 3   45
56. *Lepus callotis*, Wagler ....pl. xxv, fig. 1   45
57. *Lepus californicus*, Gray...   47
58. *Lepus sylvaticus*, Bachman.
59. *Lepus artemisia*, Bachman ....pl. xxv, fig. 2   48
60. *Lepus Bachmani*, Waterhouse.
61. *Dasypus novem-cinctus*, Linnæus...pl. xxvi   48
62. *Dicotyle Torquatus*, Cuv ....pl. xxvii, figg. 1, 2   50
63. *Cervus virginianus*, Boddaert.
64. *Cervus mexicanus*, Gmelin.
65. *Cervus macrotis*, Say...   51
66. *Antilocapra americana*, Ord.
    Ovis montana   52
67. *Bos americanus*, Gmelin.

87.

1859. BAIRD, SPENCER F.   United States and Mexican | Boundary Survey, | under the order of | Lieut. Col. W. H. Emory, | Major First Cavalry, and United States Commissioner. | = | <Birds | of the Boundary, | by | Spencer F. Baird, | Assistant Secretary of the Smithsonian Institution. | With notes by the Naturalists of the Survey.   (34th Congress, 1st Session.) House of Representatives.

1859. BAIRD, SPENCER F.—Continued.

{ Ex. Doc. No. 135. } | — | Report | on the | United States and Mexican Boundary Survey, | made under | the direction of the Secretary of the Interior, | by William H. Emory, | Major First Cavalry and United States Commissioner. | — | Volume II. | — | Washington: | Cornelius Wendell, Printer. | 1859. 4to. pp. (2) 3–33, (1) pll. xxv.

REPORT ON THE BIRDS.

1. *Cathartes aura*, Ill. .... p. 3
2. *Falco columbarius*, L.
3. *Falco aurantius*, Gm.
4. *Falco femoralis*, Temm.
5. *Falco sparverius*, Linn.
6. *Accipiter Cooperi*, Bon.
7. *Accipiter fuscus*, Gmelin.
8. *Buteo Swainsonii*, Bon.
9. *Buteo calurus*, Cassin.
10. *Buteo borealis*, Gm.
11. *Buteo montanus*, Nutt.
12. *Buteo elegans*, Cassin.
13. *Archibuteo ferrugineus*, Gray 4
14. *Asturina nitida*, Bonap.
15. *Circus Hudsonius*, Vieillot.
16. *Pandion carolinensis*, Bon.
17. *Polyborus tharus*, Cassin.
18. *Craxirex unicinctus*, Cassin.
19. *Strix pratincola*, Bon.
20. *Bubo virginianus*, Bon.
21. *Scops McCalli*, Cassin .... pl. i
22. *Athene hypugœa*, Bon. 5
23. *Athene cunicularia*, Molina.
24. *Rhynchopsitta pachyrhyncha*, Bon.
25. *Trogon mexicanus*, Swains. .... pll. i, ii
26. *Geococcyx californianus*, Baird.
27. *Picus Harrisii*, Aud.
28. *Picus scalaris*, Wagler .... pl. iii
29. *Centurus flaviventris*, Sw .... pl. iv
30. *Centurus uropygialis*, Baird 6
31. *Melanerpes formicivorus*, Bonap.
32. *Colaptes mexicanus*, Swains.
33. *Colaptes chrysoides*, Malh.
34. *Trochilus colubris*, Linn.
35. *Trochilus Alexandri*, Bourc .... pl. v, fig. 3
36. *Selasphorus rufus*, Swains.
37. *Selasphorus platycercus*, Gould .... pl. v, figg. 1 and 2
38. *Antrostomus Nuttalli*, Cassin.
39. *Chordeiles Henryi*, Baird 7
40. *Chordeiles texensis*, Lawrence .... pl. vi
41. *Ceryle alcyon*, Boie.
42. *Ceryle americana*, Boie .... pl. vii
43. *Momotus cœruliceps*, Gould .... pl. viii
44. *Pachyrhamphus aglaiae*, Lafres .... pl. ix, fig. 1
45. *Lathmidurus major*, Cabanis .... pl. ix, fig. 2
46. *Milvulus forficatus*, Sw.
47. *Tyrannus vociferans*, Swains .... pl. x ... 8
48. *Tyrannus Couchii*, Baird .... pl. xi, fig. 1
49. *Myiarchus mexicanus*, Baird.
50. *Myiarchus Lawrencii*, Baird .... pl. ix, fig. 3
51. *Sayornis nigricans*, Bonap.
52. *Sayornis fuscus*, Baird.
53. *Sayornis Sayus*, Baird 9
54. *Contopus Richardsonii*, Baird.
55. *Empidonax pusillus*, Cab.

1859. BAIRD, SPENCER F.—Continued.

1859. BAIRD, SPENCER F.—Continued.

120. *Melospiza Lincolnii*, Baird.
121. *Peucaea Cassinii*, Baird.
122. *Embernagra rufivirgata*, Lawrence ....pl. xvii, fig. 3
123. *Calamospiza bicolor*, Bonap.
124. *Guiraca cœrulea*, Swainson.
125. *Cyanospiza parellina*, Baird...pl. xviii, fig. 1 p. 17
126. *Cyanospiza versicolor*, Baird ....pl. xvii, fig. 2
127. *Cyanospiza ciris*, Baird.
128. *Spermophila Moreletii*, Pucheran....................................pl. xvi, figg. 2, 3
129. *Pyrrhuloxia sinuata*, Bonap.
130. *Cardinalis virginianus*, Bonap.
131. *Pipilo megalonyx*, Baird.
132. *Pipilo Abertii*, Baird 18
133. *Pipilo mesoleucus*, Baird.
134. *Pipilo chlorura*, Baird.
135. *Molothrus pecoris*, Swains.
136. *Agelaius phoeniceus*, Vieillot.
137. *Agelaius gubernator*, Bon.
138. *Agelaius tricolor*, Bon.
139. *Xanthocepalus icterocephalus*, Baird.
140. *Sturnella neglecta*, Aud... 19
141. *Icterus Audubonii*, Girard.
142. *Icterus parisorum*, Bonap.....,.....................................pl. xix, fig. 1
143. *Icterus Wagleri*, Sclater...pl. xix, fig. 2
144. *Icterus cucullatus*, Swains.
145. *Icterus spurius*, Bon.
146. *Icterus baltimore*, Daudin.
147. *Icterius Bullockii*, Bon....................... 20
148. *Scolecophagus cyanocephalus*, Cab.
149. *Quiscalus macroura*, Swains..pl. xx
150. *Quiscalus major*, Vieill.
151. *Corvus carnivorus*, Bartram.
152. *Corvus cryptoleucus*, Couch. .
153. *Cyanocitta californica*, Strickl.
154. *Cyanocitta Woodhousii*, Baird.. pl. xxi
155. *Cyanocitta sordida*, Baird ....pl. xxii, fig. 1 21
156. *Cyanocitta ultramarina*, Strickl.....................................pl. xxii, fig. 2
157. *Xanthoura luxuosa*, Bonap.
158. *Psilorhinus morio*, Gray.
159. *Columba fasciata*, Say.
160. *Columba flavirostris*, Wagler.......................................pl. xxiii
161. *Melopelia leucoptera*, Bonap.
162. *Zenaidura carolinensis*, Bonap.
163. *Scardafella squamosa*, Bonap...................................\.............. 22
164. *Chamaepelia passerina*, Swain.
165. *Ortalida McCalli*, Baird.
166. *Ortyx virginianus*, Bon.
167. *Ortyx texanus*, Lawrence ....pl. xxiv
168. *Lophortyx californicus*, Bonap.
169. *Lophortyx Gambelii*, Nutt.. 23
170. *Callipepla squamata*, Gray.
171. *Cyrtonyx massena*, Gould.
172. *Grus canadensis*, Temm... 24
173. *Demiegretta rufa*, Baird.
174. *Garzetta candidissima*, Bonap.
175. *Herodias egretta* var. *californica*, Baird.
176. *Ardea herodias*, Linn.
177. *Florida caerulea*, Baird.
178. *Botaurus lentiginosus*, Stephens.
179. *Butorides virescens*, Bonap.
180. *Nyctiardea Gardeni*, Baird.
181. *Nyctherodius violaceus*, Reich.
182. *Tantalus loculator*, Linn.
183. *Ibis alba*, Vieillot.

1859. BAIRD, SPENCER F.—Continued.

184. *Ibis Ordii*, Bonap.
185. *Platalea ajaja*, Linn .... p. 25
186. *Oharadrius virginicus*, Borck.
187. *Aegialitis vociferus*, Cassin.
188. *Strepsilas interpres*, ——.
189. *Recurvirostra americana*, Gm.
190. *Himantopus nigricollis*, Vieill.
191. *Gallinago Wilsonii*, Bon.
192. *Macrorhamphus griseus*, Leach.
193. *Tringa canutus*, Linn.
194. *Tringa maculata*, Vieill.
195. *Gambetta melanoleuca*, Bon.
196. *Gambetta flavipes*, Bon.
197. *Numenius longirostris*, Wilson.
198. *Rallus virginianus*, Linn. 26
199. *Porzana carolina*, Vieill.
200. *Fulica americana*, Gm.
201. *Gallinula galeata*, Bon.
202. *Anser Gambelii*, Hartlaub.
203. *Bernicla canadensis*, Boie.
204. *Dendrocygna autumnalis*, Eyton -- :pl. xxv
205. *Anas boschas*, Linn.
206. *Dafila acuta*, Jenyens.
207. *Nettion carolinensis*, Baird.
208. *Querquedula discors*, Steph.
210. *Querquedula cyanoptera*, Baird.
211. *Spatula clypeata*, Boie 27
212. *Chaulelasmus streperus*, Gray.
213. *Mareca americana*, Steph.
214. *Fulix collaris*, Baird.
215. *Aythya americana*, Bon.
216. *Bucephala albeola*, Baird.
217. *Erismatura rubida*, Bonap.
218. *Mergus americanus*, Cassin.
219. *Larus delawarensis*, Ord.
220. *Chroicocephalus atricilla*, Linn.
221. *Sterna acuflavida*, Cabot.
222. *Rhynchops nigra*, Linn. 28
223. *Pelecanus erythrorhynchus*, Gmelin.
224. *Graculus mexicanus*, Bon.
225. *Podiceps Clarkii*, Lawrence.
226. *Podiceps dominicus*, Lath.

88.

1859. BAIRD, SPENCER F.   United States and Mexican | Boundary Survey, | under the order of | Lieut. Col. W. H. Emory, | Major First Cavalry, and United States Commissioner. | — | Reptiles | of the Boundary, | by | Spencer F. Baird, | Assistant Secretary of the Smithsonian Institution. | With notes by the Naturalists of the Survey. | = | 4to.   pp. (2) 3-35, pll. xli (for title of volume see under 80 and 81).

CHELONIA.

1. *Aspidonectes Emoryi*, Agassiz .... p. 3
2. *Gypochelys lacertina*, Agassiz.
3. *Ozotheca tristycha*, Agassiz.
4. *Thyrosternum sonoriense*, Agassiz.
5. *Platythyra flavescens*, Agassiz.
6. *Ptychemys mobilensis*, Agassiz.
7. *Trachemys elegans*, Agassiz.
8. *Chrysemys oregonensis*, Agassiz 4
9. *Xerobates Berlandieri*, Agassiz.

1859. BAIRD, SPENCER F.—Continued.

1859. BAIRD, SPENCER F.—Continued.

89.

1859. BAIRD, SPENCER F. Report for 1858 of the Assistant Secretary of the Smithsonian Institution. < *Ann. Rep. Smiths. Inst. for the year* 1858, 1859, pp. 44-62.

Present condition of the Museum.

[Materials contained in the Museum of the Smithsonian Institution], pp. 52-55.

An enumeration of 49 collections, which chiefly make up the Museum.

90.

1859. BAIRD, SPENCER F. Smithsonian Miscellaneous Collections. | — | Directions | for | collecting, preserving, and transporting | specimens of natural history. | Prepared for the use of | the Smithsonian Institution. | [Seal of Smithsonian Institution.] | [Third edition.] | Washington: Smithsonian Institution. | March, 1859. 8vo. pp. 40, 6 cuts.

No. 34, S. I. In Smithsonian Miscellaneous Collections, Vol. II, Art. VII.

91.

1859. BAIRD, SPENCER F. Smithsonian Miscellaneous Collections. | — | Catalogue | of | North American Birds | chiefly in the Museum of the | Smithsonian Institution. | By | Spencer F. Baird. | [First octavo edition.] | [Seal of Smithsonian Institution.] | Washington: | Smithsonian Institution. | 1859. 8vo. 2 p. ll., pp. 19 + 2.

This catalogue is a reprint, with some changes from the one in quarto forming a portion of the Report on North American Birds in Vol. IX of the Reports of the Pacific Railroad Survey, and published as a separate paper by the Smithsonian Institution in October, 1858. Its object was to facilitate the labeling of the specimens of birds and eggs in the Museum of the Institution, as also to serve the purpose of a check-list of the species.

A special edition was printed on one side of the paper only for labeling, also an edition on thin blue paper for mailing.

In the octavo edition the note on habitat and the references to the pages of the report, which were included in the quarto edition, are omitted, the serial number of the species, the scientific name, and the common name alone being given.

This publication was issued as No. 108 of the Smithsonian series, and was included in Vol. II of the Miscellaneous Collections. Several editions have in subsequent years been struck off from the same plates, and probably 10,000 copies of the catalogue have been distributed.

This is the most usual form of the "Smithsonian Catalogue" of North American birds, which has become a classical work among ornithologists, and in accordance with which the majority of American collections of skins and eggs are labeled. A revision of this, from the hand of Robert Ridgway, is now in press.

92.

1859. BAIRD, SPENCER F. Explorations and surveys for a railroad route from the Mississippi River to the Pacific Ocean. | War Department. | — | Reptiles: | By Spencer F. Baird, | Assistant Secretary of the Smithsonian Institution. | — | Washington, D. C. | 1859. < 23d Congress, 2d Session. } House of Representatives. { Ex. Doc. No. 91. | — | Reports | of | explorations and surveys, | to | ascertain the most practicable and economical route for a railroad | from the | Mississippi River to the Pacific Ocean. | Made under the direction of the Secretary of War, in | 1853-6. | According to acts of Congress of March 3, 1853, May 31, 1854, and August 5, 1854. | — | Volume X. | — | Washington: | A. O. P. Nicholson, printer. | 1859. Parts iii and iv. 4to. First article. No text, pll. xxiv-xxvi.*

This report consists entirely of plates, the text being omitted.

"As the general report on the reptiles of Western North America, observed by the different exploring parties, has been excluded from the series for want of room, all that can be given here is an explanation of the plates prepared for this report. These represent the details of external form in different species of North American Serpents. * * * The figures have, as far as possible, been taken from the type-specimens of the species, especially those described in the Catalogue of Serpents in the Museum of the Smithsonian Institution (1853), to which the page-column refers."

---

* "The make-up of the tenth volume of the Pacific Railroad Reports is such that it might be styled the 'Bibliographer's Despair'; it contains about 20 different title-pages and a corresponding number of different paginations."—GILL & COUES.

1859. BAIRD, SPENCER F.—Continued.

Explanation of the plates.

Plate.	Fig.	Name.	Page.	Locality.	Details given.
XXIV...	1	*Crotalus durissus*, L. P.............	1	Carlisle, Pa...............	All.
	2	*Crotalus adamanteus*, Beauv......	3	Pensacola, Fla	All.
	3	*Crotalus atrox*, B. and G........	5	Indianola, Tex............	U. L. S. F.
	4	*Crotalus confluentus*, Say	8	Red River	All.
	5	*Crotalus molossus*, B. and G.......	10	Sonora.....................	U. L. S. F. SC.
	6	*Crotalus oregonus*, Holb..........	145	Oregon ?..............	U. SC.
	7	*Crotalophorus miliarius*, Holb.....	11	Charleston, S. C	All.
	8	*Crotalophorus consors*, B. and G...	12	Indianola, Tex...........	U. SC.
XXV....	9	*Crotalophorus tergeminus*, Holb...	14	Racine, Wis	All.
	10	*Crotalophorus Edwardsii*, B. and G	15	Rio Grande, Tex	All.
	11	*Crotalophorus Kirtlandii*, Holb ...	16	Ohio	All.
	11	Bis *Crotalophorus Kirtlandii*, (young).	16	Ohio	S.
	12	*Ancistrodon contortrix*, B. and G...	17	Near San Antonio, Tex ...	All.
	13	*Toxicophis piscivorus*, B. and G....	19	Red River, La.	All.
	14	*Toxicophis pugnax*, B. and G	20	Indianola	All.
	15	*Elaps fulvius*, Cuv	21	Charleston, S. C	All.
	16	*Elaps tener*, B. and G..............	22	San Felipe, Tex........	All.
	17	*Elaps tristis*, B. and G.............	23	Kemper Co., Miss	All.
	18	*Dipsas septentrionalis*, Kennicott[1]	......	Brownsville, Tex	All.
XXVI...	19	*Euatenia saurita*, B. and G	24	Carlisle, Pa	All.
	20	*Euatenia Faireyi*, B. and G........	25	Red River, La.	All.
	21	*Euatenia proxima?*, B. and G	25	Indianola, Tex............	All.
	22	*Euatenia ornata*, B. and G[2]........	28	San Antonio to El Paso...	All.
	23	*Euatenia sirtalis*, B. and G........	30	Westport, N. Y	All.
	24	*Euatenia ordinata*, B. and G	32	Riceboro', Ga............	All.
	25	*Euatenia radix*, B. and G..........	34	Racine, Wis	All.
	26	*Euatenia marciana*, B. and G......	36	Red River	U. L. S. F. SC.
	1	*Nerodia transversa*, B. and G......	148	Red River	All.
	2	*Eutaenia dorsalis*, B. and G	31	Texas	All.
	3	*Eutaenia ordinoides*, B. and G.....	33	California.............	All.[3]
XXVII..	27	*Nerodia sipedon*, B. and G.........	38	Carlisle, Pa	All.
	28	*Nerodia erythrogaster*, B. and G....	40	Red River, La.	U. L. S. F. SC.
	29	*Nerodia taxispilota*, B. and G......	43	Liberty Co., Ga	All.
	30	*Nerodia Holbrookii*, B. and G.....	43	Red River, La.	All.
	31	*Nerodia niger*, B. and G	147	Massachusetts............	All.
	32	*Regina leberis*, B. and G	45	Carlisle, Pa	All.
	33	*Regina rigida*, B. and G............	46	Pennsylvania	All.
	34	*Regina Grahamii*, B. and G........	47	Texas	All.
	, 35	*Regina Clarkii*, B. and G	48	Indianola	U. L. S. F. SC.
	36	*Regina Kirtlandii*, Kenn[4]	......	Illinois	All.
	37	*Ninia diademata*, B. and G........	49	Orizaba, Mexico	S. U.
XXVIII.	38	*Heterodon platyrhinos*, Latr.......	51	Carlisle, Pa	All.
	39	*Heterodon cognatus*, B. and G....	54	Indianola	All.
	40	*Heterodon niger*, Troost..........	55	Carlisle, Pa.............	All, Nasal Pl.
	41	*Heterodon atmodes*, B. and G......	57	Charleston, S. C...........	All, N. P.
	42	*Heterodon simus*, Holb............	59	Charleston, S. C...........	All.
	42	Bis *Heterodon simus* (second spec.).	59	Charleston, S. C...........	U.
	43	*Heterodon nasicus*, B. and G.......	61	Eagle Pass, Tex..........	All, N. P.
XXIX...	44	*Pityophis melanoleucus*, Holb	65	Carolina	U. L. S. F. SC.
	45	*Pityophis Sayi*, B. and G	151	Fort Snelling	All.
	46	*Pityophis bellona*, B. and G........	66	Rio Grande	All.
	47	*Pityophis McClellanii*, B. and G ...	68	Red River	All.
	48	*Pityophis annectens*, B. and G.....	72	San Diego, Cal...........	All.

[1] U. S. Boundary Report, ii, 1859 ; Reptiles, p. 16.
[2] U.S. Boundary Report, ii, 1859; Reptiles, p. 16: *Eutaenia parietalis*, Cat. Serpents, 28.
[3] A, head from above; B, head from side; C, head from below ; D, anal region ; E, side scales.
[4] Kennicott, Pr. A. N. Sc., viii, April, 1856, p. 95.

1859. BAIRD, SPENCER F.—Continued.

Explanation of the plates—Continued.

Plate.	Fig.	Name.	Page.	Locality.	Details given.
XXIX...	49	*Scotophis alleghaniensis*, B. and G..	73	Carlisle, Pa	U. L. S. F. SC.
	50	*Scotophis Lindheimeri*, B. and G ..	74	New Braunfels, Tex	All.
	51	*Scotophis vulpinus*, B. and G	75	Racine, Wis	All.
XXX....	52	*Scotophis confinis*, B. and G	76	Anderson, S. C	U. S.
	53	*Scotophis laetus*, B. and G	77	Red River, Ark	All.
	54	*Scotophis guttatus*, B. and G	78	Kemper Co., Miss	All.
	55	*Scotophis 4-vitattus*, B. and G	80	Florida	U. L. S. F. SC.
	56	*Scotophis Emoryi*, B. and G	157	Howard Springs, Tex	All.
	57	*Ophibolus Boylii*, B. and G	82	Eldorado Co., Cal	U. L. S. F. SC.
	58	*Ophibolus splendidus*, B. and G ...	83	Sonora	All.
	59	*Ophibolus Sayi*, B. and G	84	Red River, Ark	U. L. S. F. SC.
	60	*Ophibolus rhombomaculatus*, B. and G.	86	Georgia	All.
	61	*Ophibolus eximius*, B. and G	87	Warren, Mass	All.
	62	*Ophibolus clericus*, B. and G	88	Clark Co., Va	All.
	63	*Ophibolus doliatus*, B. and G	89	Kemper Co., Va	All.
	64	*Ophibolus gentilis*, B. and G	90	Red River, Ark	All.
XXXI...	65	*Ophibolus getulus*, B. and G	85	Charleston, S. C	All.
	66	*Georgia obsoleta*, B. and G	158	Brownsville, Tex	All.
	67	*Bascanion constrictor*, B. and G ...	93	Carlisle	All.
	68	*Bascanion Fremontii*, B. and G....	95	California	All.
	69	*Bascanion Foxii*, B. and G	96	Michigan. (Both sides of head drawn to show difference in labials.)	All.
	70	*Bascanion flaviventris*, B. and G...	96	Texas?	All.
	71	*Masticophis flageliformis*, B. and G. (old).	98	Pensacola, Fla	All.
XXXII..	72	*Masticophis flageliformis* (young) .	149	Georgia	U. L. S. F.
	73	*Masticophis flavigularis*, B. and G. (young).	99	Indianola, Tex	U. L. S. F.
	74	*Masticophis mormon*, B. and G....	101	Salt Lake	All.
	75	*Masticophis ornatus*, B. and G.....	102	San Antonio to El Paso...	All.
	76	*Masticophis taeniatus*, B. and G ...	103	California	All.
	77	*Masticophis Schottii*, B. and G.....	160	Eagle Pass	All.
	78	*Salvadora Grahamiæ*, B. and G....	104	Sonora	U. L. S. F. SC.
	79	*Leptophis æstivus*, Bell	106	Anderson, S. C	U. L. S. F. SC.
	80	*Leptophis majalis*, B. and G	107	Indianola	U. S.
	81	*Chlorosoma vernalis*, B. and G	108	Westport, N. Y. (2 diam.)	S. A.
	1	*? Diadophis docilis*, B. and G	114	Tucson. (2 diam.[1])	All.
	2	*Diadophis*		Santa Magdalena. (2 diam.*)	All.
	3	*Wenona*		Oregon*	All.
XXXIII.	82	*Diadophis punctatus*, B. and G....	112	Carlisle, Pa	U. S.
	83	*Diadophis amabilis*, B. and G.....	113	San José, Cal	All.
	84	*Diadophis docilis*, B. and G	114	San Pedro, Comanche Spr'g	All.
	85	*Diadophis pulchellus*, B. and G....	115	Eldorado Co., Cal	All.
	86	*Diadophis regalis*, B. and G	116	Sonora, Mex	All.
	87	*Taeniophis imperialis*, B. and G.[2]		Brownsville, Tex	All.
	88	*Sonora semi-annulata*, B. and G...	117	Sonora	S. U.
	89	*Rhinostoma coccinea*, Holb	118	Riceboro', Ga	All.
	90	*Rhinocheilus Lecontei*, B. and G..	120	California	U. L. S. F. SC.
	91	*Haldea striatula*, B. and G	122	Richmond, Va. (2 diam.).	U. L. S. F. SC.
	92	*Farancia abacura*, B. and G	123	Red River, La	U. L. S. F. SC.
	93	*Abastor erythrogrammus*, Gray...	125	Southern States	U. L. S. F. SC.
	94	*Virginia valeria*, B. and G	127	Maryland. (2 diam.)	U. L. S. F. SC.
	95	*Celuta amoena*, B. and G	129	Carlisle. (2 diam.)	U. L. S. F. SC.

[1] References as in pl. xxvi, figg. 1–3.

[2] *Taeniophis imperialis*, B. and G., Mexican Boundary Report, ii, 1859; Reptiles, **23**.

1859. BAIRD, SPENCER F.—Continued.

Explanation of the plates—Continued.

Plate.	Fig.	Name.	Page.	Locality.	Details given.
XXXIII.	96	*Tantilla coronata*, B. and G	131	Kemper Co., Miss. (2 diam.)	U. S.
	97	*Osceola elapsoidea*, B. and G.......	133	Charleston, S. C. (2 diam.)	U. S.
	98	*Storeria DeKayi*, B. and G........	135	Framingham, Mass. (2 diam.)	U. S.
	99	*Storeria occipito-maculata*, B. and G.	137	Madrid, N. Y. (2 diam.)..	U. S.
	100	*Rena dulcis*, B. and G	142	San Pedro, Can. Sp., Tex. (3 diam.)	U. L. S.
XXXIV.	1	*Georgia Couperi*, B. and G........	92	Georgia....................	A. B. C. D. E.[1]
	2	*Nerodia rhombifer*, B. and G	147	Arkansas	As in fig. 1.
	3	*Nerodia Woodhousii*, B. and G	42	Texas.....................	As in fig. 1.
	4	*Nerodia fasciata*, B. and G	39	South ?....................	As in fig. 1.[2]
	5	*Eutaenia radix*, B. and G..........	34	Wisconsin?...............	As in fig. 1.
	6	*? Microps lineatus*, Hall[3]	......	Fort Chadbourne, Tex	As in fig. 1.
	7	——?....................	......	....	....
XXXV ..	1	*Crotalus tigris*, Kennicott[4]	......	....	As in fig. 1.
	2	*Crotalus*	......	Colorado Bottom..........	As in fig. 1.
	3	*Crotalus*	......	Sierra Verde, California...	As in fig. 1.
	4	*Crotalus cerastes*, Hallow[5]	......	....	As in fig. 1.
	5	*Crotalus*	......	Sierra Verde, Cal	As in fig. 1.
	6	*Lamprosoma occipitale*, Hall[6] ...	......	Colorado Desert..........	As in fig. 1.
	7	*Lamprosoma occipitale*, Hall[7]	......	California	As in fig. 1.
	8	*Toluca lineata*, Kenn.[8]...............	......	Valley of Mexico..........	As in fig. 1.
XXXVI.	1	*Crotalus lucifer*, B. and G	6	Oregon....................	As in fig. 1.
	2	*Eutaenia leptocephala*, B. and G ...	29	Oregon....................	As in fig. 1.
	3	*Eutaenia Pickeringii*, B. and G	27	Oregon....................	As in fig. 1.
	4	*Pityophis catenifer*, B. and G......	69	California.................	As in fig. 1.
	5	*Pityophis Wilkesii*, B. and G	71	Oregon....................	As in fig. 1.
	6	*Bascanion vetustus*, B. and G......	97	Oregon....................	As in fig. 1.
	7	*Contia mitis*, B. and G	110	Oregon. (2 diam.)........	As in fig. 1.
	8	*Lodia tenuis*, B. and G.............	116	Oregon. (2 diam.)........	As in fig. 1.

Pp. 14, 15, 16, pll. xxiv–xxxvi.

[1] All, except head from front.
[2] Add F, dorsal scales; G, side view of ditto, showing a peculiar serration of the carination.
[3] Hallowell, Pr. A. N. Sc., viii, 1856, 240.
[4] Mex. Bound. Rep., ii, 1839; Reptiles, 14.
[5] Hallowell, Pr. A. N. Sc., vii, June, 1854, 95.
[6] U. S. Mex. Bound., ii, 1859; Reptiles, 21.
[7] Hallowell, Pr. A. N. Sc., *ut supra*.
[8] Kennicott, *ut supra*, 23.

93.

1859. BAIRD, SPENCER F. No. 1. Report upon Mammals collected on the Survey. < Explorations and surveys for a railroad route from the Mississippi River to the Pacific Ocean. | War Department. | — | Report | of | Lieut. E. G. Beckwith, | Third Artillery, | upon | Explorations for a Railroad Route, | near | the 38th and 39th parallels of north latitude, | by | Captain J. W. Gunnison, | Corps of Topographical Engineers, | and near | the forty-first parallel of north latitude, | by | Lieut. E. G. Beckwith, | Third Artillery. | — | 1854. Zoölogical Report, xx, 1857, in Report P. R. R. Surv., vol. x, 1859. *Third* Article. pp. (1) 7–9 (1), pll. v–x.

Sciurus Fremontii, Towns ....pl. xi p. 7
Tamias quadrivittatus, Wagner.
Cynomis Gunnisonii, Baird...pl. iv, fig. 2   8
Geomys castanops, Leconte ....pl. x, fig. 2
Thomomys rufescens, Maxim...pl. x, fig. 1
Dipodomys Ordii, Woodhouse.

1859. BAIRD, SPENCER F.—Continued.
    Perognathus flavus, Baird.
    Jaculus Hudsonius.
    Reithrodon montanus, Baird .... p. 9
    Arvicola modesta, Baird.

94.

1859. BAIRD, SPENCER F. No. 2. Report on Birds collected on the Survey. < Explorations and Surveys for a Railroad Route from the Mississippi River to the Pacific Ocean. | War Department. | — | Report | of | Lieut. E. G. Beckwith, | Third Artillery, | upon | Explorations for a Railroad Route, | near | the 38th and 39th parallels of north latitude, | by | Captain J. W. Gunnison, | Corps of Topographical Engineers, | and near | the forty-first parallel of north latitude, | by | Lieut. E. G. Beckwith, | Third Artillery. | — | 1854. pp. 11-16, pll. xii, xiii, xiv, xv, xvii, xxxii, xxxv, in Rep. P. R. R. Surv., vol. x. Third article.

  Buteo Swainsoni, Bonaparte..........................pll. xii and xiii p. 11
  Buteo calurus, Cassin ....pl. xiv
  Buteo oxypterus, Cassin...pl. xv
  Buteo montanus; Nuttall 12
  Circus Hudsonius, Linnæus.
  Tinnunculus sparverius,
  Otus Wilsonianus, Lesson.
  Athene cunicularia, Molina.. 13
  Chordeiles Henryi, Cassin ....pl. xvii
  Sialia arctica, Swainson ....pl. xxxv
  Eremophila cornuta, Boie ....pl. xxxii
  Xanthocephalus icterocephalus.
  Corvus carnivorus, Bartram 14
  Pica Hudsonica, Bonap.
  Perisoreus canadiensis, Bonap.
  Centrocercus urophasianus, Swainson.
  Grus canadensis, Temm.
  Symphemia semipalmata, Hartlaub 15
  Numenius longirostris, Wilson 15
  Fulica americana, Gmelin.
  Cygnus americanus, Sharpless.
  Anas boschas, L.
  Aythya americana, Bon 16
  Nettion carolinensis, Baird.
  Bucephala americana, Baird.

95.

1859. BAIRD, SPENCER F. No. 3. Report on Reptiles collected on the Survey. < Explorations and Surveys for a Railroad Route from the Mississippi River to the Pacific Ocean. | War Department. | — | Report | of | Lieut. E. G. Beckwith, | Third Artillery, | upon | Explorations for a Railroad Route, | near | the 38th and 39th parallels of north latitude, | by | Captain J. W. Gunnison, | Corps of Topographical Engineers, | and near | the forty-first parallel of north latitude, | by | Lieut. E. G. Beckwith, | Third Artillery. | — | 1854. pp. 17-20 [in Rep. P. R. R. Surv., vol. x, third article], pll. xvii, xviii, xxiii, xxiv.

  Sceloporus graciosus, B. and G... p. 17
  ? *Sceloporus occidentalis*, B. and G.
  ? *Sceloporus longipes*, Baird.
  Crotaphytus collaris, Holbrook.............................pl. xxiv, fig. 1
  Crotaphytus Wislizenii, B. and G.
  Callisaurus ventra is, Baird.
  Holbrookia maculata, Girard... 18
  Tapaya brevirostris, Girard.
  Tapaya Douglassii, Girard.
  Doliosaurus platyrhinos, Girard.
  Cnemidophorus tesselatus, Baird.

1859. BAIRD, SPENCER F.—Continued.

96.

1859. BAIRD, SPENCER F. No. 4. Report upon the Reptiles of the Route. ⟨Explorations and Surveys for a Railroad Route from the Mississippi River to the Pacific Ocean. | War Department. | — | Route near the Thirty-fifth Parallel, explored by Lieutenant A. W. Whipple, Topographical Engineers, in 1853 and 1854. | — | Zoological Report. | — | Washington, D. C. | 1859. | — | [In vol. x, Report P. R. R. Surv., part vi, No. 4.] pp. 37–45, pll. xxv, xxvi, xxvii.

1859. BAIRD, SPENCER F.—Continued.

Bufo cognatus, Say ....pl. xxvi
Acris crepitans, Baird.
Rana Oatesbiana, Shaw...p. 45
Rana clamitans, Daud.
Rana halecina, Kalm.
Rana Berlandieri, Baird.
Necturus lateralis, Baird.

97.

1859. BAIRD, SPENCER F. Report on Mammals collected on the Survey. No. 3. < Explorations and Surveys for a Railroad Route from the Mississippi River to the Pacific Ocean. | War Department. | — | Routes in California, to connect with the routes near the Thirty-fifth and Thirty-second Parallels, explored by Lieut. R. S. Williamson, Corps of Top. Eng., in 1853. | — | Zoological Report. | — | Washington, D. C. | 1859. [In vol. x, Report P. R. R. Surv., part iv, art. 3.] pp. 81, 82.

Vespertilis pallidus, Leconte ....p. 81
Lynx rufus var. *maculatus*.
Sciurus fossor, Peale.
Spermophilus Beecheyi, Rich.
Spermophilus Harrisii, Aud. and Bach 82
Thomomys bulbivorus, Baird.
Dipodomys Phillippii, Gray.
Perognathus parvus, Leconte.
Hesperomys Gambelii, Baird.

98.

1859. "EDITORIAL. [S. F. Baird's Résumé of Ornithological Field Operations in progress in America, etc.] < *Ibis*, i, 1859, pp. 334, 335.

99.

1859. BAIRD, SPENCER F. Notes on a collection of Birds made by Mr. John Xantus, at Cape St. Lucas, Lower California, and now in the Museum of the Smithsonian Institution. < *Proc. Acad. Nat. Sci. Phila.*, xi, 1859 (1860), pp. 299-306. Presented for publication Nov. 8, ordered printed Nov. 29, 1859.

The paper preceding this one is "Descriptions of supposed new species of Birds from Cape St. Lucas, Lower California. By John Xantus," *op. cit.*, pp. 297-299, in which are described 4 new species, included in the list now under consideration.

Peculiarities of the Zoology of Cape St. Lucas ....p. 290
  Bird fauna of Cape St. Lucas.
  Mammal fauna of Cape St. Lucas.
  Reptile fauna of Cape St. Lucas.
Zoology of the Gulf of California... 300
Causes of peculiar distribution of animal life in Cape St. Lucas.
Physical features of Cape St. Lucas.
Laws of distribution, migration, size, exemplified in the Cape St. Lucas collection.
Table illustrating geographical distribution of species found at Cape St. Lucas 301
Relations of land fauna of Mexico.
Relations of Marine Invertebrates of Cape St. Lucas to those of Mexico.............. 302
Note by William Stimpson on the Crustaceans of Cape St. Lucas.

42 species are enumerated, "collected from the middle of April to the middle of July, 1859," 7 of which, all new, "may as yet be considered as peculiar to Cape St. Lucas."

*1. *Tinnunculus sparverius*, Vieillot... 302
*2. *Bubo virginianus*, Bonap.
 3. *Picus lucasanus*, Xantus
 4. *Centurus uropygialis*, Baird.
 5. *Colaptes chrysoides*, Malherbe.
    (Ref. to *Colaptes* collected by Mr. Schott, P. R. R. Rep., ix, p. 125.)
*6. *Geococcyx californicus*, Baird 303
*7. *Chordeiles texensis*, Lawrence.
 8. *Myiarchus mexicanus*, Baird.

* Name only. Others have descriptive or critical remarks.

1859. BAIRD, SPENCER F.—Continued.

*9. *Sayornis nigricans*, Bonaparte.
*10. *Empidonax obscurus*, Baird.
11. *Hirundo thalassina*, Swainson.
*12. *Progne purpurea*, Boie.
*13. *Phainopepla nitens*, Sclater.
14. *Mimus polyglottus*, Boie.
15. *Harporhynchus cinereus*, Kantus...p. 303–304
16. *Campylorhynchus affinis*, Xantus.
17. *Polioptila melanura*, Lawrence.. 304
18. *Paroides flaviceps*, Baird.
19. *Carpodacus frontalis*, Gray.
*20. *Chondestes grammaca*, Bonap.
21. *Zonotrichia leucophrys*, Swains.
22. *Calamospiza bicolor*, Bonap.
*23. *Guizaca melanocephala*, Swains.
24. *Cyanospiza versicolor*, Baird.
25. *Pyrrhuloxia sinuata*, Bonap.
26. *Cardinalis igneus*, Baird, n. s...................................... 305
      Cape St. Lucas. J. Xantus.
27. *Pipilo albigula*, Baird, n. s.
      Cape St. Lucas. J. Xantus.
28. *Agelaius* ——.
29. *Icterus parisorum*, Bonap.
*30. *Icterus cucullatus*, Swainson.
31. *Cyanocitta californica*, Strickland.
32. *Melopelia leucoptera*, Bonap.
33. *Chamæpelia passerina?* var. *pallescens*, Baird, n. s.
*34. *Lophortyx californicus*, Bonap.
35. *Garzetta thula*, Bonap.?
36. *Aegialitis vociferus*, Cassin 306
*37. *Calidris arenaria*, Illiger.
*38. *Fulica americana*, Gmel.
39. *Graculus dilophus?* Gray.
40. *Thalassidroma melania*, Bonap.
41. *Blasipus Heermanni*, Bonap.
42. *Brachyrhamphus hypoleucus*, Xantus.

100.

1859. BAIRD, SPENCER F. Description of New Genera and Species of North American Lizards in the Museum of the Smithsonian Institution. < *Proc. Acad. Nat. Sci. Phila.*, x, 1858 (1859), pp. 253–256. Presented for publication Dec. 21 (p. 222), ordered printed Dec. 28.

Euphryne, Baird, n. g. (*Iguanidæ*.)
Euphryne obesus, Baird, n. s.
      Cañons of Colorado and California.
Crotaphytus reticulatus, Baird, n. s.
      Laredo and Ringgold Barracks, Texas.
Uta symmetrica, Baird, n. s.
      Fort Yuma, Cal.
Uta Schottii, Baird, n. s.
      Santa Madelina, Cal.
Uma, Baird, n. g. (*Iguanidæ*.)
Uma notata, Baird, n. s.
      Mohave Desert.
Holbrookia approximans, Baird, n. s.
      Lower Rio Grande.
Sceloporus floridanus, Baird, n. s .... p. 254
      Pensacola, Fla.
Sceloporus ornatus, Baird, n. s.
      Patos, Coahuila.
Sceloporus longipes, Baird, n. s.
      Fort Tejon, Cal.
Sceloporus Couchii, Baird, n. s.
      Santa Calenna. N. Leon.

1859. BAIRD, SPENCER F.—Continued.

Anolis Cooperi, Baird, n. s.
California.
Sphæriodactylus notatus, Baird, n. s.
Key West, Fla.
Stenodactylus variegatus, Baird, n. s.
Rio Grande and Gila Valleys.
Xantusidæ, Baird, new family.
Xantusia, Baird, n. g .... p. 255
Xantusia vigilis, Baird, n. s.
Fort Tejon, Cal.
Cnemidophorus inornatus, Baird, n. s.
New Leon.
Cnemidophorus octolineatus, Baird, n. s.
New Leon.
Gerrhonotus Webbi, Baird, n. s.
Near San Diego, Cal.
Gerrhonotus infernalis, Baird, n. s.
Devil's River, Texas.
Gerrhonotus olivaceus, Baird, n. s.
Near San Diego.
Lepidosternum floridanum, Baird, n. s.
Micanopy, Fla.
Plestiodon leptogrammus, Baird, n. s.. 256
Platte River Valley.
Plestiodon inornatus, Baird, n. s.
Sand Hills of Platte.
Plestiodon tetragrammus, Baird, n. s.
Lower Rio Grande.
Plestiodon egregius, Baird, n. s.
Indian Key, Fla.
Plestiodon septentrionalis, Baird, n. s.
Minnesota and Nebraska.

101.

1859. BAIRD, SPENCER F.   Mammals | of | North America; | the descriptions of species based chiefly on the collections | in the | Museum of the Smithsonian Institution. | By Spencer F. Baird, | Assistant Secretary of the Smithsonian Institution. | With Eighty-seven Plates of Original figures, illustrating the genera and species, and including details of external forms, | and osteology. | — | Philadelphia: | *J. B. Lippincott & Co. | 1859.   4to.   pp. xxxiv, 1–735 +1–55 + 737–764, pll. i–lxxxvi (Mammals).

This special edition is made up from the reports on the Mammals in Vol. VIII of the Pacific Railroad Reports and in Vol. II, Part II, of the Reports of the Mexican Boundary Survey.

Part I "is a reprint of the General Report on the Mammals of the Pacific Railroad surveying parties, which, for reasons explained on page xxvi of the preface, embraces all the known species (excepting of *Cheiroptera, Pinnipedia,* and *Cetacea*) north of Mexico." pp. xi–xxxiv of this edition correspond to pp. xxv–xlviii of the original work, while pp. 1–735 are the same.   To this part is prefixed a special title-page, as follows:

= | Part I. | — | General Report | upon the Mammals | of | the several Pacific Railroad Routes. | By | Spencer F. Baird, | Assistant Secretary of the Smithsonian Institution. | — | Washington, D. C. | July, 1857. | = |

"To this is added, as Part II, the Special Report on the Mammals of the United States and Mexican Boundary Survey, as in it the species found along the boundary-line are treated of more fully than in Part I."

To this part is prefixed a special title-page, as follows:

=′ | Part II. | — | Special Report | upon the Mammals of the Mexican Boundary. | By | Spencer F. Baird, | Assistant Secretary of the Smithsonian Institution; | with notes by the naturalists of the survey. | — | Washington, D. C. | January, 1859. | = |

* Some copies have the imprint of D. Appleton & Co.

1859. BAIRD, SPENCER F.—Continued.

Part II embraces pp. 3-55 of the Mexican Boundary Mammals.

Part III.—References to the figures is made up anew for this edition. "To the 43 plates accompanying the General Report on Mammals in Volume VIII of the Pacific Railroad series have been added 17 others, scattered through the different volumes, as well as 27 accompanying the Mexican Boundary Report. Such of the figures as require are colored, in this edition, for the first time. They represent the external form of 47 species, with details of structure, and the skulls and teeth of 106 others, making 161 species illustrated in some way."

SYSTEMATIC LIST OF ILLUSTRATIONS.

CHEIROPTERA.

Vespertilio pallidus, Leconte.—Animal, pl. lxi, fig. 1.
Macrotus californicus, Baird.—Animal, pl. lxi, fig. 2.

SORICINAE.

Neosorex navigator, Cooper.—Animal and skull, pl. xxvi, No. 629.
Sorex Trowbridgii, Baird.—Animal and skull, pl. xxvi.
Sorex vagrans, Cooper.—Details of external form, pl. xviii, figg. 5 and 6. Animal and skull, pl. xxvi, No. 1675.
Sorex Suckleyi, Baird.—Animal and skull, pl. xxvii, No. 1677.
Sorex pachyura, Baird.—Animal and skull, pl. xxvii, No. 1674.
Sorex Fosteri, Rich.—Details, pl. xxx, fig. 4.
Sorex platyrhinus, Wagner.—Animal and skull, pl. xxviii, No. 1699.
Sorex Cooperi, Bach.—Animal and skull, pl. xxvi, No. 2047.
Sorex Haydeni, Baird.—Animal and skull, pl. xxvi, No. 1685.
Sorex Hoyi, Baird.—Animal and skull, pl. xxviii, No. 1688.
Sorax Thomsoni, Baird.—Animal and skull, pl. xxvii, No. 1686.
Blarina talpoides, Gray.—Details of external form, pl. xviii, fig. 4, and pl. xxx, fig. 6.
Blarina brevicauda, Gray.—Details, pl. xxx, fig. 5.
Blarina carolinensis, Baird.—Details, pl. xxx, fig. 8.
Blarina angusticeps, Baird.—Details, pl. xxx, fig. 7.
Blarina cinerea, Baird.—Details, pl. xxx, figg. 9 and 10.
Blarina exilipes, Baird.—Animal and skull, pl. xxvii, No. 2157.
Blarina Berlandieri, Baird.—Animal and skull, pl. xxviii, No. 2159.

TALPINAE.

Scalops aquaticus.—Details of external form, pl. xvii, fig. 1.
Scalops Townsendii, Bach.—Details of external form, pl. xvii, fig. 5; pl. xxx, fig. 1.
Scalops Townsendii, Bach., var. *californicus.*—Details of external form, pl. xvii, figg. 2 and 6; pl. xxx, fig. 3.
Scalops Breweri, Bach.—Details of external form, pl. xvii, figg. 3-4; pl. xxx, fig. 2.
Talpa europaea, L.—Muzzle, pl. xvii, fig. 7.
Condylura cristata, Ill.—Details of external form, pl. xviii, figg. 1, 2.
Urotrichus Gibbsii, Baird.—Details of external form, pl. xviii, fig. 3. Animal and skull, pl. xxviii, No. 662.

FELINAE.

? Felis concolor, Linn.—Animal very young, pl. ii, fig. 2. Skull, adult, pl. lxxi, fig. 1; young, pl. lxxi, fig. 2.
Lynx fasciatus, Raf.—Animal, pl. ii, fig. 1.
Lynx maculatus.—Skull, adult, pl. lxxv, fig. 1; young, pl. lxxv, fig. 2.
Felis yaguarundi, Desm.—Skull, pl. lxxiv, fig. 1.
Felis eyra, Desm.—Animal, pl. lxii, fig. 1. Skull, pl. lxxiii, fig. 2.
Felis pardalis, Linn.—Skull, adult, pl. lxxii; young, pl. lxxiii, fig. 1.

LUPINAE.

Canis occidentalis var. *griseo-albus.*—Skull, pl. xxxi.
Canis latrans, Say.—Skull, pl. lxxvi.

VULPINAE.

Vulpes fulvus var. *fulvus.*—Skull, pl. xxxii.
Vulpes macroura, Baird.—Skull, pl. xxxiii.
Vulpes velox, Aud. and Bach.—Skull, pl. xxxiv.
Vulpes virginianus, Rich.—Skull, pl. xxxv, fig. 1.
Vulpes littoralis, Baird.—Animal, pl. i. Skull, pl. xxxv, fig. 2.

VINERRINAE.

Bassaris astuta, Licht.—Skull, pl. lxxiv, fig. 2.

1859. BAIRD, SPENCER F.—Continued.

MARTINAE.

Mustela Pennantii, Erxl.—Skull and gum folds, pl. xxxvi, fig. 1.
Mustela americana, Turton.—Skull, pl. xxxvi, fig. 2, and pl. xxxvii, fig. 1.
Putorius cicognanii.—Details, pl. xix, fig. 4.
Putorius Richardsonii, Bp.—Details, pl. xix, figg. 2–6.
Putorius noveboracensis, De Kay.—Skull, pl. xxxvi, fig. 3.
Putorius frenatus, Aud. and Bach.—Details, pl. xix, fig. 5. Details of young, pl. lxii, fig. 2.
Skull of adult, pl. lxxvii, fig. 1; of young, fig. 2.
Putorius xanthogenys.—Animal, pl. iii, fig. 1. Details, pl. xix, fig. 3.
Putorius vison, Rich.—Skull, pl. xxxvii, fig. 2 adult, and fig. 3 young.

LUTRINAE.

Lutra canadensis, Sab.—Details, pl. xix, fig. 7. Skull, pl. xxxviii.
Lutra californica, Gray.—Details, pl. xix, fig. 8.

MELINAE.

Mephitis varians, Gray.—Skull, pl. lx, fig. 2.
Mephitis mesoleuca, Licht.—Details, pl. xix, fig. 1. Skull, pl. xxxix, fig. 3.
Mephitis mephitica, Shaw.—Pl. lx, fig. 1.
Mephitis bicolor, Gray.—Animal, pl. xxix. Skull, pl. lx, fig. 3; pl. lxxvii, fig. 3.
Taxidea americana, Waterh.—Upper jaw, pl. xxxix, fig. 2.
Taxidea berlandieri, Baird.—Skull, pl. xxxix, fig. 7.

SUBURSINAE.

Procyon Hernandezii, Wagler.—Skull, rather young, pl. xl. Adult, pl. lxxvii.

URSINAE.

Ursus horribilis, Ord.—Skull, very old, pl. xli. Skull, rather young, pl. xlii.
Ursus horribilis var. *horiaeus*, Baird.—Skull, pl. lxxx.
Ursus arctos, L.—Skull, rather immature, pl. xliii, figg. 1–9.
Ursus americanus, Pallas.—Skull, pl. xliii, figg. 10–13.
Ursus maritimus, Linn.—Skull, pl. xliv.
Ursus cinnamoneus, Aud., Bach.—Skull, pl. lxxix.

MARSUPIALIA.

Didelphys californica, Bennett.—Animal, pl. lxiii. Fig. 1, adult; fig. 2, young.

SCIURINAE.

Sciurus castononotus, Baird.—Animal, pl. lxv. Skull, pl. lxxxi, fig. 2.
Sciurus cinereus, Linn.—Skull, pl. xlviii, fig. 2.
Sciurus limitis, Baird.—Animal, pl. lxiv. Skull, pl. lxxxi, fig. 1.
Sciurus carolinensis, Gm.—Skull, pl. xlv, fig. 2.
Sciurus Hudsonius, Pallas.—Skull, pl. xlvi, fig. 1.
Sciurus Fremontii, Towns.—Animal, pl. vi.
Sciurus Douglassii, Bach.—Details, pl. xx, fig. 1. Skull, pl. xlv, fig. 3.
Sciurus Douglassii, Bach., var. *Suckleyi.*—Animal, pl. vii.
Tamias striatus.—Skull, pl. xlvi, fig. 2.
Tamias quadrivittatus, Rich.—Details, pl. xx, fig. 2.
Tamias dorsalis, Baird.—Animal, pl. lxvi, fig. 1.
Tamias Townsendii, Bach.—Skull, pl. xlv, fig. 4.
Tamias Townsendii, Bach., var. *Cooperi.*—Animal, pl. v, fig. 2.
Spermophilus Beecheyi, Rich.—Animal, pl. iii, fig. 2. Skull, pl. xlvi, fig. 3.
Spermophilus Douglassii, Rich.—Skull, pl. xlv, fig. 1.
Spermophilus mexicanus, Licht.—Skull, pl. lxxxii, fig. 2.
Spermophilus grammura, Bachman.—Animal, pl. iv, fig. 1. Details, pl. lxvii, fig. 1. Skull, pl. lxxxii, fig. 2.
Spermophilus Couchii, Baird.—Skull, pl. lxxxi, fig. 3.
Spermophilus lateralis, Rich.—Details, pl. xx, fig. 3. Skull, pl. xlv, fig. 5.
Spermophilus spilosoma, Bennett.—Details, pl. lxvii, fig. 3. Skull, pl. lxxxix, fig. 3.
Spermophilus Harrisii, Aud. and Bach.—Skull, pl. xlviii, fig. 3.
Spermophilus Franklini, Rich.—Skull, pl. xlvi, fig. 4.
Spermophilus tereticauda, Baird.—Details, pl. lxvii, fig. 2. Skull, pl. lxxxi, fig. 4.
Cynomys lodovicianus.—Skull and teeth, pl. xlvii, figg. 2, 3.
Cynomys Gunnisonii, Baird.—Animal, pl. iv, fig. 2. Skull, pl. xlvii, fig. 4.
Arctomys monax, Gmelin.—Skull, pl. xlix, fig. 1.
Arctomys flaviventer, Bachman.—Skull, pl. xlvii, fig. 1.

1859. BAIRD, SPENCER F.—Continued.

CASTORINAE.

Aplodontia leporina, Rich.—Details, pl. xx, fig. 4.   Skull, pl. xlix, fig. 2.
Castor canadensis, Kuhl.—Skull, pl. xlviii, fig. 1.

SACCOMYINAE.

Geomys bursarius, Rich.—Details, pl. xxii, fig. 1.   Skull, pl. l, fig. 2.
Geomys breviceps, Baird.—Skull, pl. lii, fig. 2.
Geomys pinetis, Rafinesque.—Details, pl. xxii, fig. 3.
Geomys castonops, Leconte.—Animal, pl. x, fig. 2.   Skull, pl. l, fig. 2.
Geomys Clarkii, Baird.—Animals and details, pl. lxix, fig. 1.   Skull, pl. lxxxiii, fig. 2.
Geomys hispidus, Leconte.—Details, pl. xxii, fig. 4.
Thomomys bulbivorus, Baird.—Animal, with details of external form, pl. xi.   Skull, pl. l, fig. 3, and pl. lii, fig. 1.
Thomomys laticeps, Baird.—Animal, with details, pl. xii, fig 1.
Thomomys borealis, Baird.—Details, pl. xxii, fig. 2.
Thomomys umbrinus, Baird.—Animal, and details, pl. lxviii. Details, pl. lxx, fig. 1. Skull, pl. lxxxiii, fig. 5.
Thomomys rufescens, Maxim.—Animal, pl. x, fig. 1.
Thomomys fulvus, Baird.—Animal, with details, pl. xii, fig. 2.
Dipodomys Ordii, Woodhouse.—Animal, pl. v, fig. 1, and pl. lxix, fig. 3. Details, pl. xxi, fig. 1. Skull, much enlarged, pl. li, fig. 1. Natural size, fig. 2 ; also, pl. lxxxiii, fig. 3.
Dipodomys Ordii var. *montanus*, Baird.—Pl. lxxxiii, fig. 4.
Dipodomys agilis, Gambel.—Animal, pl. ix, fig. 1.   Skull, pl. lxxxiii, fig. 2.
Perognathus penicillatus, Woodhouse.—Details, pl. xx, fig. 5.
Perognathus hispidus, Baird.—Skull, pl. li, fig. 4, and pl. lxxxiii, fig. 6. Animal, pl. lxix, fig. 2.
Perognathus monticola, Baird.—Skull, pl. li, fig. 3.
Perognathus flavius, Baird.—Animal, pl. viii, fig. 2. Details, pl. xxi, fig. 3, and pl. lxx, figg. 4 and 5.

DIPODINAE.

Jaculus Hudsonius.—Details, pl. xxi, fig. 5.

MURINAE.

Mus tectorum, Savi.—Skull, pl. lii, fig. 6.
Reithrodon megalotis, Baird.—Skull, pl. lxxxiv, fig. 4.
Reithrodon montanus, Baird.—Teeth, pl. liv, No. 1306.
Reithrodon.—Details, pl. lxvii, fig. 4, and pl. lxx, fig. 6.
Hesperomys texanus, Woodhouse.—Animal, pl. viii, fig. 1.   Skull, pl. lii, fig. 5.
Hesperomys Boylii, Baird.—Animal, pl. viii, fig. 3.   Skull, pl. lii, fig. 3.
Hesperomys palustris, Wagner.—Skull, pl. lii, fig. 4.
Neotoma floridana, Say and Ord.—Skull, pl. lii, fig. 2.
Neotoma micropus, Baird.—Skull, pl. lxxxiv, fig. 2.
Neotoma mexicana, Baird.—Teeth, pl. liv, No. —, and No. 1674. Details, pl. lxx, fig. 3. Skull, pll. lxxxiv.
Neotoma fuscipes, Cooper.—Skull, pl. liii, fig. 1. Teeth, pl. liv, No. 936.
Neotoma occidentalis, Cooper.—Animal, pl. ix, fig. 2. Details, pl. xxi, fig. 4. Skull, pl. liii, fig. 3.
Neotoma cinerea, Baird.—Skull, pl. liii, fig. 5. Teeth, pl. liv, No. 1694.
Neotoma magister, Baird.—Lower jaw, pl. liii, fig. 4.
Sigmodon berlandieri, Baird.—Skull, pl. liii, figg. 6 and 7, and pl. lxxxiv, fig. 3. Animal, pl. lxvi, fig. 2. Details, pl. lxx, fig. 2.

ARVICOLINAE.

Arvicola Townsendii, Bachman.—Teeth, pl. liv, No. 1595.
Arvicola montana, Peale.—Details, pl. xxi, fig. 2.
Arvicola austera, Lec.—Teeth, pl. liv, No. 1587.
Arvicola cinnamonea, Baird.—Teeth, pl. liv, No. 1714.
Arvicola pinetorum, Leconte.—Teeth, pl. liv, No. 1719.
Fiber zibethicus, Cuv.—Teeth, pl. liv, No. 626.

HYSTRICINAE.

Erethizon dorsatus.—Upper view of skull, pl. lv, fig. 3.
Erethizon epixanthus, Brandt.—Skull, pl. lv, figg. 1-2.

1859. BAIRD, SPENCER F.—Continued.

LEPORINAE.

Lepus glacialis, Leach.—Skull, pl. lvi, fig. 1.
Lepus Washingtonii, Baird.--Animal, pl. xv.
Lepus campestris, Bach.—Skull, pl. lvi, fig. 2.
Lepus callotis, Wagler.—Skull, pl. lvii, fig. 1.
Lepus callotis var. *flavigularis.*—Skull, pl. lxxxv, fig. 1.
Lepus californicus, Gray.—Skull, pl. lvii, fig. 2.
Lepus sylvaticus, Bach.—Skull, pl. lviii, fig. 1.
Lepus Audubonii, Baird.—Animals, pl. xiii. Skull, pl. lviii, fig. 2.
Lepus Trowbridgii, Baird.—Animal, pl. xiv.
Lepus aquaticus, Bach.—Skull, pl. lix, fig. 1.
Lepus palustris, Bach.—Skull, pl. lix, fig. 2.
Lepus artemisia, Bach.—Skull, pl. lxxxv, fig. 2.

EFFODIENTA.

Dasypus novem cinctus, Linn.—Skull, pl. lxxxvi.

SUINAE.

Dicotyles torquatus, Cuv.—Skull, pl. lxxxvii, figg. 1, 2.

CERVINAE.

Alce americanus, Jardine.—Horns, adult, woodcut, fig. 1; young, fig. 2, p. 632.
Rangifer caribou, Aud. and Bach.—Horns, adult, woodcut, fig. 3; young, figg. 4, 5, and 6, p. 634.
Rangifer groenlandicus.—Horns, adult male, woodcut, fig. 7; adult female?, fig. 8, p. 635.
Cervus canadensis, Erxl.—Muzzle, woodcut, fig. 9; hoof, fig. 10, p. 639; horns, fig. 11, p. 641.
Cervus virginianus, Boddaert.—Feet, pl. xxiv, fig. 1; muzzle, woodcut, fig. 12, p. 644; horns, fig. 13, p. 648.
Cervus leucura, Douglass.—Horn, adult, wood-cut, figg. 14, 15, 17, p. 651; fig. 18, p. 652. ? Young, fig. 16, p. 651.
Cervus mexicanus, Gmelin.—Feet, pl. xxiv, fig. 2.
Cervus macrotis, Say.—Feet, pl. xxiii, fig. 1. Horns, wood-cut, figg. 19, 20, p. 657.
Cervus columbianus. Rich.—Feet, pl. xxiii, fig. 2. Horns, wood-cut, figg. 21, 22, p. 660.

ANTILOPINAE.

Antilocapra americana, Ord.—Animal, pl. xvi. Various horns, pl. xxx. Muzzle and hoof, wood-cut, figg. 23–24, p. 668.
Capella rupicapra.—Horns, pl. xxv, figg. 1854 and 882.

OVINAE.

Ovis montana, Cuvier.—Muzzle and hoof, wood-cut, figg. 24, 25, p. 674. Horns, male and female, wood-cut, figg. 26–29, p. 675. Horns of male, figg. 30–32, p. 675.

BOVINAE.

Ovibos moschatus, Blainville.—Muzzle, wood-cut, fig. 33, p. 681.
Bos americanus, Gmelin.—Muzzle and hoof, wood-cut, figg. 34, 35, p. 683.

102.

1860. BAIRD, SPENCER F. [Report for 1859 of the Assistant Secretary of the Smithsonian Institution.] < *Ann. Rep. Smiths. Inst. for the year* 1859, 1860, pp. 54–78.

American Explorations of the year .... pp. 63–68
Continuation of the enumeration of the collections making up the Museum, Nos. 50–72.. 71

103.

1860. BAIRD, SPENCER F. The Birds | of | North America; | the descriptions of species based chiefly on the collections | in the | Museum of the Smithsonian Institution. | By | Spencer F. Baird, | Assistant Secretary of the Smithsonian Institution, | with the co-operation of | John Cassin, | of the Academy of Natural Sciences of Philadelphia, | and George N. Lawrence, | of the Lyceum of Natural History of New York. | With an Atlas of One Hundred Plates. | Text. | — | Philadelphia: | J. B. Lippincott & Co. | 1860. 4to. pp. lvi, 1005.

1860. BAIRD, SPENCER F.—Continued.

"ADVERTISEMENT.

"The present work is, in part, a reprint of the General Report on North American Birds presented to the Department of War, and published in October, 1858, as one of the series of 'Reports of Explorations and Surveys of a Railroad Route to the Pacific Ocean.' In this volume, however, will be found many important additions and corrections, including detailed lists of plates, both numerical and systematic, descriptions of newly-discovered species, &c., not in the original edition.

"The Atlas contains one hundred plates, representing one hundred and forty-eight new or unfigured species of North American birds. Of these plates about fifty appear for the first time, having been prepared expressly for this work. The remainder form the ornithological illustrations of the Reports of the Pacific Railroad Survey, and of the United States and Mexican Boundary Survey under Major Emory, and are distributed throughout the numerous volumes composing those series. All have, however, been carefully retouched and lettered for this edition, and quite a number redrawn entirely from better and more characteristic specimens. In fact, the plates of the Atlas have been prepared expressly for the present edition with the utmost care and attention.

"In the volume of text will be found a complete account of the birds of North America, brought down to the present time, including accurate descriptions of all known species; their arrangement in the genera and families recognized by modern zoologists; their geographical distribution; and, as far as possible, all other information necessary to a complete summary or manual of North American ornithology. No other work extant gives a complete ornithology of our country; and it has been the especial object of the authors and publishers to adapt it to the wants of the student and lover of nature, and to present in a condensed form, and at a price within the reach of all, a reliable text-book in this favorite department of natural history. Extended bibliographical notices, embracing full references to very nearly all authors on American ornithology, have been added, and will be found to be of high interest to the student and naturalist."

The only difference between this volume and Vol. IX of the Pacific R. R. Survey consists in the addition of preface and different lists of plates. The first few pages are arranged differently, as the following table will show:

	In this volume.	In Vol. IX, P. R. R. Surv.
Advertisement	p. i	absent.
Preface	iii	absent.
Explanation of the plates	vii	absent.
Systematic list of illustrations	xiii	absent.
Systematic list of higher groups	xvii	p. xvii.

The remaining pages appear to be precisely the same.

104.

1860. BAIRD, SPENCER F. The Birds | of | North America; | the descriptions of species based chiefly on the collections | in the | Museum of the Smithsonian Institution. | By | Spencer F. Baird, | Assistant Secretary of the Smithsonian Institution, | with the co-operation of | John Cassin | of the Academy of Natural Sciences of Philadelphia, | and George N. Lawrence, | of the Lyceum of Natural History of New York. | With an Atlas of One Hundred Plates. | Atlas. | — | Philadelphia: | J. B. Lippincott & Co. | 1860. | 4to. p. (2) i–xi (1), 100 plates.

"The present Atlas has been prepared for the two-fold purpose of completing the series of illustrations of the Birds of North America, and to give accurate and easily recognized figures of the numerous hitherto unknown birds described in the first volume. It contains figures of all birds inhabiting the United States which have not been given by former American authors, in connection with whose works it continues and concludes, as far as possible, to the present time the pictorial representation of all North American birds. In the accompanying volume of text will be found descriptions of all the known birds of the United States; their arrangement in the genera and families of modern zoologists; their geographical distribution; and, it is believed, everything necessary to a complete and thorough knowledge of this favorite department of the natural history of our country.

"In 1843 the distinguished ornithologist, Mr. Audubon, brought to a completion the second and last edition of his great work on the Birds of North America, in which are given faithful and accurate representations of nearly five hundred species. This elaborate work included all the birds known to that celebrated author as inhabiting the continent of America north of Mexico.

1860. BAIRD, SPENCER F.—Continued.

"In 1853 Mr. Cassin commenced the publication of a work entitled 'Illustrations of the Birds of California, Texas, Oregon, and British and Russian America: intended to contain descriptions and figures of all North American birds not given by former American authors; and a general synopsis of North American Ornithology.' Philadelphia: J. B. Lippincott & Co. The first series, containing plates of fifty species not given by Audubon, was completed in 1855, and has not been extended, having been superseded by the present work.

"Many of the birds of the United States, not included in the works of the preceding or other American authors, having been collected by the several parties for the Survey of a Railroad Route to the Pacific Ocean, and of the Boundary between the United States and Mexico, as mentioned in the preface to volume I of this work; they were figured in the reports of these expeditions published by Congress under the direction of the War and Interior Departments. All of these birds appear in the present volume, but, in almost every instance, redrawn from better and more characteristic specimens. Of the one hundred plates, however, of this volume, about one-half appear for the first time, having been prepared expressly for the present work. Many of the latter represent birds of Eastern North America.

"As already stated, the work of Mr. Audubon contains figures of somewhat less than five hundred species of North American birds; that of Mr. Cassin contains fifty species. In the present volume will be found one hundred and forty-eight species, nearly all of which are now represented for the first time in any work on American Ornithology. The three works together include illustrations of very nearly all the known birds of North America. A few species only are wanting, chiefly of Russian America and other remote localities, of which no specimens are preserved in American museums. All are carefully described, however, in the preceding volume."

EXPLANATION OF PLATES.

[NOTE.—Where not otherwise mentioned the specimens figured are to be considered as in the Museum of the Smithsonian Institution, and the numbers refer to the Smithsonian record of birds. The original of each figure is indicated as far as can now be ascertained. The numbers in parenthesis refer to the numbers of the plates in the Mexican Boundary series.]

1859. BAIRD, SPENCER F.—Continued.

SYSTEMATIC LIST OF ILLUSTRATIONS.*

* The figures to the left of the name refer to the systematic list of North American birds in the first

1860. BAIRD, SPENCER F.—Continued.

1860. Baird, Spencer F.—Continued.

105.

1860. [Baird, Spencer F.] (*translator.*) On the principal plants used as food by man.—Sketch of the plants chiefly used as food by man, in different parts of the world and at various periods.—By Dr. F. Unger. | (Translated from the German for this Report.) < *Report of the Commissioner of Patents for the year* 1859.—*Agriculture*, 1860, pp. 299–362.

106.

1861. [Baird, Spencer F.] Smithsonian Miscellaneous Collections.—Instructions in reference to collecting Nests and Eggs of North American Birds. 8vo. pp. 82. (No title-page.)

Issued as No. 139 of the Smithsonian Series, and in Vol. II of the Miscellaneous Collections. The instructions for the preparation and preservation of öological specimens are quoted from Dr. T. M. Brewer, and in the appendix, pp. 10–22, is printed "Suggestions for forming collections of birds' eggs, by Alfred Newton."

107.

1861. Baird, Spencer F. [Report for 1860 of the Assistant Secretary of the Smithsonian Institution.] < *Ann. Rep. Smithsonian Institution for the year* 1860, 1861, pp. 55–86.

American explorations of the year...pp. 66–72
Continuation of enumeration of collections in the Museum, Nos. 73–94............. 75

108.

1861. Baird, Spencer F. Report | upon the | Colorado River of the West, | Explored in 1857 and 1858 by | Lieutenant Joseph C. Ives, | Corps of Topographical Engineers, | Under the direction of the Office of Explorations and Surveys, | A. A. Humphreys, Captain Topographical Engineers, in charge. | — | By order of the | Secretary of War. | — | Washington : | Government Printing Office. | 1861. 1 vol., 4to.

Zoology. By Professor S. F. Baird. pp. 1–6. (This part dated 1860.)
List of birds collected on the expedition, pp. 5, 6; 65 spp., with localities.

109.

1862. Baird, Spencer F. [Report for 1861 of the Assistant Secretary of the Smithsonian Institution.] < *Ann. Rep. Smithsonian Institution for the year* 1861, 1860, pp. 48–67.

Explorations of the year ....pp. 58–61

110.

1863. BAIRD, SPENCER F. [Report for 1862 of the Assistant Secretary of the Smithsonian Institution.] < *Ann. Rep. Smithsonian Institution for the year* 1862, 1863, pp. 46–59.

111.

1863. BAIRD, SPENCER F. [Notice of R. Kennicott's and J. Xantus's Movements in North America.] < *Ibis*, v, 1863, pp. 238, 239.

112.

1863. BAIRD, SPENCER F. [Letter on J. Xantus's collections at Colima, Mexico. < *Ibis*, v, 1863, p. 476.

113.

1864. BAIRD, SPENCER F. [Report for 1863 of the Assistant Secretary of the Smithsonian Institution.] < *Ann. Rep. Smithsonian Institution for the year* 1863, 1864, pp. 44–63.

Explorations of the year.. pp. 52–56

114.

1863. BAIRD, SPENCER F. (*editor*). Notes on the Birds of Jamaica. By W. T. March, with remarks by S. F. Baird. < *Proc. Acad. Nat. Sci. Phila.*, 1863, i, pp. 150–154 (May); ii, pp. 283–304 (Nov.); iii, 1864, pp. 62–72 (March, 1864).

Remarks chiefly critical.

115.

[1864–66.] BAIRD, SPENCER F. Smithsonian Miscellaneous Collections. | — | Review | of | American Birds, in the Museum of the | Smithsonian Institution. | By | S. F. Baird. | — | Part I. North and Middle America. | — | [Medallion.] | Washington : | Smithsonian Institution. | [No date on title : June, 1864, to p. 33; July, 1864, to p. 81; Aug., 1864, to p. 129; Sept., 1864, to p. 145; Oct., 1864, to p. 161; Nov., 1864, to p. 177; Apr., 1865, to p. 241; May, 1865, to p. 321; May, 1866, to p. 417; June, 1866, to end.] 1 vol. 8vo. Originally issued in sheets as successively printed, at above dates. pp. iv, 450.

"The present work is intended as a catalogue of the birds of Northern and Middle America in the Museum of the Smithsonian Institution, with such critical notices of the same as appear to be called for, and a list of the specimens, or of such of them as best show the geographical distribution of the species. Species not in the Smithsonian collection, but which I have had the opportunity of personally examining and comparing, are also included. Species mentioned by authors, but which I have not seen, will be mentioned at the end of the genera or families to which they are supposed to belong."

TABLE OF CONTENTS.

OSCINES.

SIGNATURES 1 AND 2, JUNE, 1864.

OSCINES.

[NOTE.—An asterisk (*) denotes analytical key; a dagger (†), analytical table; a double dagger (‡), arrangement of genera; and a section mark (§), key to subgenera.]

7 BD

[1864-66.] BAIRD, SPENCER F.—Continued.

SIGNATURES 3, 4, 5, JULY, 1864.

[1864–66.] BAIRD, SPENCER F.—Continued.

Harporhynchus Lecontei (Lawr.), Baird. R....................................... p. 47
Harporhynchus crissalis, Henry.
Harporhynchus redivivus (Gambel), Cabanis 48
Mimus, Boie.
Mimus polyglottus, Linn.
Mimus orpheus (Linn.), Sclater... 50
Mimus bahamensis, Bryant 52
Mimus Hillii, March.
Mimus gracilis, Cabanis.. 54
Galeoscoptes, Cabanis.
Galeoscoptes carolinensis (Linn.), Cabanis.
Melanoptila, Sclater.. 55
Melanoptila glabrirostris, Sclater.
Melanotis, Bonap... 56
Melanotis cærulescens (Swainson), Bonaparte.
Melanotis hypoleucus, Hartlaub 57
Donacobius, Swainson.
Donacobius atricapillus (Linn.), Bonap.
Donacobius albo-vittatus, d'Orbigny.
Turdus pinicola, Sclater....................................... 58
Turdus plebeius, Cabanis.
Turdus nigrescens, Cab.
Margarops densirostris (Vieill.), Sclater................................... 59
Margarops montanus (Lafr.), Sclater.
Cichlerminia Bonapartii (Lafr.), Sclater.
Cinclocerthia ruficauda (Gould), Sclater 59
Cinclocerthia gutturalis, Sclater.
Mimus dominicus, Sclater.
Mimus Gundlachi, Sclater.
Harporhynchus ocellatus, Sclater.
Turdus phæopygus, Sclater.
Turdus gymnopthalmus (Cab.), Scl.
Turdus xanthoscelis, Jardine.
Mimus melanopterus, Lawr.
Family Cinclidæ.
Cinclus, Bechst.
Cinclus mexicanus, Swainson.. 60
Family Saxicolidæ.
Saxicola, Bechst 61
Saxicola œnanthe (Linn.), Bechst.
Sialia, Swainson.. 62
Sialia sialis (Linn.), Latham.
Sialia azurea, Swainson.
Sialia mexicana, Swainson.. 63
Sialia arctica (Swainson), Nuttall... 64
Family Sylviidæ.
†Regulinæ 65
†Polioptilinæ.. 65
Regulus, Cuv.
Regulus satrapa, Licht.
Regulus Cuvieri, Aud 66
Regulus calendula (Linn.), Licht.
Polioptila, Sclater.. 67
Polioptila melanura, Lawrence.. 68
Polioptila nigriceps, Baird, n. s 69
      Mazatlan.
Polioptila leucogastra (Maxim.), Sclater.
Polioptila Buffoni, Sclater 70
Polioptila albiloris, Salvin.
Polioptila superciliaris, Lawrence 71
Polioptila bilineata (Bonaparte), Sclater.................................. 72
Polioptila dumicola (Vieill.), Sclater 73
Polioptila plumbea, Baird 74
Polioptila cærulea (Linn.), Sclater.

[1864–66.] BAIRD, SPENCER F.—Continued.

1864–66.] BAIRD, SPENCER F.—Continued.
   7' 7a. *Helminthophaga celata.*
   8, 8a. *Teretristis forusii.*
   9, 9a. *Vireo barbatula.*
  (B.) *Telmatodytes.*
    Cistothorus palustris (Wilson), Baird .... p. 147
    Cistothorus palustris var. *paludicola*, Baird, n. var 148
      Pacific Coast, United States.
    Thryothorus albinucha, Cabot 149
      (Note assigning to this species *Thryothorus petenicus*, Salvin.)
Family *Motacillidæ* 150
  **Motacilla.*
    *Motacilla 151
   **Anthus.*
    **Anthus.*
    **Neocorys.*
    *Notiocorys, Baird, n. subg.
    *Pediocorys, Baird, n. subg.
   Motacilla, Linn.
    Motacilla alba, Linn 152
  (A.) *Anthus*, Bechst.
    Anthus ludovicianus (Gmel.), Licht 153
    Anthus pratensis (Linn.), Bechst 155
  (B.) *Neocorys*, Sclater.
    Anthus Spraguei (Aud.), Baird. R.
  (C.) *Notiocorys*, Baird 156
    Anthus rufus (Gmel.), Burm.
  (D.) *Pediocorys*, Baird 157
    Anthus bogotensis, Sclater.
    Anthus, —— 158
Family *Sylvicolidæ* 160
  *Sylvicolinae 166
   **Mniotilteae.*
   **Vermivoreae.*
   **Sylvicoleae.*
  **Geothlypinae.*
   **Seuireae.*
   **Geothlypeae.*
  **Icterianae.*
   **Icterieae.*
   **Teretristeae.*
  *Setophaginae 167
Subfamily *Sylvicolinæ.*
  Mniotilta, Vieill.
    Mniotilta varia (Linn.), Vieill.
  Parula, Bon.
   *Parula americana 160
    **Parula, pityayumi.*
    **Parula inornata.*
    **Parula superciliosa.*
    **Parula gutturalis.*
    Parula americana (Linn.), Bon.
    Parula pitiayumi (Vieill.), Sclater 170, 260
    Parula inornata, Baird, n. s 171
      Guatemala and Costa Rica.
    Parula superciliosa (Hartlaub), Sclater.
    Parula gutturalis (Cabanis), Baird. R 172
  Protonotaria, Baird 173
    Protonotaria citrea (Bodd.), Baird.
  Helminthophaga, Cabanis.
    Helminthophaga pinus (Linn.), Baird 174
    Helminthophaga chrysoptera (Linn.), Cab 175
    Helminthophaga Bachmani (Aud.), Cabanis.
    Helminthophaga ruficapilla (Wilson), Baird.
    Helminthophaga celata (Say), Baird 176

1864–66.] BAIRD, SPENCER F.—Continued.

[1864-66.] BAIRD, SPENCER F.—Continued.

MAY, 1865.

[1864-66.] BAIRD, SPENCER F.—Continued.

1864-66.] BAIRD, SPENCER F.—Continued.

Progne, Boie.

[1864-66.] BAIRD, SPENCER F.—Continued.

116.

1865. BAIRD, SPENCER F. Report of the Assistant Secretary (of the Smithsonian Institution for the year 1864). < *Ann. Rep. Smithsonian Institution for the year* 1864, 1865, pp. 74–100.

Explorations of the year .... pp. 81–84

117.

1866. BAIRD, SPENCER F. Report of the Assistant Secretary, Spencer F. Baird, relative to exchanges, collections of Natural History, &c. < *Ann. Rep. Smithsonian Institution for the year* 1865, pp. 75–88.

118.

1866. BAIRD, SPENCER F. The Distribution and Migrations of North American Birds; by Spencer F. Baird, Asst. Sec. Smithsonian Institution. (Abstract of a memoir presented to the National Academy of Sciences, Jan., 1865.) < *Amer. Journ. Sci. and Arts* (2), xli, pp. 78–90, Jan., 1866, art. xii.

119.

1866. BAIRD, SPENCER F. The Distribution and Migrations of North American Birds; by Spencer F. Baird, Asst. Sec. Smithsonian Institution. (Abstract of a memoir presented to the National Academy of Sciences, Jan., 1865.) [Continued from p. 90.] < *Amer. Journ. Sci. and Arts* (2), xli, pp. 184–192, March, 1866, art. xxv.

120.

1866. BAIRD, SPENCER F. The Distribution and Migrations of North American Birds; by Spencer F. Baird, Asst. Sec. Smithsonian Institution. (Abstract of a memoir presented to the National Academy of Sciences, Jan., 1865.) [Concluded from p. 192.] < *Amer. Journ. Sci. and Arts* (2), xli, pp. 337–347, May, 1866, art. xli.

Notices.—*Annals and Magazine of Natural History,* xviii, 1866, pp. 141-144; *Ibis,* iii, 1867, pp. 257-293.

These papers were stitched together and issued as an excerpt, without title or repaging.

121.

1866. [BAIRD, SPENCER F.] Smithsonian Miscellaneous Collections—210.—Arrangement of Families of Birds. [Adopted provisionally by the Smithsonian Institution.] 8vo. pp. 8.

A.—American, Nòs. 1-81, pp. 1-6.

B.—Old World exclusively, Nos. 1-30, pp. 7, 8.

Dated Smithsonian Institution, June, 1866.

"The classification of birds here presented is based essentially upon that of Prof. Lilljeborg, of Upsala."

An edition without heading "Smithsonian Misc. Coll." was also published.

122.

1866. BAIRD, SPENCER F. The origin of the domestic Turkey. < *Report of the Commissioner of Agriculture for* 1866, pp. 288–290.

123.

1866. BAIRD, SPENCER F. Prof. Spencer F. Baird. Die Verbreitung und Wanderungen der Vögel Nord-Amefika's. <*Journal für Ornithologie,* xiv, 1866, pp. 244–269, 338–352.
Translated from *American Journal of Science and Arts,* vol. xli, 1866.

124.

1867. BAIRD, SPENCER F. The Distribution and Migrations of North American Birds. < *Ibis,* 1867, 2d ser., iii, pp. 257–293.
Reprinted from *Am. Journ. Sci. and Arts,* xli, Jan., Mar., May, 1866.

125.

1867. BAIRD, SPENCER F. [Note on the Pelican] in "The Pelican in Cayuga County, N. Y., by W. J. Beal." < *American Naturalist,* i, pp. 323–4, Aug., 1867.

126.

1868. BAIRD, SPENCER F. Report of Prof. S. F. Baird (Assistant Secretary of the Smithsonian Institution, for the year 1867.) < *Ann. Rep. Smithsonian Institution for the year* 1867, 1868, pp. 64–78.
The report on the additions to the Museum and on the explorations, hitherto given in the report of the Assistant Secretary, are this year and afterwards incorporated in the report of the Secretary.

127.

1868. BAIRD, SPENCER F. The Basking Shark—The "Great Sea Monster." <*American Agriculturist,* xxvii, 1868, p. 130.
Notes on *Cetorhinus maximus.*

128.

1869. BAIRD, SPENCER F. Explorations and Collections in Natural History (etc.). < *Ann. Rep. Smithsonian Institution for the year* 1868, 1869 (*in Rep. of the Secretary,* pp. 22–41), and pp. 54–67.

129.

1869. BAIRD, SPENCER F. On additions to the Bird-Fauna of North America, made by the Scientific Corps of the Russo-American Telegraph Expedition. <*Trans. Chicago Acad. Sciences,* i, pt. ii, 1869, pp. 311–325, pll. xxvii–xxxiv.

Scops Kennicotti, Elliot, n. s...pl. xxvii p. 311
    Sitka, Alaska.
Budytes flava (Lin.), Bon ....pl. xxx, fig. 1   312
Phyllopneuste Kennicotti, Baird, n. s.....................:.......pl. xxx, fig. 2
    St. Michael's, Alaska.
Troglodytes alascensis, Baird, n. s..pl. xxx, fig. 3   315
    St. George's Island, Bering Sea.
Pyrrhula coccinea, De Selys, var. *Cassinii,* Baird, n. var...............pl. xxix, fig. 1   316
Leucosticte griscinucha (Brandt), Baird ....pl. xxviii, fig. 2
Leucosticte littoralis, Baird, n. s.....................................pl. xxviii, fig. 1   318
    Sitka, Alaska; Ft. Simpson, British Columbia.
Melospiza insignis, Baird, n. s. ....pl. xxix, fig. 2   319
    Kodiak.
Spermophila badiiventris, Lawrence....................................pl. xxviii, fig. 3
Limosa uropygialis, Gould ....pl. xxxii   320
Sterna aleutica, Baird, n. s ....pl. xxxi, fig. 1   321
    Kodiak.

1869. BAIRD, SPENCER F.—Continued.

Graculus bicristatus, Pallas ....pl. xxxiii p. 321

Puffinus tenuirostris, Temm...pl. xxxiv, fig. 2   322

Fulmarus Rodgersi, Cassin..   323

Larus borealis, Brandt...   324

Simorhynchus Cassini, Coues ....pl. xxxi, fig. 2

130.

1870. BAIRD, SPENCER F.   Fossil Birds of the United States.   *<Harper's New Monthly Mag.*, xl, 1870, pp. 467, 469, and 470.

Brief notices of current discoveries.

131.

1871. BAIRD, SPENCER F.   New link between Reptiles and Birds.   *< Harper's New Monthly Mag.*, xl, 1870, p. 628.   See, also, p. 469.

132.

1871. BAIRD, SPENCER F.   Spawning of the Goose Fish (*Lophius Americanus*).   *<American Naturalist*, v, 1871, p. 785.

133.

1871. BAIRD, SPENCER F.   [Letter addressed to fishermen and others living on the shores of Lake Michigan, and announcing that fishes with metallic tags had been liberated at twenty points, and asking for information about their subsequent capture.]   8vo.   1 page.   [U. S. F. C. 1.]   Dated Oct. 30, 1871.

134.

1871. BAIRD, SPENCER F.   Rocky Mountain Explorations.   *<Annual Record of Science and Industry*,* 1871–72, p. 131.

135.

1871. BAIRD, SPENCER F.   Explorations of Professor Powell.   *<Ann. Rec.*, 1871–72, pp. 132, 133.

---

*Nearly all the articles on natural history included in the *Annual Record of Science and Industry* were published during 1871 either in *Harper's New Monthly Magazine* or in *Harper's Weekly, a Journal of Civilization*, and are entitled to individual record, but it is more convenient, both for compilation and reference, to mention them collectively under the title of the *Annual Record* for each year.

I have been somewhat perplexed to decide in what manner it is best to cite the articles contributed by Professor Baird to the scientific columns of *Harper's Monthly* and *Harper's Weekly* and subsequently each year collected together in the *Annual Record of Science and Industry*. Many of these are original contributions to knowledge, never elsewhere published. Others are critical reviews or notes upon the current literature of science. Others are abstracts of scientific papers, with the addition of explanatory or illustrative remarks. Others still are abstracts of papers, for the most part in the words of the authors of the papers or of some other reviewer. Those in the latter category are the most numerous, and these it has not been thought necessary to cite in this list. I have endeavored to refer to all those articles, however brief, which contain original matter or critical opinions. Where these are reviews or abstracts, this fact is indicated in the appended explanatory remark.

A strict adherence to the usage of bibliographers would necessitate references to the original places of publication. This would have made necessary a great amount of extra labor, which would hardly have been profitable, since close questions of priority of authorship are rarely, if ever, involved. I rest contented, therefore, with the citation of pages in the *Annual Record*, following the example of Mr. S. H. Scudder and other bibliographers in treating this as a scientific serial, to be regarded in the same light as the proceedings of a society. The headings are those used in the *Annual Record*, the originally published articles having been printed as paragraphs without titles. Persons who may need to use the original publication will find this list an aid.

I have been obliged to use my personal judgment in culling the articles. No rules could be applied.

136.

BAIRD, SPENCER F. · Explorations of Professor Cope. < *Ann. Rec.*, 1871–72, pp. 133, 134.

137.

BAIRD, SPENCER F. Headwaters of the Yellowstone. < *Ann. Rec.*, 1871–72, p. 137.

138.

BAIRD, SPENCER F. Raymond's Report on the Yukon. < *Ann. Rec.*, 1871–72, pp. 138, 139.

139.

BAIRD, SPENCER F. Explorations in Vineyard Sound [by Prof. Verrill]. < *Ann. Rec.*, 1871–72, p. 140.

140.

BAIRD, SPENCER F. Explorations in the West Indies. < *Ann. Rec.*, 1871–72, pp. 141, 142.

141.

BAIRD, SPENCER F. Explorations of Dr. Habel (in South America). < *Ann. Rec.*, 1871, pp. 142, 143.

142.

BAIRD, SPENCER F. Explorations of Professor Hartt (in Brazil). < *Ann. Rec.*, 1871–72, pp. 146, 147.

143.

BAIRD, SPENCER F. Faunal peculiarities of the Azores. < *Ann. Rec.*, 1871–72, pp. 149, 150.
Review of work of F. D. Godman.

144.

BAIRD, SPENCER F. Faunal provinces of the west coast of America. < *Ann. Rec.*, 1871–72, p. 152.
Review of paper by Prof. Verrill in *Transactions, Connecticut Academy of Sciences*, 1871.

145.

BAIRD, SPENCER F. Darwin on The Descent of Man. < *Ann. Rec.*, 1871–72, p. 156.
Notice of a review in *The Academy* by A. R. Wallace, from which the main points of Darwin's book are stated. Remarks upon the general acceptance of the doctrine of evolution by naturalists.

146.

BAIRD, SPENCER F. Shell-heaps in New Brunswick. < *Ann. Rec.*, 1871–72, p. 182.

147.

BAIRD, SPENCER F. Variation of color in Birds with the locality. < *Ann. Rec.*, 1871–72, pp. 190, 191.

148.

BAIRD, SPENCER F. Catalogue of Fishes in the British Museum. < *Ann. Rec.*, 1871–72, p. 206.
Alludes to the collection of fishes in the Museum of Comp. Zoology, which is compared with that in the British Museum.

149.

1871. BAIRD, SPENCER F.   British Museum Fishes. < *Ann. Rec.*, 1871–72, p. 206.
  Notice of the lack of facilities for obtaining specimens complained of by the officers of the
  British Museum.

150.

1871. BAIRD, SPENCER F.   Spawning of Herring. < *Ann. Rec.*, 1871–72, p. 207.
  Statement of writer in *Land and Water.*

151.

1871. BAIRD, SPENCER F.   The Food of the Sea-Herring. < *Ann. Rec.*, 1871–72, pp.
  208–210.
  Including a review of Axel Boeck's investigations.

152.

1871. BAIRD, SPENCER F.   Fishes of Cuba. < *Ann. Rec.*, 1871–72, p. 211.
  Notice of Poey's recent paper on the *Percoids* of Cuba.

153.

1871. BAIRD, SPENCER F.   Phosphorescence of Dead Fish. < *Ann. Rec.*, 1871–72, p.
  211.
  Notice of Panceri's theory.

154.

1871. BAIRD, SPENCER F.   Fresh-Water Fishes of Algeria. < *Ann. Rec.*, 1871–72, p.
  211.
  Review of Plairfair's paper on the "Hydrographical System and Fresh-Water Fish of
  Algeria."

155.

1871. BAIRD, SPENCER F.   Confusion of names of Fishes. < *Ann. Rec.*, 1871–72, p.
  207.
  Explanation of error in *Land and Water* by which "horse-mackerel" and "blue-fish" are
  confounded.

156.

1871. BAIRD, SPENCER F.   Tame Codfish. < *Ann. Rec.*, 1871–72, p. 212.
  Notice of tne codfish confined at Port Logan, England. Stones in the stomach of codfish.

157.

1871. BAIRD, SPENCER F.   Teeth of the Sturgeon. < *Ann. Rec.*, 1871–72, p. 213.

158.

1871. BAIRD, SPENCER F.   Development of the Lamprey. < *Ann. Rec.*, 1871–72, p
  213.
  A review of Owsjannikow's investigation on *Petromyzon.*

159.

1871. BAIRD, SPENCER F.   Lütken on Ganoid Fishes. < *Ann. Rec.*, 1871–72, p. 214.

160.

1871. BAIRD, SPENCER F.   Gourami Fish. < *Ann. Rec.*, 1871–72, p. 214.
  Note on the facility with which this fish may be transported.

161.

1871. BAIRD, SPENCER F.   A new Lophioid Fish. < *Ann. Rec.*, 1871–72, p. 214.
  The dorsal fin ray of *Oneirodes*, with suggestion as to the use of this organ in Lophioids.

162.

BAIRD, SPENCER F. Peculiarities of Salmon Kelts. <*Ann. Rec.*, 1871–72, p. 215.

Frank Buckland's observation upon the thickening of the skin in male and female kelts.

163.

BAIRD, SPENCER F. Salmon-Fishing in Loch Tay. <*Ann. Rec.*, 1871–72, p. 215.

Review of article by Frank Buckland in *Land and Water*.

164.

BAIRD, SPENCER F. "Landlocked Salmon." <*Ann. Rec.*, 1871–72, p. 216.

A notice of Livingston Stone's observations on Maine land-locked salmon.

165.

BAIRD, SPENCER F. Food for Young Trout. <*Ann. Rec.*, 1871–72, p. 217.

166.

BAIRD, SPENCER F. Tailless Trout in Scotland. <*Ann. Rec.*, 1871–72, p. 217.

167.

BAIRD, SPENCER F. Hermit Crabs Climbing Trees. <*Ann. Rec.*, 1871–72, p. 229.

168.

BAIRD, SPENCER F. Ancient City in New Mexico. <*Ann. Rec.*, 1871–72, p. 241.

169.

BAIRD, SPENCER F. Fossil Fishes of Wyoming. <*Ann. Rec.*, 1871–72, p. 248.

Notice of recent investigations by Prof. Cope.

170.

BAIRD SPENCER F. Cephalaspis in America. <*Ann. Rec.*, 1871–72, p. 248.

Notice of *Cephalaspis Dawsoni*, E. R. Lankester, from the Siluro-Devonian beds of Gaspé.

171.

BAIRD, SPENCER F. Port Kennedy Bone-Cave. <*Ann. Rec.*, 1871–72, p. 249.

172.

BAIRD, SPENCER F. Immunity of the Pig from injury by serpent bite. <*Ann. Rec.*, 1871–72, pp. 255, 256.

173.

BAIRD, SPENCER F. Peculiarities of the Florida Wild Turkey. <*Ann. Rec.*, 1871–72, p. 257.

174.

BAIRD, SPENCER F. Existing specimens of the Great Auk. <*Ann. Rec.*, 1871–72, pp. 258, 259.

175.

BAIRD, SPENCER F. Relation of weight to length in Crocodiles and Alligators. <*Ann. Rec.*, 1871–72, p. 259.

Calling attention to Prof. Phillips's circular.

176.

1871. BAIRD, SPENCER F.   Cod-fisheries of Alaska.   < *Ann. Rec.*, 1871–72, p. 259.
    Suggests that with the increasing importance of the Alaska cod-fisheries they may yet
    replace those of Newfoundland.

177.

1871. BAIRD, SPENCER F.   Occurrence of the Pompano (and Spanish Mackerel) north-
    ward.   < *Ann. Rec.*, 1871–72, p. 260.

178.

1871. BAIRD, SPENCER F.   Increase of Salmon in the British Provinces.   < *Ann. Rec.*,
    1871–72, p. 260.

179.

1871. BAIRD, SPENCER F.   Use of the pectoral fins of fish.   < *Ann. Rec.*, 1871–72, p. 261.
    Notice of Mr. Hausen's discussion of this subject.

180.

1871. BAIRD, SPENCER F.   Relations of Ganoids to Plagiostomes.   < *Ann. Rec.*, 1871–
    72, pp. 261–263.
    Review of a communication by Dr. Gunther in *Nature*.

181.

1871. BAIRD, SPENCER F.   Theory of the Salmon Fly.   < *Ann. Rec.*, 1871–72, p. 263.
    Dr. Gunther's theory of the resemblance of the artificial fly to the prawn.

182.

1871. BAIRD, SPENCER F.   Capture of Horse Mackerel in Buzzard's Bay.   < *Ann.
    Rec.*, 1871–72, p. 263.

183.

1871. BAIRD, SPENCER F.   Did Hendrik Hudson find Salmon in the Hudson River?
    < *Ann. Rec.*, 1871–72, p. 264.

184.

1871. BAIRD, SPENCER F.   Black Bass in the Potomac.   < *Ann. Rec.*, 1871–72, p. 264.
    The manner of introduction of Black Bass into the Potomac.

185.

1871. BAIRD, SPENCER F.   Cause of Death of Fresh-water Fish in Salt Water.   < *Ann.
    Rec.*, 1871–72, p. 265.
    Review of M. Beit's communication to the Academy of Sciences, Paris.

186.

1871. BAIRD, SPENCER F.   Proper Fish for Stocking Rivers.   < *Ann. Rec.*, 1871–72,
    p. 265.

187.

1871. BAIRD, SPENCER F.   Living Eyeless Fish.   < *Ann. Rec.*, 1871–2, p. 266.
    Notice of the eyeless fish from Mammoth Cave in the Dublin Zoological Gardens.

188.

1871. BAIRD, SPENCER F.   Stocking waters of New York with Fish.   < *Ann. Rec.*,
    1871–72, p. 266.

189.

1871. BAIRD, SPENCER F. Killing Fish with Torpedoes in Florida. <*Ann. Rec.*, 1871–72, p. 267.
Comment on the practice of killing fish for manure with torpedoes at New Smyrna, Florida.

190.

1871. BAIRD, SPENCER F. Fungus growth on Fish and their Eggs. <*Ann. Rec.*, 1871–72, p. 267.
Review of the investigation of growth of mould on fish by Professor Willcome, of Tharandt, Germany.

191.

1871. BAIRD, SPENCER F. Allen on the Birds of East Florida. <*Ann. Rec.*, 1871–72, p. 2.

192.

1871. BAIRD, SPENCER F. Zoological Stations in the Gulf of Naples. <*Ann. Rec.*, 1871–72, pp. 274, 275.
Alludes to the proposed establishment of a station at Eastport.

193.

1871. BAIRD, SPENCER F. Verrill's Exploration in New Jersey. <*Ann. Rec.*, 1871–72, p. 276·
Professor Verrill's exploration at Beesley's Point.

194.

1871. BAIRD, SPENCER F. Dr. Stimpson's Exploration in Florida. <*Ann. Rec.*, 1871–72, p. 377.

195.

1871. BAIRD, SPENCER F. Schools of Young Bluefish. <*Ann. Rec.*, 1871–72, p. 278.
The occurrence of large schools of Bluefish, four inches and less in length, in Beaufort Harbor, N. C., Dec., 1871.

196.

1871. BAIRD, SPENCER F. American Tapirs. <*Ann. Rec.*, 1871–72, p. 278.
Skeleton of *Tapirus Roulini* obtained by the Smithsonian.

197.

1871. BAIRD, SPENCER F. Fish-Guano Flour from Loffoden. <*Ann. Rec.*, 1871–72, pp. 342, 343.

198.

1871. BAIRD, SPENCER F. Report of the Connecticut Fish Commissioners. <*Ann. Rec.*, 1871–72, p. 348.
Review.

199.

1871. BAIRD, SPENCER F. Nutrition of Young Fish in Hatching Establishments. <*Ann. Rec.*, 1871–72, p. 350.
Notice of Dr. Hartmann's communication to the German Fishery Society in regard to the age at which artificially hatched fish should be turned out.

200.

1871. BAIRD, SPENCER F. Irish Oyster Fisheries. <*Ann. Rec.*, 1871–72, pp. 352, 353.
Review of Report of Commission.

201.

1871. BAIRD, SPENCER F.   Artificial Ice in Packing Fish. *< Ann. Rec.*, 1871–72, p. 355.

202.

1871. BAIRD, SPENCER F.   Preservation of Dead Salmon for an indefinite time.
< Ann. Rec., 1871–72, p. 356.

203.

1871. BAIRD, SPENCER F.   Importance of Killing freshly-captured Fish. *< Ann.
Rec.*, 1871–72, p. 387.

204.

1871. BAIRD, SPENCER F.   Ship-Canal across Cape Cod. *< Ann. Rec.*, 1871–72, pp.
422, 423.

205.

1871. BAIRD, SPENCER F.   Ship-Canal across New Jersey. *< Ann. Rec.*, 1871–72, pp.
423, 424.

206.

1871. BAIRD, SPENCER F. . Oil from Birds. *< Ann. Rec.*, 1871–72, pp. 466, 467.

207.

1871. BAIRD, SPENCER F.   Pegging Lobster Claws. *< Ann. Rec.*, 1871–72, pp. 552, 553.

208.

1871. BAIRD, SPENCER F.   Fayrer on Snake Bites. *< Ann. Rec.*, 1871–72, p. 577.

209.

1871. BAIRD, SPENCER F.   "Archives of Science." *< Ann. Rec.*, 1871–72, p. 604.
Notice of a new periodical, and reminiscence of Mr. C. C. Frost, the botanist.

210.

1871. BAIRD, SPENCER F.   Commissioner of Fisheries. *< Ann. Rec.*, 1871–72, pp.
605, 606.

211.

1871. BAIRD, SPENCER F.   Fishing Steamer. *< Ann. Rec.*, 1871–72, p. 606.
Notice of a steamer devised in England for sea-fishing.

212.

1871. BAIRD, SPENCER F.   International Exchanges of Holland [and of United States
by Smithsonian Institution]. *< Ann Rec.*, 1871–72, p. 607.

213.

1871. BAIRD, SPENCER F.   Return of Mr. Gwyn Jeffreys to England. *< Ann. Rec.*,
1871–72, p. 608.

214.

1871. BAIRD, SPENCER F.   Destruction of the Chicago Academy of Sciences by Fire.
< Ann. Rec., 1871–72, p. 609.

215.

1871. BAIRD, SPENCER F.   Necrology of Science for 1871. *< Ann Rec.*, 1871–72, pp.
611–613.

216.

[BAIRD, SPENCER F.] Memoranda of Inquiry | relative to the | food-fishes of the United States. [Washington: Government Printing Office. 1872.] 8vo. 5 pp., no title-page. [U. S. F. C., No. 2.]

Included in Report of Commissioners, Part I, pp. 1-3.

217.

[BAIRD, SPENCER F.] Questions relative | to the | Food-Fishes of the United States. [Washington: Government Printing Office, 1872.] 8vo. 7 pp., no title-page. [U. S. F. C., No. 3.]

218.

BAIRD, SPENCER F. [Letter to accompany No. 217 inviting information concerning Food-Fishes.] 1 p., letter size. [U. S. F. C., No. 4.]

219.

BAIRD, SPENCER F. Explorations and Collections (of the Smithsonian Institution in the year 1871). < Ann. Rep. Smithsonian Institution for the year 1871, 1872, in Report of Secretary, pp. 26–34 + and 42–62.

220.

BAIRD, SPENCER F. [Instructions to Capt. C. F. Hall for collecting objects of Natural History on the Expedition toward the North Pole.] < Ann. Rep. Smithsonian Institution for the year 1871, 1872, pp. 379–381.

221.

BAIRD, SPENCER F. Annual Record | of | Science and Industry | for 1871. | Edited by | Spencer F. Baird, | with the assistance of eminent men of science. | [Cut.] | New York: | Harper & Brothers, publishers, | Franklin Square. | 1872. [8vo. pp. xxxii, 634.]

222.

BAIRD, SPENCER F. General summary of scientific and industrial progress for the year 1871. < Ann. Rec., 1871–72, pp. xvii–xxxii.

Explorations, pp. xxiv-xxvi.

223.

BAIRD, SPENCER F. Yellowstone Park. < Ann. Rec., 1872–73, p. 125.

Notice of passage of law establishing National Park.

224.

BAIRD, SPENCER F. Second Report of the Geological Survey of Indiana for 1870. < Ann. Rec., 1872–73, p. 125.

225.

BAIRD, SPENCER F. Report of the Geological Survey of Ohio for 1870. < Ann. Rec., 1872–73, p. 125.

226.

BAIRD, SPENCER F. Report of Professor Hayden's Explorations. < Ann. Rec., 1872–73, p. 126.

227.

1872. BAIRD, SPENCER F. Report of Mr. Clarence King—Vol. 5. < *Ann. Rec.*, 1872–73, p. 127.

228.

1872. BAIRD, SPENCER F. Progress of the Geological Survey of California. < *Ann. Rec.*, 1872–73, p. 128.

229.

1872. BAIRD, SPENCER F. Geology of the Bermudas. < *Ann. Rec.*, 1872–73, p. 137.

230.

1872. BAIRD, SPENCER F. Explorations of Lieutenant Wheeler in 1871. < *Ann. Rec.*, 1872–73, p. 152.

231.

1872. BAIRD, SPENCER F. Explorations of Major Powell in 1871. < *Ann. Rec.*, 1872–73, p. 152.

232.

1872. BAIRD, SPENCER F. Professor Marsh's Explorations in 1871. < *Ann. Rec.*, 1872–73, p. 153.

233.

1872. BAIRD, SPENCER F. Explorations of Dr. Stimpson. < *Ann. Rec.*, 1872–73, p. 154.

234.

1872. BAIRD, SPENCER F. Explorations of Prof. Hartt in Brazil. < *Ann. Rec.*, 1872–73, p. 157.

235.

1872. BAIRD, SPENCER F. Recent Explorations in the United States. < *Ann. Rec.*, 1872–73, p. 173.

236.

1872. BAIRD, SPENCER F. Explorations of William H. Dall. < *Ann. Rec.*, 1872–73, p. 175.

237.

1872. BAIRD, SPENCER F. Explorations of the Navy Department in the North Pacific. < *Ann. Rec.*, 1872–73, p. 180.

238.

1872. BAIRD, SPENCER F. Explorations of the Challenger. < *Ann. Rec.*, 1872–73, p. 181.

239.

1872. BAIRD, SPENCER F. Explorations of Professor Powell in 1872. < *Ann. Rec.*, 1872–73, p. 192.

240.

1872. BAIRD, SPENCER F. Report of the Circumnavigating Committee of the Royal Society. < *Ann. Rec.*, 1872–73, p. 193.

241.

1872. BAIRD, SPENCER F. Explorations of the Portsmouth. < *Ann. Rec.*, 1872–73, p. 196.

242.

1572. BAIRD, SPENCER F. Explorations by Professor Hayden in 1872. $<$ *Ann. Rec.*, 1872–73, p. 197.

243.

1872. BAIRD, SPENCER F. Marine Zoology of the Bay of Fundy. $<$ *Ann. Reo.*, 1872–73, p. 201.

244.

1872. BAIRD, SPENCER F. The Voyage of the Hassler. $<$ *Ann. Rec.*, 1872–73, p. 204.

245.

1872. BAIRD, SPENCER F. Exploration in Central America. $<$ *Ann. Rec.*, 1872–73, p. 211.

246.

1872. BAIRD, SPENCER F. Powell's Report. $<$ *Ann. Rec.*, 1872–73, p. 212.

247.

1872. BAIRD, SPENCER F. Grandidier on the Zoology of Madagascar. $<$ *Ann. Rec.*, 1872–73, p. 213.

248.

1872. BAIRD, SPENCER F. Visit of Abbé David to Thibet. $<$ *Ann. Reo.*, 1872–73, p. 213.

249.

1872. BAIRD, SPENCER F. Fulfillment of the Predictions of Professor Agassiz. $<$ *Ann. Rec.*, 1872–73, p. 214.

250.

1872. BAIRD, SPENCER F. Professor Agassiz's Prophecies. $<$ *Ann. Rec.*, 1872–73, p. 215.

251.

1872. BAIRD, SPENCER F. Smith on Tomocaris Peircei. $<$ *Ann. Rec.*, 1872–73, p. 216.

252.

1872. BAIRD, SPENCER F. Memoirs of the Cambridge Museum. $<$ *Ann. Reo.*, 1872–73, p. 217.

253.

1872. BAIRD, SPENCER F. Prehistoric Beads. $<$ *Ann. Rec.*, 1872–73, p. 221.

254.

1872. BAIRD, SPENCER F. The Tanis Stone. $<$ *Ann. Rec.*, 1872–73, p. 230.

255.

1872. BAIRD, SPENCER F. German Central Museum of Ethnology. $<$ *Ann. Rec.*, 1872–73, p. 232.

256.

1872. BAIRD, SPENCER F. Stranding of a Japanese Junk on the Aleutian Islands. $<$ *Ann. Reo.*, 1872–73, p. 232.

257.

1872. BAIRD, SPENCER F. Use of the Boomerang by American Indians. < *Ann. Rec.,* 1872–73, p. 235.

258.

1872. BAIRD, SPENCER F. Dwarfed Human Head. < *Ann. Rec.,* 1872–73, p. 235.

259.

1872. BAIRD, SPENCER F. Shell Mound near Newburyport. < *Ann. Rec.,* 1872–73, p. 236.

260.

1872. BAIRD, SPENCER F. Journal of the Anthropological Institute of New York. < *Ann. Rec.,* 1872–73, p. 236.

261.

1872. BAIRD, SPENCER F. Characteristics of higher groups of Mammals. < *Ann. Rec.,* 1872–73, p. 238.

262.

1872. BAIRD, SPENCER F. Fur-bearing Animals of New Jersey. < *Ann. Rec.,* 1872–73, p. 240.

263.

1872. BAIRD, SPENCER F. Capture of Bassaris in Ohio. < *Ann. Rec.,* 1872–73, p. 240.

264.

1872. BAIRD, SPENCER F. Law in regard to Killing Buffalo. < *Ann. Rec.,* 1872–73, p. 241.

265.

1872. BAIRD, SPENCER F. New American Mastodon. < *Ann. Rec.,* 1872–73, p. 242.

266.

1872. BAIRD, SPENCER F. American Birds in Europe. < *Ann. Rec.,* 1872–73, p. 247.

267.

1872. BAIRD, SPENCER F. Antagonism of Harmless Serpents to Poisonous ones. < *Ann. Rec.,* 1872–73, p. 255.

268.

1872. BAIRD, SPENCER F. Blood from the Eye of the Horned Toad. < *Ann. Rec.,* 1872–73, p. 256.

269.

1872. BAIRD, SPENCER F. Horned Frogs Viviparous. < *Ann. Rec.,* 1872–73, p. 258.

270.

1872. BAIRD, SPENCER F. Cope on the Fossil Fish of the Kansas Cretaceous. < *Ann. Rec.,* 1872–73, p. 258.

271.

1872. BAIRD, SPENCER F. Nest-building Fish. < *Ann. Rec.,* 1872–73, p. 259.
Notice of Agassiz's observation on nest of *Pterophryne.*

272.

BAIRD, SPENCER F. Another Pelagic Fish-Nest. < *Ann. Rec.*, 1872–73, p. 260.
Notice of discovery of nest of *Pterophryne* at the Bermudas by J. Matthew Jones.

273.

BAIRD, SPENCER F. Respiration in Fish. < *Ann. Rec.*, 1872–73, p. 261.
Notice of a lecture by M. Gréhant.

274.

BAIRD, SPENCER F. Genesis of Hippocampus. <*Ann. Rec.*, 1872–73, p. 261.
Notice of Canestrinis' observation of rudimentary caudal fin in *Hippocampus*.

275.

BAIRD, SPENCER F. Chinese Cyprinidæ. <*Ann. Rec.*, 1872–73, p. 262.
Notice of Bleeker's paper on the *Cyprinoids* of China.

276.

BAIRD, SPENCER F. Death of an Aged Carp. < *Ann. Rec.*, 1872–73, p. 262.

277.

BAIRD, SPENCER F. Salmon Fly-fishing on the Northwest Coast of America.
< *Ann. Rec.*, 1872–73, p. 262.

278.

BAIRD, SPENCER F. Venomous Fish in the Mauritius. < *Ann. Rec.*, 1872–73,
p. 263.
Notices of venomous fish.

279.

BAIRD, SPENCER F. Teeth in Young Sturgeons. < *Ann. Rec.*, 1872–73, p. 264.

280.

BAIRD, SPENCER F. Monster Cod. <*Ann. Rec.*, 1872–73, p. 264.

281.

BAIRD, SPENCER F. Stones in the Stomachs of Codfish. <*Ann. Rec.*, 1872–73,
p. 264.

282.

BAIRD, SPENCER F. Bluefish on the Southern Coast. <*Ann. Rec.*, 1872–73, p.
265.

283.

BAIRD, SPENCER F. Edward's Work on North American Butterflies. <*Ann.
Rec.*, 1872–73, p. 267.

284.

BAIRD, SPENCER F. Have Trilobites Legs? < *Ann. Rec.*, 1872–73, p. 268.

285.

BAIRD, SPENCER F. Early stages of the American Lobster. <*Ann. Rec.*, 1872–
73, p. 269.
Review of paper by S. I. Smith.

286.

1872. BAIRD, SPENCER F.   Professor Gill's Arrangement of Mollusks. <*Ann. Reo.*, 1872–73, p. 269.

287.

1872. BAIRD, SPENCER F.   Embryology of Terebratulina and Ascidia, and Protective Coloration of Mollusca. <*Ann. Rec.*, 1872–73, p. 271.

288.

1872. BAIRD, SPENCER F.   Parasites and Commensals of Fish. <*Ann. Reo.*, 1872–73, p. 272.
Review of Van Beneden's Memoirs on this subject.

289.

1872. BAIRD, SPENCER F.   Worms in the Trout of Yellowstone Lake. <*Ann. Reo.*, 1872–73, pp. 274, 275.

290.

1872. BAIRD, SPENCER F.   Prehistoric (?) Man in America. <*Ann. Rec.*, 1872–73, p. 295.

291.

1872. BAIRD, SPENCER F.   A Sea-Serpent in a Highland Loch. <*Ann. Rec.*, 1872–73, p. 296.

292.

1872. BAIRD, SPENCER F.   Generation of Eels. <*Ann. Rec.*, 1872–73, p. 299.
Ercolanis claims that the eel is hemaphrodite.

293.

1872. BAIRD, SPENCER F.   Alleged Gigantic Pike. <*Ann. Rec.*, 1872–73, p. 301.
Notice of the big pike mentioned by Walton.

294.

1872. BAIRD, SPENCER F.   Nature of the Blue Coloring Matter in Fishes. <*Ann. Rec.*, 1872–73, p. 302.
Notice of Pouchet's investigation.

295.

1872. BAIRD, SPENCER F.   New American Fossil Vertebrates. <*Ann. Reo.*, 1872–73, p. 305.
Discoveries of Prof. O. C. Marsh.

296.

1872. BAIRD, SPENCER F.   The Proboscidians of the American Eocene. <*Ann. Reo.*, 1872–73, p. 307.
Discoveries of Prof. E. D. Cope.

297.

1872. BAIRD, SPENCER F.   The Armed Metalophodon. <*Ann. Rec.*, 1872–73, p. 308.

298.

1872. BAIRD, SPENCER F.   Fossil Fishes and Insects from the Nevada Shales. <*Ann. Rec.*, 1872–73, p. 308.

299.

BAIRD, SPENCER F. Report of the Museum of Comparative Zoology. <*Ann. Rec.*, 1872–73, p. 309.

300.

BAIRD, SPENCER F. "The Lens," a new Scientific Journal. <*Ann. Rec.*, 1872–73, p. 310.

301.

BAIRD, SPENCER F. Coues's Work on American Birds. <*Ann. Rec.*, 1872–73, p. 310.
Review of "A Key to North American Birds."

302.

BAIRD, SPENCER F. New Ornithological Periodical. <*Ann. Rec.*, 1872–73, p. 311.
"American Ornithology."

303

BAIRD, SPENCER F. Edwards on North American Butterflies. <*Ann. Rec.*, 1872–73, p. 311.
Review.

304.

BAIRD, SPENCER F. Scammon on the West Coast Cetaceans. <*Ann. Rec.*, 1872–73, p. 312.
Review.

305.

BAIRD, SPENCER F. People using the Boomerang. <*Ann. Rec.*, 1872–73, p. 316.

306.

BAIRD, SPENCER F. Origin of the Domestic Dog. <*Ann. Rec.*, 1872–73, p. 317.

307.

BAIRD, SPENCER F. The Mammals of Thibet. <*Ann. Rec.*, 1872–73, p. 318.
Researches of Abbé David.

308.

BAIRD, SPENCER F. Change of Color in Fishes. <*Ann. Rec.*, 1872–73, p. 319.

309.

BAIRD, SPENCER F. Maynard on the Birds of Florida. <*Ann. Rec.*, 1872–73, p. 320.
Review.

310.

BAIRD, SPENCER F. Allen on the Birds of Kansas, etc. <*Ann. Rec.*, 1872–73, p. 321.
Review.

311.

BAIRD, SPENCER F. Filaria in the Brain of the Water-Turkey. <*Ann. Rec.*, 1872–73, p. 324.
Investigations of Jeffries Wyman.

312.

BAIRD, SPENCER F. Use of the Bill of the Huia Bird. <*Ann. Rec.*, 1872–73, p. 324.

313.

1872. BAIRD, SPENCER F. A Fossil Lemuroid from the Eocene of Wyoming. <*Ann. Rec.*, 1872-73, p. 326.

314.

1872. BAIRD, SPENCER F. Prehistoric Remains in Unalashka. <*Ann. Rec.*, 1872-73, p. 327.
Discoveries of W. H. Dall.

315.

1872. BAIRD, SPENCER F. Archæology in America. <*Ann. Rec.*, 1872-73, p. 329
Researches of Dr. Schmidt, of Essen, Germany.

316.

1872. BAIRD, SPENCER F. Prehistoric Remains in Wyoming. <*Ann. Rec.*, 1872-73, p. 330.
Discoveries of Prof. Leidy.

317.

1872. BAIRD, SPENCER F. Peculiar Mound Crania. <*Ann. Rec.*, 1872-73, p. 321.
Discoveries of Dr. G. W. Foster.

318.

1872. BAIRD, SPENCER F. Migrations of the California Gray Whale. <*Ann. Rec.*, 1872-73, p. 332.

319.

1872. BAIRD, SPENCER F. Objects from the Florida Mounds. <*Ann. Rec.*, 1872-73, p. 332.
Discoveries of Prof. Wyman.

320.

1872. BAIRD, SPENCER F. Fossil Elephant in Alaska. <*Ann. Rec.*, 1872-73, p. 333.
Discoveries of M. Pinart.

321.

1872. BAIRD, SPENCER F. Enumeration of American Serpents. <*Ann. Rec.*, 1872-73, p. 334.

322.

1872. BAIRD, SPENCER F. Carpal and Tarsal Bones of Birds. <*Ann. Rec.*, 1872-73, p. 336.
Views of Prof. E. C. Morse.

323.

1872. BAIRD, SPENCER F. Coues on the Birds of the United States. <*Ann. Rec.*, 1872-73, p. 336.

324.

1872. BAIRD, SPENCER F. Fossil Mammals from the West. <*Ann. Rec.*, 1872-73, p. 337.
Discoveries of Prof. O. C. Marsh

325.

1872. BAIRD, SPENCER F. Algæ of Rhode Island. <*Ann. Rec.*, 1872-73, p. 342.
Olney's list.

326.

BAIRD, SPENCER F. Relation of Recent North American Flora to Ancient. <*Ann. Rec.*, 1872–73, p. 352.
Review of Lesquereux's work.

327.

BAIRD, SPENCER F. Report of C. V. Riley, State Entomologist of Missouri, for 1871. <*Ann. Rec.*, 1872–73, p. 378.

328.

BAIRD, SPENCER F. Report of the United States Agricultural Department of Cattle Diseases. <*Ann. Rec.*, 1872–73, p. 381.

329.

BAIRD, SPENCER F. R. D. Cutts on Sea Fisheries. <*Ann. Rec.*, 1872–73, p. 396.

330.

BAIRD, SPENCER F. French Fisheries for 1870. <*Ann. Rec.*, 1872–73, p. 397.

331.

BAIRD, SPENCER F. Comparison of American and French Fisheries. <*Ann. Rec.*, 1872–73, p. 398.

332.

BAIRD, SPENCER F. German Fishery Association. <*Ann. Rec.*, 1872–73, p. 399.

333.

BAIRD, SPENCER F. Fisheries of the Gulf of Naples. <*Ann. Rec.*, 1872–73, p. 399.

334.

BAIRD, SPENCER F. French Fish-Breeding Establishment. <*Ann. Rec.*, 1872–73, p. 399.
Notice of the establishment at Montbeliard.

335.

BAIRD, SPENCER F. Fishery Exposition at Gothenburg, (Sweden.) <*Ann. Rec.*, 1872–73, p. 400.

336.

BAIRD, SPENCER F. Fish-Culturists' Association at Albany. <*Ann. Rec.*, 1872–73, p. 400.
Second annual meeting of the American Fish-Culturists' Association.

337.

BAIRD, SPENCER F. U. S. Appropriation for the Propagation of Fish. <*Ann. Rec.*, 1872–73, p. 401.

338.

BAIRD, SPENCER F. Report of Maine Fish Commissioners for 1871. <*Ann. Rec.*, 1872–73, p. 403.

339.

1872. BAIRD, SPENCER F. Fish Culture in New Hampshire. $<$ *Ann. Rec.*, 1872–73, p. 405.
   Review of Report of Commissioners of Fisheries for 1872.

340.

1872. BAIRD, SPENCER F. Prizes of the Massachusetts Agricultural Society for Fish-Culture. $<$ *Ann. Rec.*, 1872–73, p. 406.

341.

1872. BAIRD, SPENCER F. Alabama Fish Commissioners. $<$ *Ann. Rec.*, 1872–73, p. 406.

342.

1872. BAIRD, SPENCER F. Fish-Culture in California. $<$ *Ann. Rec.*, 1872–73, p. 407.
   Review of Biennial Report of Commissioners of Fisheries of the State of California for the years 1870 and 1871.

343.

1872. BAIRD, SPENCER F. Report of California Fish Commissioners. $<$ *Ann. Rec.*, 1872–73, p. 408.

344.

1872. BAIRD, SPENCER F. Stocking California Waters with Trout. $<$ *Ann. Rec.*, 1872–73, p. 409.

345.

1872. BAIRD, SPENCER F. Transporting Black Bass to California. $<$ *Ann. Rec.*, 1872–73, p. 409.

346.

1872. BAIRD, SPENCER F. Report of Connecticut Fish Commissioners for 1871. $<$ *Ann. Rec.*, 1872–73, p. 409.

347.

1872. BAIRD, SPENCER F. Planting of Shad in the Valley of the Mississippi and the Lakes. $<$ *Ann. Rec.*, 1872–73, p. 410.

348.

1872. BAIRD, SPENCER F. Report of the New York Fish Commissioners for 1871. $<$ *Ann. Rec.*, 1872–73, p. 412.

349.

1872. BAIRD, SPENCER F. Second Report of New Jersey Fish Commissioners. $<$ *Ann. Rec.*, 1872–73, p. 413.

350.

1872. BAIRD, SPENCER F. Transportation of Black Bass to England. $<$ *Ann. Rec.*, 1872–73, p. 414.

351.

1872. BAIRD, SPENCER F. Fisheries of the North Carolina Coast. $<$ *Ann. Rec.*, 1872–73, p. 414.

352.

1872. BAIRD, SPENCER F. Consumption of Bluefish in New York. $<$ *Ann. Rec.*, 1872–73, p. 414.

353.

1872. BAIRD, SPENCER F. Breeding Salmon and Trout in Inclosures. $<$ *Ann. Reo.*, 1872–73, p. 415.

Experiments in the North Sea by Professor Rasch.

354.

1872. BAIRD, SPENCER F. Capture of Rhine Salmon in Holland. $<$ *Ann. Reo.*, 1872–73, p. 416.

355.

1872. BAIRD, SPENCER F. Cost of Salmon-Eggs in Europe. $<$ *Ann. Reo.*, 1872–73, p. 418.

356.

1872. BAIRD, SPENCER F. Artificial Breeding of Salmon. $<$ *Ann. Reo.*, 1872–73, p. 418.

Arrangements proposed by Frank Buckland for salmon-breeding in the Aberdeenshire Dee.

357.

1872. BAIRD, SPENCER F. Do Salmon need to Reside in Salt Water? $<$ *Ann. Reo.*, 1872–73, p. 418.

Notice of the claims of a writer in *The Field*.

358.

1872. BAIRD, SPENCER F. Profitable result of Salmon Planting in Germany. $<$ *Ann. Rec.*, 1872–73, p. 419.

359.

1872. BAIRD, SPENCER F. Trout-Breeding in France. $<$ *Ann. Reo.*, 1872–73, p. 420.

360.

1872. BAIRD, SPENCER F. Renewal of Salmon-Planting in the Delaware. $<$ *Ann. Rec.*, 1872–73, p. 421.

361.

1872. BAIRD, SPENCER F. Salmon and Trout in Australia. $<$ *Ann. Rec.*, 1872–73, p. 422.

362.

1872. BAIRD, SPENCER F. Spawning of Herring. $<$ *Ann. Reo.*, 1872–73, p. 422.
Notice of Matthew Dunn's observations on herring-eggs.

363.

1872. BAIRD, SPENCER F. Breeding of Smelt in Europe. $<$ *Ann. Rec.*, 1872–73, p. 423.
Notice of statement by correspondent of *Land and Water*.

364.

1872. BAIRD, SPENCER F. Raising Otsego Bass. $<$ *Ann. Rec.*, 1872–73, p. 423.
Notice of Seth Green's experiments.

365.

1872. BAIRD, SPENCER F. Fisheries on the Coast of Norway. $<$ *Ann. Reo.*, 1872–73, p. 423.

9 BD

366.

1872. BAIRD, SPENCER F.   Loffoden Codfishery.   <*Ann. Rec.*, 1872–73, p. 424.

367.

1872. BAIRD, SPENCER F.   Spawning of Menhaden.   <*Ann. Rec.*, 1872–73, p. 425.

368.

1872. BAIRD, SPENCER F.   Herring-Fishery in Great Britain.   <*Ann. Rec.*, 1872–73, p. 425.

369.

1872. BAIRD, SPENCER F.   Reappearance of a Peculiar Herring on the Norway Coast.   <*Ann. Rec.*, 1872–73, p. 426.
   Notice of article in *Land and Water.*

370.

1872. BAIRD, SPENCER F.   Use of Fishes as Manure in England.   <*Ann. Rec.*, 1872–73, p. 426.
   Notice of article in *Land and Water.*

371.

1872. BAIRD, SPENCER F.   Food of Shad.   <*Ann. Rec.*, 1872–73, p. 426.
   Notice of observation by Professor Leidy.

372.

1872. BAIRD, SPENCER F.   Bryan on the Decrease of Shad.   <*Ann. Rec.*, 1872–73, p. 427.

373.

1872. BAIRD, SPENCER F.   Shad in Alabama.   <*Ann. Rec.*. 1872–73, p. 427.

374.

1872. BAIRD, SPENCER F.   Shad in Red River, Arkansas.   <*Ann. Rec.*, 1872–73, p. 428.

375.

1872. BAIRD, SPENCER F.   Shad Hatching in the Hudson River.   <*Ann. Rec.*, 1872–73, p. 428.

376.

1872. BAIRD, SPENCER F.   Planting of Shad in the Genesee River.   <*Ann. Rec.*, 1872–73, p. 429.

377.

1872. BAIRD, SPENCER F.   Planting of Shad in Lake Champlain.   <*Ann. Rec.*, 1872–73, p. 429.

378.

1872. BAIRD, SPENCER F.   Stocking California with Shad.   <*Ann. Rec.*, 1872–73, p. 430.

379.

1872. BAIRD, SPENCER F.   Transferring Shad to the Sacramento River.   <*Ann. Rec.*, 1872–73, p. 430.

380.

1872. BAIRD, SPENCER F. *Cyprinus orfus* as an Ornamental and Food Fish. <*Ann. Rec.*, 1872–73, p. 431.

381.

1872. BAIRD, SPENCER F. Prize Essay on the Reproduction of Eels. <*Ann. Rec.*, 1872–73, p. 434.
Notice of prize offered by the Belgium Academy of Sciences, with a statement of the commonly accepted theory on this subject.

382.

1872. BAIRD, SPENCER F. Marking Whitefish. <*Ann. Rec.*, 1872–73, p. 435.
Experiments of George Clark, of Ecorse, Mich.

383.

1872. BAIRD, SPENCER F. Catch of Fur Seals in 1872. <*Ann. Rec.*, 1872–73, p. 435.

384. ·

1872. BAIRD, SPENCER F. Oil Works on Unalaschka. <*Ann. Rec.*, 1872–73, p. 436.

385.

1872. BAIRD, SPENCER F. Spawning of Codfish in Alaska. <*Ann. Rec.*, 1872–73, p. 436.
Observations of Mr. W. H. Dall.

386.

1872. BAIRD, SPENCER F. Codfishing in the Shumagin Islands. <*Ann. Rec.*, 1872–73, p. 436.
Statement of a correspondent of the *Alaska Herald*.

387.

1872. BAIRD, SPENCER F. American Whale Fishery in 1871. <*Ann. Rec.*, 1872–73, p. 437.

388.

1872. BAIRD, SPENCER F. Salmon Fisheries in the Columbia River. <*Ann. Rec.*, 1872–73, p. 440.

389.

1872. BAIRD, SPENCER F. Capture of Sacramento Salmon with the Hook. <*Ann. Rec.*, 1872–73, p. 441.

390.

1872. BAIRD, SPENCER F. Breeding of Leeches. <*Ann. Rec.*, 1872–73, p. 441.

391.

1872. BAIRD, SPENCER F. Spawning of the Sterlet. <*Ann. Rec.*, 1872–73, p. 442.
Observations in the Volga, by Prof. Owsjannikow.

392.

1872. BAIRD, SPENCER F. Tunny Fisheries on the South Shore of the Mediterranean. <*Ann. Rec.*, 1872–73, p. 442.

393.

1872. BAIRD, SPENCER F. Fishing Statistics of Great Britain for 1869. <*Ann. Rec.*, 1872–73, p. 443.

394.

1872. BAIRD, SPENCER F.   Fisheries of the Shumagin Islands. $<$ *Ann. Rec.*, 1872–73, p. 444.

Notices of letter by correspondent of the *Alaska Herald.*

395.

1872. BAIRD, SPENCER F.   Utilization of Refuse Fish. $<$ *Ann. Rec.*, 1872–73, p. 444.

Schacht Brothers' establishment for utilizing the products of the sturgeon at Sandusky. Ohio.

396.

1872. BAIRD, SPENCER F.   Peculiarities of Reproduction of California Salmon. $<$ *Ann. Rec.*, 1872–73, p. 445.

Observations of Livingstone Stone.

397.

1872. BAIRD, SPENCER F.   Winter-Quarters of Nova Scotia Salmon. $<$ *Ann. Rec.*, 1872–73, p. 446.

Observations by Dr. Gilpin, of Halifax.

398.

1872. BAIRD, SPENCER F.   Growth of Salmon. $<$ *Ann. Rec.*, 1872–73, p. 446.

Observations at Hameln, on the Weser.

399.

1872. BAIRD, SPENCER F.   Best kind of Water for Salmon Hatching. $<$ *Ann. Rec.*, 1872–73, p. 446.

Controversy between Dr. Hetting, of Norway, and Herr Von der Wengen.

400.

1872. BAIRD, SPENCER F.   Alleged Discovery of Young Shad in the Sacramento River. $<$ *Ann. Rec.*, 1872–73, p. 447.

401.

1872. BAIRD, SPENCER F.   Report of Fish Commissioners of Vermont (for 1871–72). $<$ *Ann. Rec.*, 1872–73, p. 447.

402.

1872. BAIRD, SPENCER F.   Rat Catching. $<$ *Ann. Rec.*, 1872–73, p. 458.

403.

1872. BAIRD, SPENCER F.   Department Report on the Preparation of Timber. $<$ *Ann. Rec.*, 1872–73, p. 485.

404.

1872. BAIRD, SPENCER F.   Report of the Zoological Society for 1871. $<$ *Ann. Rec.*, 1872–73, p. 601.

405.

1872. BAIRD, SPENCER F.   Washington Meeting of the Natural Academy of Sciences. $<$ *Ann. Rec.*, 1872–73, p. 604.

406.

1872. BAIRD, SPENCER F.   Twenty-first Meeting of the American Association for the Advancement of Science. $<$ *Ann. Rec.*, 1872–73, p. 607.

407.

1872. BAIRD, SPENCER F. Fifth Report of the Peabody Museum, Cambridge. <*Ann. Rec.*, 1872–73, p. 608.

408.

1872. BAIRD, SPENCER F. Report of the Peabody Academy of Science for 1871. <*Ann. Rec.*, 1872–73, p. 608.

409.

1872. BAIRD, SPENCER F. Bloomington Scientific Association. <*Ann. Rec.*, 1872–73, p. 609.

410.

1872. BAIRD, SPENCER F. Meeting of the American Philological Society. <*Ann. Rec.*, 1872–73, p. 609.

411.

1872. BAIRD, SPENCER F. Circular of the Chicago Academy of Sciences. <*Ann. Rec.*, 1872–73, p. 610.

412.

1872. BAIRD, SPENCER F. Regulations of the New York Museum of Natural History. <*Ann. Rec.*, 1872–73, p. 611.

413.

1872. BAIRD, SPENCER F. American Journal of Conchology. <*Ann. Rec.*, 1872–73, p. 611.

414.

1872. BAIRD, SPENCER F. Gold Medal to Professor Dana. <*Ann. Rec.*, 1872–73, p. 617.

415.

1872. BAIRD, SPENCER F. Renewed Activity of the St. Louis Academy of Sciences. <*Ann. Rec.*, 1872–73, p. 612.

416.

1873. BAIRD, SPENCER F. Necrology of Science for 1872. <*Ann. Rec.*, 1872–73, p. 619.

417.

1873. BAIRD, SPENCER F. Annual Record | of | Science and Industry | for 1872. | Edited by | Spencer F. Baird, | with the assistance of eminent men of science. | [Cut.] | New York: | Harper & Brothers, Publishers, | Franklin Square. | 1873. 8vo. pp. lxviii, 650.
See above, Nos. 224–415.

418.

1873. [BAIRD, SPENCER F.] A bill to Regulate the use of Stationary Apparatus in the Capture of Fish. fol. bill-form. 6 pp.
Reproduced in Report of Commissioner Part I. See No. 424.

419.

1873. BAIRD, SPENCER F. Statistics of the Menhaden Fisheries, etc. [Questions addressed to fishermen, etc.]—Washington, December 20, 1873. 4to. letter form, 2 l. [U. S. F. C., 5.]

420.

1873. BAIRD, SPENCER F. "Account of the additions to the National Museum, and the various operations connected with it during 1872." <*Ann. Rep. Smithsonian Institution for the year* 1872, 1873, pp. 43–52 and 55–62.

421.

1873. BAIRD, SPENCER F. 42d Congress, } 2d session. } Senate. { Mis. Doc. | No. 61. | — | United States Commission of Fish and Fisheries. | —Part 1.— | Report | on the | condition of the sea fisheries | of the south coast of New England | in | 1871 and 1872, | by | Spencer F. Baird, Commissioner. | — | Washington : | Government Printing Office. | 1873. 8vo, pp. (5) 6–41 (1). [U. S. F. C., 6.]

The report of the Commissioner, without supplementary papers, pp. xlvii, was issued separately in advance, paged in Arabic numbers.

422.

1873. BAIRD, SPENCER F. 42d Congress, } 2d session,' } Senate. { Mis. Doc. | No. 61. | — | United States Commission of Fish and Fisheries. | —Part 1.— | Report | on the | condition of the sea fisheries | of the south coast of New England | in | 1871 and 1872, | by | Spencer F. Baird, Commissioner. | — | With supplementary papers. | — | Washington : | Government Printing Office. | 1873. 8vo, pp. xlvii, 852, plates xxxviii, with 38 leaves explanatory to plates, 2 maps. [U. S. F. C., 7.]

CONTENTS.

REPORT OF THE COMMISSIONER.

423.

1873. [BAIRD, SPENCER F.] Memoranda of inquiry relative to the Food-Fishes of the United States. < Rep. U. S. Comm. Fish and Fisheries, part i, 1873, pp. 1–3.

A general plan of investigation, "for the purpose of securing greater precision in the inquiries prosecuted in reference to the natural history of the fishes and the influences exerted upon their multiplication." Prepared with the assistance of Prof. Gill.

Also issued separately. See above. No. 216.

424.

1873. [BAIRD, SPENCER F.] Questions relative to the Food-Fishes of the United States. < Rep. U. S. Comm. Fish and Fisheries, part i, 1873, pp. 3–6.

A schedule of 88 questions, embodying the points of investigation included in the plan of the memoranda of inquiry.

Also issued separately in advance. See above. No. 217.

425.

[BAIRD, SPENCER F.] Testimony in regard to the present condition of the Fisheries, taken in 1871. < *Rep. U. S. Comm. Fish and Fisheries*, part i, 1873, pp. 7–72.

Testimony of the fishermen of Southern New England, gathered by Prof. Baird, with the aid of Mr. H. E. Rockwell, phonographic reporter, in August and September, 1871.

426.

[BAIRD, SPENCER F.] Report of conference held [by the U. S. Commissioner of Fish and Fisheries] at Boston, October 5, 1871, with the Fishery Commissioners of Massachusetts and Rhode Island. < *Rep. U. S. Comm. Fish and Fisheries*, 1873, part i, pp. 125–131.

427.

[BAIRD, SPENCER F.] Draught of law proposed for the consideration of and enactment by the Legislatures of Massachusetts, Rhode Island, and Connecticut. < *Rep. U. S. Comm. Fish and Fisheries*, part i, 1873, pp. 132–134.

"The draught, as originally prepared, was first discussed at the conference with the commissioners of Massachusetts and Rhode Island and then submitted to several eminent legal gentlemen for consideration: among others to Mr. Henry Williams and Mr. George H. Palmer, of Boston, from whom criticisms were received." See No. 416.

428.

[BAIRD, SPENCER F.] Notices in regard to the Abundance of Fish on the New England Coast in former times. < *Rep. U. S. Comm. Fish and Fisheries*, part i, 1873, pp. 148–172.

A collection of extracts from early colonial accounts of the abundance of fishes.

429.

[BAIRD, SPENCER F.] Statistics of Fish and Fisheries on the South Shore of New England. < *Rep. U. S. Comm. Fish and Fisheries*, part i, 1873, pp. 172–181.

430.

[BAIRD, SPENCER F.] Supplementary Testimony and Information relative to the Condition of the Fisheries of the South Side of New England, taken in 1872. < *Rep. U. S. Comm. Fish and Fisheries*, part i, 1873, pp. 182–195.

Memoranda obtained at Wood's Hole, Newport, Hyannis, Nantucket, and Martha's Vineyard by the Commissioner and Mr. Vinal N. Edwards.

431.

[BAIRD, SPENCER F.] Natural History of some of the more important Food-fishes of the South Shore of New England.—I. The Scup, *Stenotomus argyrops*, (Linn.,) Gill. < *Rep. U. S. Com. Fish and Fishcries*, part i, 1873, pp. 228–235.

432.

[BAIRD, SPENCER F.] [Natural History of some of the more important Food-fishes of the South Shore of New England.]—II. The Blue-Fish, *Pomatomus saltatrix*, (Linn.,) Gill. < *Rep. U. S. Comm. Fish and Fisheries*, part i, 1873, pp. 235–252.

433.

[BAIRD, SPENCER F.] Description of Apparatus used in capturing Fish on the Sea-coast and Lakes of the United States. < *Rep. U. S. Comm. Fish and Fisheries*, part i, 1873, pp. 253–274.

434.

1873. [BAIRD, SPENCER F.] List of Patents granted by the United States to the end of 1872, for inventions connected with the Capture, Utilization or Cultivation of Fishes and Marine Invertebrates. < *Rep. U. S. Comm. Fish and Fisheries*, part i, 1873, pp. 275–280.

435.

1873. BAIRD, SPENCER F. List of Fishes collected at Wood's Hole. < *Rep. U. S. Comm. Fish and Fisheries*, part i, 1873, pp. 823–827.

A name-list of 121 species collected by the U. S. Commission in 1871, with the exception of a few taken there the following year by Vinal N. Edwards and of *Lactophrys trigonus* and *Hippocampus Hudsonius*, given by Dr. Storer as taken at Vineyard Haven.

436.

1873. BAIRD, SPENCER F. Salmon in the Hudson. < *Forest and Stream*, i, 1873, p. 233.

Stocking the Hudson with salmon by the U. S. Fish Commission.

437.

1873. BAIRD, SPENCER F. Planting California Salmon at Fort Edward. < *Forest and Stream*, i, 1873, p. 298.

Letter to editor.

438.

1873. [BAIRD, SPENCER F.] The New England Fisheries. < *American Sportsman*, iii, 1873, p. 23.

Quotations in editions of *Scientific Miscellany* from letter to Massachusetts commissioners.

439.

1873. BAIRD, SPENCER F. Signal Telegraphy and the Herring-Fishery. < *Ann. Rec.*, 1873–75, p. 73.*

Establishment of a signal station at Eastport, Me.

440.

1873. BAIRD, SPENCER F. Third and Fourth Annual Report of the Geological Survey of Indiana for 1871, 1872. < *Ann. Rec.*, 1873–75, p. 202.

441.

1873. BAIRD, SPENCER F. Report for 1872 on the Geology of New Jersey. < *Ann. Rec.*, 1873–75, p. 203.

442.

1873. BAIRD, SPENCER F. Geological Survey of Canada for 1871–72. < *Ann. Rec.*, 1873–75, p. 202.

443.

1873. BAIRD, SPENCER F. Fourth Annual Report of Mining Statistics for 1872. < *Ann. Rec.*, 1873–75, p. 203.

Report of R. W. Raymond.

444.

1873. BAIRD, SPENCER F. Geological Survey of Ohio. < *Ann. Rec.*, 1873–75, p. 204.

445.

1873. BAIRD, SPENCER F. Final Report of the Geological Survey of Ohio. < *Ann. Rec.*, 1873–75, p. 205.

---

* For full title of the volume of the Annual Record for 1873, see under 1875.

446.

BAIRD, SPENCER F. Lesquereux on the Fossil Plants of the Northern Hemisphere. < *Ann. Rec.*, 1873–75, p. 210.

447.

BAIRD, SPENCER F. Explorations in the Gulf of St. Lawrence in 1872. < *Ann. Rec.*, 1873–75, p. 216.
Explorations of J. F. Whiteaves.

448.

BAIRD, SPENCER F. Fauna of the St. George's Bank and adjacent waters. < *Ann. Rec.*, 1873–75, p. 218.
Verrill on the discoveries of the steamer "Bache."

449.

BAIRD, SPENCER F. Report of the German North Polar Expedition. < *Ann. Rec.*, 1873–75, p. 220.

450.

BAIRD, SPENCER F. Fictitious Account of Pary's Explorations. < *Ann. Rec.*, 1873–75, p. 221.

451.

BAIRD, SPENCER F. Report on the Yellowstone Park. < *Ann. Rec.*, 1873–75, p. 222.
Report of Gov. Langford.

452.

BAIRD, SPENCER F. Explorations of Lieut. Wheeler in 1871. < *Ann. Rec.*, 1873–75, p. 223.

453.

BAIRD, SPENCER F. Dr. Hayden's Surveys. < *Ann. Rec.*, 1873–75, p. 226.

454.

BAIRD, SPENCER F. Sixth Annual Report of Dr. Hayden's Explorations. < *Ann. Rec.*, 1873–75, p. 232.

455.

BAIRD, SPENCER F. Report of Major J. W. Barlow. < *Ann. Rec.*, 1873–75, p. 232.
Expedition down the Yellowstone River.

456.

BAIRD, SPENCER F. Final Report of Dr. Hayden's Explorations. < *Ann. Rec.*, 1873–75, p. 236.

457.

BAIRD, SPENCER F. The History of the Polaris. < *Ann. Rec.*, 1873–75, p. 237.

458.

BAIRD, SPENCER F. The Cruise of the "Challenger" in 1873. < *Ann. Rec.*, 1873–75, p. 243.

459.

BAIRD, SPENCER F. Explorations of W. H. Dall in the Aleutian Islands. < *Ann. Rec.*, 1873–75, p. 246.

460.

1873. BAIRD, SPENCER F.   Dr. Hayden's Geological Explorations in 1873.   <*Ann. Rec.*, 1873–75, p. 249.

461.

1873. BAIRD, SPENCER F.   Lieutenant Wheeler's Exploration in 1873.   <*Ann. Rec.*, 1873–75, p. 251.

462.

1873. BAIRD, SPENCER F.   Explorations of Captain William A. Jones.   <*Ann. Rec.*, 1873–75, p. 254.

463.

1873. BAIRD, SPENCER F.   Interoceanic Canal Explorations by the United States Navy.   <*Ann. Rec.*, 1873–75, p. 255.

464.

1873. BAIRD, SPENCER F.   Natural-History Explorations of the Northern Boundary Survey.   <*Ann. Rec.*, 1873–75, p. 257.

465.

1873. BAIRD, SPENCER F.   Explorations of Professor Powell.   <*Ann. Rec.*, 1873–75, p. 258.

466.

1873. BAIRD, SPENCER F.   Yellowstone Expedition.   <*Ann. Rec.*, 1873–75, p. 261.

467.

1873. BAIRD, SPENCER F.   Recent Explorations in Spitzenberg.   <*Ann. Rec.*, 1873–75, p. 262.
Explorations of Dr. Richard Von Drasche.

468.

1873. BAIRD, SPENCER F.   Effects of Seasons on the Distribution of Animals and Plants.   <*Ann. Rec.*, 1873–75, p. 263.
Observations of Prof. Shales.

469.

1873. BAIRD, SPENCER F.   The Oldest Zoological Museum in America.   <*Ann. Rec.*, 1873–75, p. 264.
The collections of M. Delacoste now in the museum of Princeton College.

470.

1873. BAIRD, SPENCER F.   Cincinnati Acclimation Society.   <*Ann. Rec.*, 1873–75, p. 265.

471.

1873. BAIRD, SPENCER F.   "Revision of the Echini," by Alexander Agassiz.   <*Ann. Rec.*, 1873–75, p. 265.

472.

1873. BAIRD, SPENCER F.   Opening of the "Anderson School of Natural History."   <*Ann. Rec.*, 1873–75, p. 266.

473.

1873. BAIRD, SPENCER F.   The Brighton Aquarium.   <*Ann. Rec.*, 1873–75, p. 267.

474.

BAIRD, SPENCER F. The Zoological Gardens of London. <*Ann. Rec.*, 1873–75, p. 267.

475.

BAIRD, SPENCER F. The Godeffroy Museum, at Hamburg. <*Ann. Rec.*, 1873–75, p. 268.

476.

BAIRD, SPENCER F. Preservation of British Prehistoric Monuments. <*Ann. Rec.*, 1873–75, p. 276.
Suggestions for United States.

477.

BAIRD, SPENCER F. Prehistoric Cannibalism in Florida. <*Ann. Rec.*, 1873–75, p. 281.
Observations of Prof. Wyman.

478.

BAIRD, SPENCER F. Working of Mica Mines in North Carolina in Prehistoric Times. <*Ann. Rec.*, 1873–75, p. 282.
Observations of Prof. Kerr.

479.

BAIRD, SPENCER F. New Fossil Carnivora. <*Ann. Rec.*, 1873–75, p. 284.
Discoveries of Leidy.

480.

BAIRD, SPENCER F. Orophippus agilis. <*Ann. Rec.*, 1873–75, p. 286.
Discoveries of Manly.

481.

BAIRD, SPENCER F. ·Maynard on the Mammals of Florida. <*Ann. Rec.*, 1873–75, p. 286.

482.

BAIRD, SPENCER F. A New Fossil Bird. <*Ann. Rec.*, 1873–75, p. 288.
Ichthyornis of Manly.

483.

BAIRD, SPENCER F. Development of a Guadeloupe Frog. <*Ann. Rec.*, 1873–75, p. 290.

484.

BAIRD, SPENCER F. Geographical Distribution of Percoid Fishes. <*Ann. Rec.*, 1873–75, p. 291.
Notice of paper by Vaillant.

485.

BAIRD, SPENCER F. Allman on Tubularian Hydroids. <*Ann. Rec.*, 1873–75, p. 296.
A monograph of the Gymnoblastic or Tubularian Hydroids.

486.

BAIRD, SPENCER F. Haeckel on the Calcareous Sponges. <*Ann. Rec.*, 1873–75, p. 298.

487.

BAIRD, SPENCER F. Blood Corpuscles of the Salmonidæ. <*Ann. Rec.*, 1873–75, p. 299.
Gulliver on blood corpuscles of fishes.

488.

1873. BAIRD, SPENCER F. Number of Glyptodonts, or Extinct Giant Armadillos.
 <*Ann. Rec.*, 1873–75, p. 302.
 Conclusions of Burmeister.

489.

1873. BAIRD, SPENCER F. International Exhibition of Horns. <*Ann. Rec.*, 1873–75,
 p. 303.
 Proposed exhibition in London, May 1, 1874.

490.

1873. BAIRD, SPENCER F. Respiration in Fishes at Different Ages. <*Ann. Rec.*, 1873–
 75, p. 304.
 Experiments by Quinquand.

491.

1873. BAIRD, SPENCER F. Absence of Fish Above the Yosemite Falls (and in the
 headwaters of the Hudson). <*Ann. Rec.*, 1873–75, p. 305.

492.

1873. BAIRD, SPENCER F. Reproduction of the Eel. <*Ann. Rec.*, 1873–75, p. 306.

493.

1873. BAIRD, SPENCER F. Influence of External Conditions on the Structure of Insects. <*Ann. Rec.*, 1873–75, p. 307.
 Observations of G. H. Horn.

494.

1873. BAIRD, SPENCER F. Number of the Red Blood Corpuscles. <*Ann. Rec.*, 1873–
 75, p. 308.

495.

1873. BAIRD, SPENCER F. New Vertebrate Fossils. <*Ann. Rec.*, 1873–75, p. 308.
 Discoveries of Cope and Stevenson.

496.

1873. BAIRD, SPENCER F. Aboriginal Monkey. <*Ann. Rec.*, 1873–75, p. 310.
 Review of paper by Stearns.

497.

1873. BAIRD, SPENCER F. The Prehistoric Races of America. <*Ann. Rec.*, 1873–75,
 p. 311.
 Review of book by Dr. J. W. Foster.

498.

1873. BAIRD, SPENCER F. The Cesnola Collection. <*Ann. Rec.*, 1873–75, p. 312.

499.

1873. BAIRD, SPENCER F. The Canstadt Race of Mankind. <*Ann. Rec.*, 1873–75, p.
 312.
 Researches of Quatrefages and Harvey.

500.

1873. BAIRD, SPENCER F. The Relationship of the Coyote to the Pointer Dog.
 <*Ann. Rec.*, 1873–75, p. 314.
 Views of Dr. Elliot Coues.

501.

1873. BAIRD, SPENCER F.   New Fossils Discovered by Professor Cope. $<$*Ann. Rec.,*
1873–75, p. 315.

502.

1873. BAIRD, SPENCER F.   Additions to Yale College Museum. $<$*Ann. Rec.,* 1873–75,
p. 315.
Pterodactyl and Zeltner Collection of Central American Antiquities.

503.

1873. BAIRD, SPENCER F.   Pterodactyl in the Cambridge Museum. $<$*Ann. Rec.,*
1873–75, p. 316.

504.

1873. BAIRD, SPENCER F.   A Large Fish. $<$*Ann. Rec.,* 1873–75, p. 317.
Promocrops guasa from the St. Johns, Fla.

505.

1873. BAIRD, SPENCER F.   Relations of the Megatheriidæ. $<$*Ann. Rec.,* 1873–75, p.
318.
Views of Paul Gervais.

506.

1873. BAIRD, SPENCER F.   Recent Explorations of Professor Cope. $<$*Ann. Rec.,* 1873–
75, p. 319.

507.

1873. BAIRD, SPENCER F.   Binney on Geographical Distribution of Mollusks. $<$*Ann.
Rec.,* 1873–75, p. 320.

508.

1873. BAIRD, SPENCER F.   Habits of the Black Bass. $<$*Ann. Rec.,* 1873–75, p. 322.

509.

1873. BAIRD, SPENCER F.   Food of the Basking Shark. $<$*Ann. Rec.,* 1873–75, p. 328.

510.

1873. BAIRD, SPENCER F.   Determining Sex in Butterflies. $<$*Ann. Rec.,* 1873–75, p.
329.
Investigations of Mary Treat.

511.

1873. BAIRD, SPENCER F.   Distribution of California Moths. $<$*Ann. Rec.,* 1873–75,
p. 330.
Paper by A. S. Packard.

512.

1873. BAIRD, SPENCER F.   Pavonaria Blakei, a New Alcyonoid Polyp. $<$*Ann. Rec.,*
1873–75, p. 332.

513.

1873. BAIRD, SPENCER F.   Terrestrial Mollusca in the Bahamas. $<$*Ann. Rec.,* 1873–
75, p. 333.
Bland "On the Physical Geography of and the Distribution of Terrestrial Mollusca in the
Bahama Islands."

10 BD

514.

1873. BAIRD, SPENCER F. On the Nature of Aptychus. <*Ann. Rec.*, 1873–75, p. 334
Views of Prof. Waager.

515.

1873. BAIRD, SPENCER F. Bird Collections in London. <*Ann. Rec.*, 1873–75, p. 336.

516.

1873. BAIRD, SPENCER F. Brighton Aquarium. <*Ann. Rec.*, 1873–75, p. 336.

517.

1873. BAIRD, SPENCER F. New Scaphirhynchus in Turkestan. <*Ann. Rec.*, 1873–75,
p. 336.
Alludes to American species.

518.

1873. BAIRD, SPENCER F. An Aquarium for Central Park. <*Ann. Rec.*, 1873–75, p.
337.

519.

1873. BAIRD, SPENCER F. Report of the Central Park Menagerie. <*Ann. Rec.*, 1873–
75, p. 338.

520.

1873. BAIRD, SPENCER F. The Gardens of the Acclimation Society of Paris. <*Ann.
Rec.*, 1873–75, p. 338.

521.

1873. BAIRD, SPENCER F. Catalogue of Rhode Island Mollusca. [By H. F. Carpen-
ter]. <*Ann. Rec.*, 1873–75, p. 339.

522.

1873. BAIRD, SPENCER F. The Mummied Heads of the Peruvian Indians. <*Ann.
Rec.*, 1873-75, p. 339.

523.

1873. BAIRD, SPENCER F. Number of American Birds. <*Ann. Rec.*, 1873–75, p. 340.

524.

1873. BAIRD, SPENCER F. Ethnology of the Peat Bogs. <*Ann. Rec.*, 1873–75, p. 341.

525.

1873. BAIRD, SPENCER F. Curious Fish. <*Ann. Rec.*, 1873–75, p. 342.
Polyodon folium in Chatauqua Lake.

526.

1873. BAIRD, SPENCER F. The Classes of Vertebrates and their Relationship. <*Ann.
Rec.*, 1873–75, p. 342.
Views of Theodore Gill.

527.

1873. BAIRD, SPENCER F. The Fossils Discovered by Professor Cope. <*Ann. Rec.*,
1873–75, p. 348.

1873. BAIRD, SPENCER F. Antiquities of the Southern Indians. <*Ann. Rec.*, 1873–75, p. 348.
Book by C. C. Jones.

1873. BAIRD, SPENCER F. Alleged Shower of Fish-Scales. <*Ann. Rec.*, 1873–75, p. 350.

1873. BAIRD, SPENCER F. Habits of the Craw-Fish. <*Ann. Rec.*, 1873–75, p. 351.
Chantran,s memoir.

1873. BAIRD, SPENCER F. The Sequoias of California, and their History. <*Ann. Rec.*, 1873–75, p. 363.
Dr. Gray's American Association address.

1873. BAIRD, SPENCER F. Forest Growth in the Wabash Valley. <*Ann. Rec.*, 1873–75, p. 367.
Investigations of Robert Ridgway.

1873. BAIRD, SPENCER F. Fish Guano. <*Ann. Rec.*, 1873–75, p. 387.

1873. BAIED, SPENCER F. Value of Sea-Weed Manure. <*Ann. Rec.*, 1873–75, p. 395.

1873. BAIRD, SPENCER F. Extermination of Field-Mice. <*Ann. Rec.*, 1873–75, p. 412.

1873. BAIRD, SPENCER F. Tenth Annual Report of the Massachusetts Agricultural College. <*Ann. Rec.*, 1873–75, p. 418.

1873. BAIRD, SPENCER F. Statistics of Canada Fisheries for 1869. <*Ann. Rec.*, 1873–75, p. 427.

1873. BAIRD, SPENCER F. British Exhibition of Fishing Products at Vienna. <*Ann. Rec.*, 1873–75, p. 427.

1873. BAIRD, SPENCER F. Exhibition of Fishery Products at Vienna. <*Ann. Rec.*, 1873–75, p. 429.

1873. BAIRD, SPENCER F. Fishery Models at the late Scandinavian Exhibition. <*Ann. Rec.*, 1873–75, p. 429.

1873. BAIRD, SPENCER F. Is Seal Oil Fish Oil? <*Ann. Rec.*, 1873–75, p. 430.

542.

1873. BAIRD, SPENCER F.   Gloucester Winter Herring Fishery.   <*Ann. Rec.*, 1873–75, p. 431.

543.

1873. BAIRD, SPENCER F.   Emden Herring-Fishery for 1872.   <*Ann. Rec.*, 1873–75, . p. 431.

544.

1873. BAIRD, SPENCER F.   Trade in Frozen Herring.   <*Ann. Rec.*, 1873–75, p. 432.

545.

1873. BAIRD, SPENCER F.   Improvement in Value of the British Salmon Fisheries.   <*Ann. Rec.*, 1873–75, p. 433.

546.

1873. BAIRD, SPENCER F.   Fishery Laws in Germany.   <*Ann. Rec.*, 1873–75, p. 433.

547.

1873. BAIRD, SPENCER F.   Shipments Eastward of California Salmon.   <*Ann. Rec.*, 1873–75, p. 433.

548.

1873. BAIRD, SPENCER F.   Meeting of the American Fish-Culturists' Association.   <*Ann. Rec.*, 1873–75, p. 434.
    First annual meeting of the American Fish-Culturists' Association, New York, Feb. 11, 1872.

549.

1873. BAIRD, SPENCER F.   Culture of Sea-Fish in Fresh Water.   <*Ann. Rec.*, 1873–75, p. 435.
    Notice of the experiments of J. B. Arnold, of Gurnsey, in 1829.

550.

1873. BAIRD, SPENCER F.   Sixth Report of the Maine Commissioners of Fisheries for 1872.   <*Ann. Rec.*, 1873–75, p. 436.

551.

1873. BAIRD, SPENCER F.   Report of the Fish Commission of Rhode Island for 1872.   <*Ann. Rec.*, 1873–75, p. 437.

552.

1873. BAIRD, SPENCER F.   Report of the Fish Commission of New York for 1872.   <*Ann. Rec.*, 1873–75, p. 438.

553.

1873. BAIRD, SPENCER F.   Ohio Fish Commission.   <*Ann. Rec.*, 1873–75, p. 441.
    Appointment of Fish Commissioners.

554.

1873. BAIRD, SPENCER F.   Michigan Fishery Bill.   <*Ann. Rec.*, 1873–75, p. 441.
    Establishment of a board of Fish Commissioners.

555.

1873. BAIRD, SPENCER F.   Hybrids of Salmon and Trout.   <*Ann. Rec.*, 1873–75, p. 442.

556.

1873. BAIRD, SPENCER F. Cultivation of Fish in Ditches and Ponds. < *Ann. Rec.,* 1873–75, p. 443.

557.

1873. BAIRD, SPENCER F. United States Salmon-Breeding Establishment at Bucksport, Maine. < *Ann. Rec.,* 1873–75, p. 443.

558.

1873. BAIRD, SPENCER F. Marked Salmon on the American Coast. < *Ann. Rec.,* 1873–75, p. 444.

559.

1873. BAIRD, SPENCER F. Transporting Salmon Eggs to New Zealand. < *Ann. Rec.,* 1873–75, p. 445.

560.

1873. BAIRD, SPENCER F. Naturalization of Trout in New Zealand. < *Ann. Rec.,* 1873–75, p. 447.
Introduction of English trout.

561.

1873. BAIRD, SPENCER F. Food for Diminutive Trout. < *Ann. Rec.,* 1873–75, p. 447.
Discovery of Fred. Mather.

562.

1873. BAIRD, SPENCER F. Alleged occurrence of Shad in the Mississippi. < *Ann. Rec.,* 1873–75, p. 448.

563.

1873. BAIRD, SPENCER F. Increase in the Growth of Trout. < *Ann. Rec.,* 1873–75, p. 448.

564.

1873. BAIRD, SPENCER F. Shad in the Sacramento River. < *Ann. Rec.,* 1873–75, p. 449.

565.

1873. BAIRD, SPENCER F. Shad in California Waters. < *Ann. Rec.,* 1873–75, p. 449.

566.

1873. BAIRD, SPENCER F. Hatching Striped Bass Artificially. < *Ann. Rec.,* 1873–75, p. 450.
Experiments of M. G. Holton, at Weldon, N. C.

567.

1873. BAIRD, SPENCER F. Shad in the Altahama River. < *Ann. Rec.,* 1873–75, p. 450.

568.

1873. BAIRD, SPENCER F. Treatment of Fish-Ponds. < *Ann. Rec.,* 1873–75, p. 452.

569.

1873. BAIRD, SPENCER F. Culture of the Sterlet. < *Ann. Rec.,* 1873–1875, p. 452.
Dr. Knoch's experiments in the Volga.

570.

1873. BAIRD, SPENCER F. Maritime Fisheries of France for 1871. < *Ann. Rec.,* 1873–75, p. 453.

571.

1873. BAIRD, SPENCER F.   Laws regulating the Newfoundland Fisheries. $<Ann.$
$Rec.$, 1873–75, p. 454.

572.

1873. BAIRD, SPENCER F.   Fish Inspection Law of Canada. $<Ann. Rec.$, 1873–75,
p. 455.
Rate of growth in trout.

573.

1873. BAIRD, SPENCER F.   Recent Fishery and Game Laws of the Ohio Legislature.
$<Ann. Rec.$, 1873–75, p. 457.

574.

1873. BAIRD, SPENCER F.   Pacific Cod-Fisheries of 1873. $<Ann Rec.$, 1873–75, p. 458.

575.

1873. BAIRD, SPENCER F.   German Report of United States Fisheries and Fish-Culture. $<Ann. Rec.$, 1873–75, p. 458.
Report of Drs. Finsch and Lindeman.

576.

1873. BAIRD, SPENCER F.   The Fish of the Caspian Sea. $<Ann. Rec.$, 1873–75, p. 459.

577.

1873. BAIRD, SPENCER F.   Prices of American Fish-Eggs and Fry in England. $<Ann.$
$Rec.$, 1873–75, p. 459.

578.

1873. BAIRD, SPENCER F.   Gloucester Halibut Fishery. $<Ann. Rec.$, 1873–75, p. 460.

579.

1873. BAIRD, SPENCER F.   Statistics of Egyptian Fisheries. $<Ann. Rec.$, 1873–75,
p. 460.
Importation of cured fish into England in 1873.

580.

1873. BAIRD, SPENCER F.   Arrival of Salmon Eggs in New Zealand. $<Ann. Rec.$,
1873–75, p. 462.

581.

1873. BAIRD, SPENCER F.   Shad in the Alleghany River. $<Ann. Rec.$, 1873–75, p. 462.

582.

1873. BAIRD, SPENCER F.   Second Annual Meeting of the American Fish-Culturists
Association. $<Ann. Rec.$, 1873–75, p. 463.

583.

1873. BAIRD, SPENCER F.   Taking California Salmon with the Hook. $<Ann. Rec.$,
1873–75, p. 464.

584.

1873. BAIRD, SPENCER F.   The Fresh-Water Fisheries of India. $<Ann. Rec.$, 1873–
75, p. 465.

585.

1873. BAIRD, SPENCER F.   Oil from Birds. $<$ *Ann. Rec.*, 1873–75, p. 566.

586.

1873. BAIRD, SPENCER F.   Influence of External Pressure in the Life of Fishes. $<$ *Ann. Rec.*, 1873–75, p. 467.

587.

1873. BAIRD, SPENCER F.   Utilization of Old Fish Pickle. $<$ *Ann. Rec.*, 1873–75, p. 502.

588.

1873. BAIRD, SPENCER F.   The Sponge Trade. $<$ *Ann. Rec.*, 1873–75, p. 569.

589.

1873. BAIRD, SPENCER F.   Action of Cod-Liver Oil. $<$ *Ann. Rec.*, 1873–75, p. 623.

590.

1873. BAIRD, SPENCER F.   Buffalo Society of Natural History. $<$ *Ann. Rec.*, 1873–75, p. 653.

591.

1873. BAIRD, SPENCER F.   The Torrey Botanical Club. $<$ *Ann. Rec.*, 1873–75, p. 654.

592.

1873. BAIRD, SPENCER F.   Minnesota Academy of Natural Science. $<$ *Ann. Rec.*, 1873–75, p. 655.

593.

1873. BAIRD, SPENCER F.   Agassiz Natural-History Club at Penikese. $<$ *Ann. Rec.*, 1873–75, p. 655.

594.

1873. BAIRD, SPENCER F.   The Sixth Annual Report of the Peabody Museum, Cambridge. $<$ *Ann. Rec.*, 1873–75, p. 656.

595.

1873. BAIRD, SPENCER F.   Condition of the Boston Natural-History Society, 1871–2. $<$ *Ann. Rec.*, 1873–75, p. 656.

596.

1873. BAIRD, SPENCER F.   Building of the New York Museum of Natural History. $<$ *Ann. Rec.*, 1873–75, p. 657.

597.

1873. BAIRD, SPENCER F.   Appropriations for the New York State Cabinet of Natural-History. $<$ *Ann. Rec.*, 1873–75, p. 657.

598.

1873. BAIRD, SPENCER F.   Report of the National Academy of Sciences for 1872. $<$ *Ann. Rec.*, 1873–75, p. 657.

599.

1873. BAIRD, SPENCER F.   Nourse's History of the U. S. Naval Observatory. $<$ *Ann. Rec.*, 1873–75, p. 657.

600.

1873. BAIRD, SPENCER F. Twenty-second Meeting of the American Association for the Advancement of Science. <*Ann. Rec.*, 1873–75, p. 659.

601.

1873. BAIRD, SPENCER F. The Centennial Exhibition. <*Ann. Rec.*, 1873–75, p. 664.

602.

1873. BAIRD, SPENCER F. American Department of the Vienna Exposition. <*Ann. Rec.*, 1873–75, p. 662.

603.

1873. BAIRD, SPENCER F. The Bache Fund. <*Ann. Rec.*, 1873–75, p. 662.

604.

1873. BAIRD, SPENCER F. Gift of Land to the California Academy of Sciences. <*Ann. Rec.*, 1873–75, p. 663.

605.

1873. BAIRD, SPENCER F. The James Lick Donation to the California Academy of Sciences. <*Ann. Rec.*, 1873–75, p. 664.

606.

1873. BAIRD, SPENCER F. Woodward's Gardens in San Francisco. <*Ann. Rec.*, 1873–75, p. 664.

607.

1873. BAIRD, SPENCER F. "Forest and Stream," a New Weekly Journal. <*Ann. Rec.*, 1873–75, p. 665.

608.

1873. BAIRD, SPENCER F. Catalogue of the Army Medical Museum. <*Ann. Rec.*, 1873–75, p. 665.

609.

1873. BAIRD, SPENCER F. National Photographic Institute. <*Ann. Rec.*, 1873–75, p. 666.

610.

1873. BAIRD, SPENCER F. National Invitation to the National Statistical Congress. <*Ann. Rec.*, 1873–75, p. 666.

611.

1873. BAIRD, SPENCER F. Benevolent Endowment in the United States Treasury. <*Ann. Rec.*, 1873–75, p. 666.

612.

1873. BAIRD, SPENCER F. Memorial to Galileo. <*Ann. Rec.*, 1873-75, p. 678.

613.

1874. BAIRD, SPENCER F. Necrology of Science for 1873. <*Ann. Rec.*, 1873–75, p. 681.

614.

1874. BAIRD, SPENCER F. Food-Fishes of the United States. <*Forest and Stream*, i, 1874, p. 330.
Letter dated Washington, Jan. 2, 1874, communicating memoranda of inquiry. See No. 218.

615.

BAIRD, SPENCER F. [The introduction of Young Salmon into New York waters.] <*Forest and Stream*, i, 1874, p. 347.

616.

BAIRD, SPENCER F. The Garfish. <*Forest and Stream*, i, 1874, p. 375.

617.

BAIRD, SPENCER F. Pisciculture and the Fisheries. <*Forest and Stream*, ii, 1874, p. 52.

A paper read before the American Fish Culturists' Association at the New York meeting, Feb. 10, 1874.

618.

BAIRD, SPENCER F. [Fly-fishing for Shad.] <*Forest and Stream*, ii, 1874, p. 155.

619.

BAIRD, SPENCER F. [Introduction of California Salmon into Australia.] <*Forest and Stream*, ii, 1874, p. 229.

620.

BAIRD, SPENCER F. [Letter concerning Fishways.] <*Forest and Stream*, ii, 1874, p. 340.

621.

BAIRD, SPENCER F. The Tarpum. <*Forest and Stream*, ii, 1874, p. 389.

622.

BAIRD, SPENCER F. The Blue-Back Trout. <*Forest and Stream*, iii, 1874, p. 277.

623.

[BAIRD, SPENCER F.] Prof. Baird's Report. Extracts. <*Forest and Stream*, iii, 1874, pp. 276, 292, 308, 324, 340, 356, 388.

624.

[BAIRD, SPENCER F.] [Extracts from Prof. Baird's Report.] <*American Sportsman*, v, 1874, pp. 148, (?), 162, 178.

625.

BAIRD, SPENCER F. "Opinion as to the probable cause of the rapid diminution of the supply of food-fishes on the coast of New England, and especially of Maine." Letter to E. M. Stilwell, Esq., Bangor, Me. <*Rep. Comm. of Inland Fisheries Mass.*, 1874, pp. 42–45.

626.

BAIRD, SPENCER F. "Report * * * of the additions to the Museum and the onerous operations connected with it during the year 1873." <*Ann. Rep. Smithsonian Institution for the year* 1873, 1874, pp. 36–53 and 58–69.

627.

BAIRD, SPENCER F. United States Commission of Fish and Fisheries. | — | Part II. | — | Report | of | the Commissioner | for | 1872 and 1873. | — | A— Inquiry into the decrease of the food-fishes. | B—The propagation of food-fishes in the waters | of the United States. | — | Washington : | Government Printing Office. | 1874. | 8vo, pp. (v) vi–viii, (1) 2–92. [U. S. F. C., No. 8.]

Report of the Commissioner without supplement.

628.

1874. BAIRD, SPENCER F.   United States Commission of Fish and Fisheries. | Part
II. | Report | of | the Commissioner | for | 1872 and 1873. | — | A.—Inquiry
into the decrease of the food-fishes. | B.—The propagation of food-fishes in
the waters | of the United States. | — | With supplementary papers. | Wash-
ington: | Government Printing Office. | 1874.   8vo, pp. cii, 808, pls. xxxvii,
4 maps.   (U. S. F. C., No. 9.)

CONTENTS.

REPORT OF THE COMMISSIONER.

A. Inquiry into the decrease of the food-fishes.

1874. BAIRD, SPENCER F.—Continued.

629.

1874. BAIRD, SPENCER F. Temperatures in the Gulf of Mexico. < *Rep. U. S. Comm. Fish and Fisheries*, part ii, 1874, pp. 745–748.

630.

1874. BAIRD, SPENCER F. Reports of Special Conferences (of the U. S. Commissioner of Fisheries) with the American Fish-Culturists' Association and State Commissioners of Fisheries. < *Rep. U. S. Comm. Fish and Fisheries*, part ii, 1874, pp. 757–773.

631.

1874. BAIRD, SPENCER F. Statistics of the Menhaden Fisheries, etc. (12). [Questions addressed to fishermen, etc.]—Washington, December 20, 1873. 4to. letter form, 2 l. [U. S. F. C., 10.] Reprint of U. S. F. C., 5.

632.

1874. BAIRD, SPENCER F. Explorations of Pinart in Alaska. <*Ann. Rec.*, 1874–75, p. 246.*

---

* For the full title of the *Annual Record* for 1874, see 1875.

633.

1874. BAIRD, SPENCER F. Explorations of W. M. Gabb in Costa Rica. <*Ann. Rec.*, 1874–75, p. 246.

634.

1874. BAIRD, SPENCER F. Professor Orton's Explorations. < *Ann. Rec.*, 1874–75, p. 248.

635.

1874. BAIRD, SPENCER F. Horetzky on the Hudson's Bay Territory. <*Ann. Rec.*, 1874–75, p. 256.

636.

1874. BAIRD, SPENCER F. Professor Stoddard's Expedition to Colorado. <*Ann. Rec.*, 1874–75, p. 257.

637.

1874. BAIRD, SPENCER F. Explorations of Professor Powell in 1874. < *Ann. Rec.*, 1874–75, p. 262.

638.

1874. BAIRD, SPENCER F. Explorations in 1874 of Lieutenant G. M. Wheeler, United States Engineers. <*Ann. Rec.*, 1874–75, p. 267.

639.

1874. BAIRD, SPENCER F. Explorations of Dr. Hayden in 1874. <*Ann. Rec.*, 1874–75, p. 275.

640.

1874. BAIRD, SPENCER F. The Godeffroy Museum at Hamburg. <*Ann. Rec.*, 1874–75, p. 282.

641.

1874. BAIRD, SPENCER F. Natural History of the Bermudas. <*Ann. Rec.*, 1874–75, p. 283.
Explorations of J. Matthew Jones.

642.

1874. BAIRD, SPENCER F. New Experiments on the Venom of East Indian Serpents. <*Ann. Rec.*, 1874–75, p. 287.
Memoirs of Drs. Fayrer and Bunton.

643.

1874. BAIRD, SPENCER F. Composition of the Cartilage of the Shark. <*Ann. Rec.*, 1874–75, p. 289.
Investigations of Petersen and Golet.

644.

1874. BAIRD, SPENCER F. Composition of the Body Fluids of Fish and Invertebrates. <*Ann. Rec.*, 1874–75, p. 291.
Inquiries of Messrs. Rateau and Papellon.

645.

1874. BAIRD, SPENCER F. Malformation of Fish Embryos. <*Ann. Rec.*, 1874–75, p. 294.

646.

BAIRD, SPENCER F. The Sea-Serpent on the Scotch Coast. $<$ *Ann. Rec.*, 1874–75, p. 296.

647.

BAIRD, SPENCER F. Prepared Heads of Macas Indians. $<$ *Ann. Rec.*, 1874–75, p. 297.

648.

BAIRD, SPENCER F. Paucity of Mammals in Cuba. $<$ *Ann. Rec.*, 1874–75, p. 300.

649.

BAIRD, SPENCER F. Embryology of the Lemurs. $<$ *Ann. Rec.*, 1874–75, p. 301
Discoveries of Milne Edwards and Cole.

650.

BAIRD, SPENCER F. Genesis of the Horse. $<$ *Ann. Rec.*, 1874–75, p. 301.
Views of O. C. Maul.

651.

BAIRD, SPENCER F. Extermination of Buffaloes. $<$ *Ann. Rec.*, 1874–75, p. 303.

652.

BAIRD, SPENCER F. The new Fossil Bird of the Sheppey Clay. $<$ *Ann. Rec.*, 1874–75, p. 305.

653.

BAIRD, SPENCER F. Affinities of Heloderma horridum. $<$ *Ann. Rec.*, 1874–75, p. 308.

654.

BAIRD, SPENCER F. Occurrence of a Cuban Crocodile in Florida. $<$ *Ann. Rec.*, 1874–75, p. 308.

655.

BAIRD, SPENCER F. Food of the Shad. $<$ *Ann. Rec.*, 1874–75, p. 310.

656.

BAIRD, SPENCER F. The Structure of the Laucelot. $<$ *Ann. Rec.*, 1874–75, p. 310.
Reviews of memoir by Dr. Stieda.

657.

BAIRD, SPENCER F. Fish Living in Dried Mud. $<$ *Ann. Rec.*, 1874–75, p. 311.
Note by M. Dareste.

658.

BAIRD, SPENCER F. The "Nerfling" Fish. $<$ *Ann. Rec.*, 1874–75, p. 311.

659.

BAIRD, SPENCER F. Longevity of Fishes. $<$ *Ann. Rec.*, 1874–75, p. 312.
Observation of Dr. Buchner, of Gressen, upon longevity of eels.

660.

1874. BAIRD, SPENCER F.   Spawning of Whiting-Pout. $<$ *Ann. Rec.*, 1874–75, p. 312.
Observation of Henry Lee upon *Gadus luscus.*

661.

1874. BAIRD, SPENCER F.   Sensibility of Fish to Poisons. $<$*Ann. Rec.*, 1874–75, p. 313.
Observations of M. Rabuteau and Papellon at Concameau.

662.

1874. BAIRD, SPENCER F.   Structure of the Embryonic Cellule in the Eggs of Bony
Fishes. $<$ *Ann. Rec.*, 1874–75, p. 314.
Investigations of Balbani.

663.

1874. BAIRD, SPENCER F.   The Embryology of Terebratulina. $<$*Ann. Rec.*, 1874–75,
p. 314.
Labors of E. S. Morse.

664.

1874. BAIRD, SPENCER F.   The Food of the Oyster, and a New Parasite. $<$*Ann.
Rec.*, 1874–75, p. 315.
Paper by John M. Crady.

665.

1874. BAIRD, SPENCER F.   Explanation of the alleged occurrence of the King-Crab
in Holland. $<$ *Ann. Rec.*, 1874–75, p. 316.

666.

1874. BAIRD, SPENCER F.   Success of the Naples Zoological Station. $<$*Ann. Rec.*,
1874–75, p. 317.

667.

1874. BAIRD, SPENCER F.   Zoological Garden of Hamburg. $<$*Ann. Rec.*, 1874–75,
p. 318.

668.

1874. BAIRD, SPENCER F.   Fossil Vertebrates in Ohio. $<$*Ann. Rec.*, 1874–75, p. 322.

669.

1874. BAIRD, SPENCER F.   Exhibition of British Ethnology. $<$*Ann. Rec.*, 1874–75,
p. 326.
Describes ethnological department of U. S. National Museum.

670.

1874. BAIRD, SPENCER F.   Footprints in Solid Rock. $<$*Ann. Rec.*, 1874–75, p. 328.
Refers to specimens in U. S. National Museum.

671.

1874. BAIRD, SPENCER F.   Ancient Stone Fort in Indiana. $<$*Ann. Rec.*, 1874–75,
p. 329.

672.

1874. BAIRD, SPENCER F.   Dall's Ethnological Explorations in Alaska. $<$*Ann. Rec.*,
1874–75, p. 329.

673.

1874. BAIRD, SPENCER F.   Trade among the Aborigines. $<$*Ann. Rec.*, 1874–75, p. 331.

674.

1874. BAIRD, SPENCER F. The Species of American Squirrels. <*Ann. Rec.*, 1874–75, p. 332.
Notice of work of J. A. Allen.

675.

1874. BAIRD, SPENCER F. Relationship of American Deer to their British anologues. <*Ann. Rec.*, 1874–75, p. 335.

676.

1874. BAIRD, SPENCER F. The Characters and Relations of the Hyopotamidæ. <*Ann. Rec.*, 1874–75, p. 336.
Review of Kowalevsky's memoir.

677.

1874. BAIRD, SPENCER F. Shad in the Gulf of Mexico. <*Ann. Rec.*, 1874–75, p. 338.

678.

1874. BAIRD, SPENCER F. History of the Pacific Coast Marine Mammals. <*Ann. Rec.*, 1874–75, p. 338.

679.

1874. BAIRD, SPENCER F. Decrease in the European Bison. <*Ann. Rec.*, 1874–75, p. 339.
Scammon's "Marine Mammals of the Pacific Coast."

680.

1874. BAIRD, SPENCER F. Eggs of the Siluridæ. <*Ann. Rec.*, 1874–75, p. 339.
Observations of Dr. Ray.

681.

1874. BAIRD, SPENCER F. Discovery of Putorius Nigripes. <*Ann. Rec.*, 1874–75, p. 339.

682.

1874. BAIRD, SPENCER F. The Fossil Hog of America. <*Ann. Rec.*, 1874–75, p. 340.

683.

1874. BAIRD, SPENCER F. Dall on the Birds of Alaska. <*Ann. Rec.*, 1874–75, p. 340.

684.

1874. BAIRD, SPENCER F. Lawrence's Birds of Northwestern Mexico. <*Ann. Rec.*, 1874–75, p. 341.

685.

1874. BAIRD, SPENCER F. Collection of Birds of Paradise. <*Ann. Rec.*, 1874–75, p. 341.

686.

1874. BAIRD, SPENCER F. Geographical Distribution of Asiatic Birds. <*Ann. Rec.*, 1874–75, p. 341.
Review of memoir by H. J. Elwes.

687.

1874. BAIRD, SPENCER F. Suggested Introduction of the Rook into the United States. <*Ann. Rec.*, 1874–75, p. 342.

688.

1874. BAIRD, SPENCER F.   Dr. Coues' Manual of Field Ornithology. <*Ann. Rec.*, 1874–75, p. 343.

689.

1874. BAIRD, SPENCER F.   Catalogue of American Birds. <*Ann. Rec.*, 1874–75, p. 343.
Sclater and Salvin's Index Avium Neotropicalium.

690.

1874. BAIRD, SPENCER F.   American King-Crab on the European Coast. <*Ann. Rec.*, 1874–75, p. 344.

691.

1874. BAIRD, SPENCER F.   An "Army Worm." <*Ann. Rec.*, 1874–75, p. 344.

692.

1874. BAIRD, SPENCER F.   Dall's Catalogue of the Shells of Behring's Strait. <*Ann. Rec.*, 1874–75, p. 345.

693.

1874. BAIRD, SPENCER F.   Discovery of the Aleut Mummies. <*Ann. Rec.*, 1874–75, p. 345.

694.

1874. BAIRD, SPENCER F.   Change of Volume of Fish in Swimming. <*Ann. Rec.*, 1874–75, p. 348.
Investigation by Harting.

695.

1874. BAIRD, SPENCER F.   The Development of Sharks and Rays. <*Ann. Rec.*, 1874–75, p. 349.
Investigations of F. M. Balfour at Naples.

696.

1874. BAIRD, SPENCER F.   Respiration in the Amphibia. <*Ann. Rec.*, 1874–75, p. 350.

697.

1874. BAIRD, SPENCER F.   The Basking Shark. <*Ann. Rec.*, 1874–75, p. 351.
Note on *Selache maxima*.

698.

1874. BAIRD, SPENCER F.   Taming the Zebra. <*Ann. Rec.*, 1874–75, p. 352.

699.

1874. BAIRD, SPENCER F.   Lieutenant Wheeler's Expedition. <*Ann. Rec.*, 1874–75, p. 353.

700.

1874. BAIRD, SPENCER F.   Sea-Weeds of the Bay of Fundy. <*Ann. Rec.*, 1874–75, p. 356.
Notice of work done by U. S. Fish Commission.

701.

1874. BAIRD, SPENCER F.   New Yearly Report of the Progress of Botany. <*Ann. Rec.*, 1874–75, p. 361.

702.

1874. BAIRD, SPENCER F. Recent Publications in Systematic Botany. < *Ann. Rec.,* 1874–75, p. 363.

703.

1874. BAIRD, SPENCER F. Introduction of Prairie Chickens into the Eastern States. < *Ann. Rec.,* 1874–75, p. 391.

704.

1874. BAIRD, SPENCER F. Destructiveness of Rodents in California. < *Ann. Rec.,* 1874–75, p. 397.

705.

1874. BAIRD, SPENCER F. Oil from Sharks' Livers. < *Ann. Rec.,* 1874–75, p. 419.

706.

1874. BAIRD, SPENCER F. Fisheries and Sea Temperatures. < *Ann. Rec.,* 1874–75, p. 419.
Observations of the Scotch Meteorological Society.

707.

1874. BAIRD, SPENCER F. Marine Fisheries of Maine in 1873. < *Ann. Rec.,* 1874–75, p. 420.

708.

1874. BAIRD, SPENCER F. Consumption of Marine Products in Washington. < *Ann. Rec.,* 1874–75, p. 420.

709.

1874. BAIRD, SPENCER F. The French Fisheries. < *Ann. Rec.,* 1874–75, p. 422.

710.

1874. BAIRD, SPENCER F. The Seal and Herring Fisheries of Newfounland. < *Ann. Rec.,* 1874–75, p. 424.

711.

1874. BAIRD, SPENCER F. Alaska Cod-Fisheries in 1873. < *Ann. Rec.,* 1874–75, p. 424.

712.

1874. BAIRD, SPENCER F. Fish-Culture in Castalia Springs. < *Ann. Rec.,* 1874–75, p. 425.
Fish-culture in mineral springs in Erie County, Ohio.

713.

1874. BAIRD, SPENCER F. Restocking Otsego Lake, N. Y., with Fish. < *Ann. Rec.,* 1874–75, p. 426.

714.

1874. BAIRD, SPENCER F. Introduction of British Fish into India. < *Ann. Rec.,* 1874–75, p. 426.

715.

1874. BAIRD, SPENCER F. Transporting Living Trout. < *Ann. Rec.,* 1874–75, p. 427.
Experience of German experts.

716.

1874. BAIRD, SPENCER F. Caution in Planting Young Salmon. <*Ann. Rec.*, 1874-75, p. 427.

Statement of Riedel.

717.

1874. BAIRD, SPENCER F. Destruction of Fish on the Oregon Coast by Nitro-glycerine. <*Ann. Rec.*, 1874-75, p. 428.

718.

1874. BAIRD, SPENCER F. Sterlet from St. Petersburg at the Brighton Aquarium. <*Ann. Rec.*, 1874-75, p. 428.

719.

1874. BAIRD, SPENCER F. Stocking a Pond in Utah with Eels. <*Ann. Rec.*, 1874-75, p. 428.

720.

1874. BAIRD, SPENCER F. Spinal Column of the Sturgeon as an Article of Food. <*Ann. Rec.*, 1874-75, p. 447.

721.

1874. BAIRD, SPENCER F. New Survey of the State of Massachusetts. <*Ann. Rec.*, 1874-75, p. 573.

722.

1874. BAIRD, SPENCER F. Cambridge Entomological Club. <*Ann. Rec.*, 1874-75, p. 574.

723.

1874. BAIRD, SPENCER F. Twenty-third Annual Meeting of the American Association for the Advancement of Science. <*Ann. Rec.*, 1874-75, p. 574.

724.

1874. BAIRD, SPENCER F. Seventh Annual Report of the Peabody Museum, Cambridge, Massachusetts. <*Ann. Rec.*, 1874-75, p. 575.

725.

1874. BAIRD, SPENCER F. First Report of the Anderson School of Natural History at Penikese. <*Ann. Rec.*, 1874-75, p. 576.

726.

1874. BAIRD, SPENCER F. The Penikese School. <*Ann. Rec.*, 1874-75, p. 578.

727.

1874. BAIRD, SPENCER F. Opening of the Anderson School at Penikese. <*Ann. Rec.*, 1874-75, p. 579.

728.

1874. BAIRD, SPENCER F. Report for 1873 of the Peabody Academy of Science, Salem. <*Ann. Rec.*, 1874-75, p. 579.

729.

1874. BAIRD, SPENCER F. Report of the Museum of Comparative Zoology for 1873. <*Ann. Rec.*, 1874-75, p. 580.

730.

1874. BAIRD, SPENCER F. Bulletin of the Museum of Comparative Zoology. <*Ann. Rec.*, 1874–75, p. 581.

731.

1874. BAIRD, SPENCER F. Catalogues of the Museum of Comparative Zoology. <*Ann. Rec.*, 1874–75, p. 581.

732.

1874. BAIRD, SPENCER F. Annual Meeting of the Trustees of the Museum of Comparative Zoology. <*Ann. Rec.*, 1874–75, p. 582.

733.

1874. BAIRD, SPENCER F. Report of the Bussey Institution. <*Ann. Rec.*, 1874–75, p. 582.

734.

1874. BAIRD, SPENCER F. The "Torrey Memorial Cabinet." <*Ann. Rec.*, 1874–75, p. 584.

735.

1874. BAIRD, SPENCER F. "Directory" of the Torrey Botanical Club. <*Ann. Rec.*, 1874–75, p. 584.

736.

1874. BAIRD, SPENCER F. The Bulletin of the Science Department of Cornell University. <*Ann. Rec.*, 1874–75, p. 585.

737.

1874. BAIRD, SPENCER F. Issue of "Proceedings" by the New York Lyceum of Natural History. <*Ann. Rec.*, 1874–75, p. 585.

738.

1874. BAIRD, SPENCER F. Publishing Fund of the Historical Society of Pennsylvania. <*Ann. Rec.*, 1874–75, p. 586.

739.

1874. BAIRD, SPENCER F. Report of the Philadelphia Academy of Natural Sciences. <*Ann. Rec.*, 1874–75, p. 586.

740.

1874. BAIRD, SPENCER F. Philadelphia National Museum. <*Ann. Rec.*, 1874–75, p. 587.

741.

1874. BAIRD, SPENCER F. Report of the Zoological Society of Philadelphia. <*Ann. Rec.*, 1874–75, p. 587.

742.

1874. BAIRD, SPENCER F. The Zoological Society of Philadelphia. <*Ann. Rec.*, 1874–75, p. 588.

743.

1874. BAIRD, SPENCER F. Reorganization of the Maryland Academy of Sciences. <*Ann. Rec.*, 1874–75, p. 589.

744.

1874. BAIRD, SPENCER F.　Botanical Conservatory of the Maryland Academy of Sciences.　< *Ann. Rec.*, 1874–75, p. 590.

745.

1874. BAIRD, SPENCER F.　Recent Publications of the Smithsonian Institution. < *Ann. Rec.*, 1874–75, p. 591.

746.

1874. BAIRD, SPENCER F.　United States Departmental Centennial Board.　< *Ann. Rec.*, 1874–75, p. 593.

747.

1874. BAIRD, SPENCER F.　Additions to the National Herbarium in 1874.　< *Ann. Rec.*, 1874–75, p. 593.

748.

1874. BAIRD, SPENCER F.　European Savans in American Institutions.　< *Ann. Rec.*, 1874–75, p. 594.

749.

1874. BAIRD, SPENCER F.　Sale of Dr. Troost's Cabinet of Minerals and Antiquities. < *Ann. Rec.*, 1874–75, p. 594.

750.

1874. BAIRD, SPENCER F.　Report of the Zoological Society of London.　< *Ann. Rec.*, 1874–75, p. 595.

751.

1874. BAIRD, SPENCER F.　Annual Return of the British Museum.　< *Ann. Rec.*, 1874–75, p. 596.

752.

1874. BAIRD, SPENCER F.　Temporary Museum at the late Meeting of the British Association.　< *Ann. Rec.*, 1874–75, p. 599.

753.

1874. BAIRD, SPENCER F.　Meeting of the French Association for the Advancement of Science.　< *Ann. Rec.*, 1874–75, p. 601.

754.

1874. BAIRD, SPENCER F.　The "American Society" of Paris.　< *Ann. Rec.*, 1874–75, p. 602.

755.

1874. BAIRD, SPENCER F.　Life-Saving Stations on the Coast of the United States. < *Ann. Rec.*, 1874–75, p. 606.

756.

1875. BAIRD, SPENCER F.　Annual Record | of | Science and Industry | for 1873. | Edited by | Spencer F. Baird, | with the assistance of Eminent Men of Science. | [Cut.] | New York: Harper & Brother, Publishers, | Franklin Square. | 1875. | (8vo.　pp. cxxxii, 714.)

757.

BAIRD, SPENCER F. Further Contributions to the Minute Anatomy of the Tæniæ, which prey on Fish. <*Ann. Rec.*, 1873-75, p. xcii.

758.

BAIRD, SPENCER F. Further Observation on the Cercariæ in the Intestines of Fish. <*Ann. Rec.*, 1873-75, p. xcvii.

759.

BAIRD, SPENCER F. Fish-way at Holyoke. <*Ann. Rec.*, 1873-75, p. cx.

760.

BAIRD, SPENCER F. Pisciculture and the Fisheries. (A general summary of progress.) <*Ann. Rec.*, 1873-75, pp. cx, cxviii.

761.

BAIRD, SPENCER F. General Summary | of | Scientific and Industrial Progress | during the year 1874. <*Ann. Rec.*, 1874-75, p. xix.

762.

BAIRD, SPENCER F. Annual Record | of | Science and Industry | for 1874. | Edited by | Spencer F. Baird, | with the assistance of Eminent Men of Science. | [Cut.] | New York: Harper & Brother, Publishers, | Franklin Square. | 1875. | (8vo. pp. cciv, 665.)

763.

BAIRD, SPENCER F. Necrology. <*Ann. Rec.*, 1874-75, p. 611.

764.

BAIRD, SPENCER F. Bibliography. <*Ann. Rec.*, 1874-75, p. 617-632.

765.

BAIRD, SPENCER F. Mr. Balfour on the Embryology of Sharks. <*Ann. Rec.*, 1874-75, p. cliv, general summary.
Dr. Lamper on the segmentary organs in the embryos of rays.
Discovery of *Ceratodus* in Queensland.

766.

BAIRD, SPENCER F. Fisheries (and Pisciculture) general summary. <*Ann. Rec.*, 1874-75, p. clxix.

767.

[BAIRD, SPENCER F.] Prof. Baird's Report. (Editorial, quoting.) <*Forest and Stream*, iii, 1875, p. 340.

768.

BAIRD, SPENCER F. Prof. Baird's Report. Comparative value of Anadromous and other fishes. (Editorial quoting.) <*Forest and Stream*, iii, 1875, p. 356.

769.

BAIRD, SPENCER F. Prof. Baird's Report. Different Methods of Multiplying Fish. (Editorial quoting.) <*Forest and Stream*, iii, 1875, p. 388.

770.

1875. BAIRD, SPENCER F. (E)$\frac{c}{1}$ | U. S. Commission of Fish and Fisheries | — | Sta-
tistics of the Fishery Marine. | — | Circular. [U. S. F. C., 12.] [Foolscap
size, 2 pp. Washington: Government Printing Office, 1875.]
    The blank tables to accompany this circular were printed in uniform style, and are regis-
tered (U. S. F. C., 11). Prepared by G. Brown Goode.

771.

1875. BAIRD, SPENCER F. "(Report on) additions to the Museum and the various
operations connected with it during the past year." <*Ann. Rep. Smithsonian
Institution for the year* 1874, 1875, pp. 27–44, 49–76.

772.

1875. BAIRD, SPENCER F. (D.) Conclusions as to decrease of Cod-Fisheries on the
New England Coast (Report of U. S. Commissioner on Fisheries). <*Rep.
Comm. Inland Fisheries Massachusetts,* 1875, pp. 38–41.
    Extract from Report, Part II, pp. xi–xiv.

773.

1875. BAIRD, SPENCER F. Soles and Turbot for American Waters. <*Rod and Gun,*
vii, 1875, p. 150.
    Letter to Frank Buckland.

774.

1875. BAIRD, SPENCER F. Protection of Salmon. <*Forest and Stream,* v, 1875, p. 166.

775.

1875. BAIRD, SPENCER F. Fish Culture in Kentucky. <*Forest and Stream,* v, 1875,
p. 243.

776.

1875. BAIRD, SPENCER F. "Prof. Baird made a brief statement as to the action of
the United States Fish Commission." [Abstract.] <*Proc. American Fish-
Culturists' Association,* 4th ann. meeting, 1875, pp. 8, 9.

777.

1875. BAIRD, SPENCER F. The Saranac Exploring Expedition. <*Ann. Rec.,* 1875–
76, p. 260.

778.

1875. BAIRD, SPENCER F. Explorations under Dr. Hayden in 1875. <*Ann. Rec.,*
1875–76, p. 263.

779.

1875. BAIRD, SPENCER F. Explorations under Major Powell in 1875. <*Ann. Rec.,*
1875–76, p. 286.

780.

1875. BAIRD, SPENCER F. Explorations and Surveys under Lieutenant George M.
Wheeler, U. S. Army, in 1875. <*Ann. Rec.,* 1875–76, p. 293.

781.

1875. BAIRD, SPENCER F. Major Powell's Final Report. <*Ann. Rec.,* 1875–76, p. 298.

782.

1875. BAIRD, SPENCER F.   Reports of the Northern Boundary Surveys.   <*Ann. Rec.*, 1875–76, p. 300.

783.

1875. BAIRD, SPENCER F.   Origin of Animal Forms.   <*Ann. Rec.*, 1875–76, p. 305.

784.

1875. BAIRD, SPENCER F.   Discovery of Animal Remains in the Lignite Beds of the Saskatchewan District.   <*Ann. Rec.*, 1875–76, p. 311.

785.

1875. BAIRD, SPENCER F.   Fauna of the Mammoth Cave.   <*Ann. Rec.*, 1875–76, p. 313.

786.

1875. BAIRD, SPENCER F.   Is Sex Distinguishable in Egg-Shells?   <*Ann. Rec.*, 1875–76, p. 320.

787.

1875. BAIRD, SPENCER F.   Mr. George Latimer's Archæological Collection from Porto Rico.   <*Ann. Rec.*, 1875–76, p. 325.

788.

1875. BAIRD, SPENCER F.   Stone Knives with Handles, from the Pai-Utes.   <*Ann. Rec.*, 1875–76, p. 326.

789.

1875. BAIRD, SPENCER F.   Archæology of the Mammoth Cave.   <*Ann. Rec.*, 1875–76, p. 327.

790.

1875. BAIRD, SPENCER F.   Discovery in Newfoundland of the Great Auk.   <*Ann. Rec.*, 1875–76, p. 339.

791.

1875. BAIRD, SPENCER F.   Habits of Kingfishers.   <*Ann. Rec.*, 1875–76, p. 339.

792.

1875. BAIRD, SPENCER F.   Professor Alfred Newton on the Migration of Birds.   <*Ann. Rec.*, 1875–76, p. 340.

793.

1875. BAIRD, SPENCER F.   The Batrachia and Reptilia of North America.   <*Ann. Rec.*, 1875–76, p. 343.

794.

1875. BAIRD, SPENCER F.   Report of the Occurrence of Large Codfish off Mazatlan.   <*Ann. Rec.*, 1875–76, p. 344.*
Criticism of a statement in *Land and Water.*

795.

1875. BAIRD, SPENCER F.   Grayling in the Au Sable River, Mich.   <*Ann. Rec.*, 1875–76, p. 344.

---

* For full title of *Annual Record* for 1875, see 1876.

12 BD

796.

1875. BAIRD, SPENCER F. Respiration of the Loach. <*Ann. Rec.*, 1875–76, p. 345.
Notices of M. Rougemont's investigations.

797.

1875. BAIRD, SPENCER F. Monograph on the Anguilliform Fish. <*Ann. Rec.*, 1875–76, p. 345.
M. Dareste's monograph.

798.

1875. BAIRD, SPENCER F. Largest Pike ever taken in England. <*Ann. Rec.*, 1875–76, p. 346.

799.

1875. BAIRD, SPENCER F. Habits of Eels. <*Ann. Rec.*, 1875–76, p. 346.
Observations of M. E. Noel at Rouen.

800.

1875. BAIRD, SPENCER F. Fossil Lepidosteus. <*Ann. Rec.*, 1875–76, p. 347.

801.

1875. BAIRD, SPENCER F. Productive Season of the Cod on the Faroe Islands. <*Ann. Rec.*, 1875–76, p. 347.

802.

1875. BAIRD, SPENCER F. Softness of Bones in Old Congers. <*Ann. Rec.*, 1875–76, p. 347.

803.

1875. BAIRD, SPENCER F. Leptocephali are Larval Forms of Congers, etc. <*Ann. Rec.*, 1875–76, p. 348.

804.

1875. BAIRD, SPENCER F. Have Jelly-Fishes a Nervous System? <*Ann. Rec.*, 1875–76, p. 348.

805.

1875. BAIRD, SPENCER F. Giant Cuttle-Fish Found on the Grand Bank, December, 1874. <*Ann. Rec.*, 1875–76, p. 351.

806.

1875. BAIRD, SPENCER F. Fauna of the Caspian. <*Ann. Rec.*, 1875–76, p. 351.

807.

1875. BAIRD, SPENCER F. Giant Cuttle-Fish found on the Grand Bank, December, 1874. <*Ann. Rec.*, 1875–76, p. 351.

808.

1875. BAIRD, SPENCER F. Introduction of the American Turkey. <*Ann. Rec.*, 1875–76, p. 354.

809.

1875. BAIRD, SPENCER F. Report of the Fish Commission of Canada for 1874. <*Ann. Rec.*, 1875–76, p. 405.

810.

1875. BAIRD, SPENCER F. Ninth Annual Report of the Massachusetts Commissioners of Fisheries. <*Ann. Rec.*, 1875–76, p. 406.

811.

BAIRD, SPENCER F. Ninth Report of the Fish Commissioners of Connecticut. < *Ann. Rec.*, 1875–76, p. 407.

812.

BAIRD, SPENCER F. First Report of the Commissioners of Fisheries of Michigan. < *Ann. Rec.*, 1875–76, p. 407.

813.

BAIRD, SPENCER F. First Annual Report of the Fish Commissioners of Minnesota. < *Ann. Rec.*, 1875–76, p. 408.

814.

BAIRD, SPENCER F. Fifth Report of the Fish Commissioners of Rhode Island. < *Ann. Rec.*, 1875–76, p. 408.

815.

BAIRD, SPENCER F. Report of the Fish Commissioners of Pennsylvania for 1874. < *Ann. Rec.*, 1875–76, p. 409.

816.

BAIRD, SPENCER F. Report of the Fish Commissioners of New Hampshire for 1874. < *Ann. Rec.*, 1875–76, p. 410.

817.

BAIRD, SPENCER F. Second Report of the Fish Commissioners of Vermont. < *Ann. Rec.*, 1875–76, p. 411.

818.

BAIRD, SPENCER F. First Report of the Fish Commissioners of Wisconsin. < *Ann. Rec.*, 1875–76, p. 411.

819.

BAIRD, SPENCER F. Third Annual Report of the American Fish-Culturists' Association. < *Ann. Rec.*, 1875–76, p. 412.

820.

BAIRD, SPENCER F. Meeting of the American Fish-Culturists' Association. < *Ann. Rec.*, 1875–76, p. 412.

821.

BAIRD, SPENCER F. Objection to the Use of Submerged Net-Weirs. < *Ann. Rec.*, 1875–76, p. 413.

822.

BAIRD, SPENCER F. Fisheries and Seal-Hunting in the White Sea and Northern Ocean. < *Ann. Rec.*, 1875–76, p. 413.

823.

BAIRD, SPENCER F. Close time for the Capture of Seals. < *Ann. Rec.*, 1875–76, p. 414.

824.

BAIRD, SPENCER F. Bad Condition of the Hair-Seal Fisheries. < *Ann. Rec.*, 1875–76, p. 414.

825.

1875. BAIRD, SPENCER F.   Fish Consumption of Washington. <*Ann. Rec.*, 1875–76, p. 415.

826.

1875. BAIRD, SPENCER F.   Effect of Polluted Water on Fishes. <*Ann. Rec.*, 1875–76, p. 416.

827.

1875. BAIRD, SPENCER F.   Menhaden Oil and Guano. <*Ann. Rec.*, 1875–76, p. 418.

828.

1875. BAIRD, SPENCER F.   Hybrid Fish. <*Ann. Rec.*, 1875–76, p. 418.

829.

1875. BAIRD, SPENCER F.   Experiments with Young Maine Salmon. <*Ann. Rec.*, 1875–76, p. 418.

830.

1875. BAIRD, SPENCER F.   Increase of English Fishes in Tasmania. <*Ann. Rec.*, 1875–76, p. 419.

831.

1875. BAIRD, SPENCER F.   Stocking the Rivers on the West Side of Lake Champlain by the United States Fish Commission. <*Ann. Rec.*, 1875–76, p. 419.

832.

1875. BAIRD, SPENCER F.   Distribution of Trout Eggs from Tasmania to the Neighboring Colonies. <*Ann. Rec.*, 1875–76, p. 420.

833.

1875. BAIRD, SPENCER F.   Importation of the Gourami into Paris. <*Ann. Rec.*, 1875–76, p. 420.

834.

1875. BAIRD, SPENCER F.   Mr. C. G. Atkins' Experiments on the Artificial Hatching of the Smelt. <*Ann. Rec.*, 1875–76, p. 421.

835.

1875. BAIRD, SPENCER F.   Seth Green's Artificial Hatching of Sturgeon. <*Ann. Rec.*, 1875–76, p. 422.

836.

1875. BAIRD, SPENCER F.   The New Westminster Aquarium. <*Ann. Rec.*, 1875–76, p. 422.

837.

1875. BAIRD, SPENCER F.   Fish at Great Depths. <*Ann. Rec.*, 1875–76, p. 425.

838.

1875. BAIRD, SPENCER F.   Piscicultural Prizes. <*Ann. Rec.*, 1875–76, p. 425.

839.

1875. BAIRD, SPENCER F.   Change of Water in Aquaria. <*Ann. Rec.*, 1875–76, p. 426.

840.

1875. BAIRD, SPENCER F.   French Prizes for American Fish. <*Ann. Rec.*, 1875–76, p. 426.

841.

1875. BAIRD, SPENCER F. Fish-Culture in China. < *Ann. Rec.*, 1875–76, p. 427.

842.

1875. BAIRD, SPENCER F. Newfoundland Fisheries in 1874–5. < *Ann. Rec.*, 1875–76, p. 427.

843.

1875. BAIRD, SPENCER F. Illumination for Attracting Fish. < *Ann. Rec.*, 1875–76, p. 428.

844.

1875. BAIRD, SPENCER F. Manufacture of Cod-Liver Oil. < *Ann. Rec.*, 1875–76, p. 428.

845.

1875. BAIRD, SPENCER F. Operations of the United States Fish Commission in 1875. < *Ann. Rec.*, 1875–76, p. 429.

846.

1875. BAIRD, SPENCER F. Salmon in the San Joaquin. < *Ann. Rec.*, 1875–76, p. 430.

847.

1875. BAIRD, SPENCER F. Salmon Trade of the Columbia River. < *Ann. Rec.*, 1875–76, p. 431.

848.

1875. BAIRD, SPENCER F. Married Salmon. < *Ann. Rec.*, 1875–76, p. 432.

849.

1875. BAIRD, SPENCER F. Salmon in the Sacramento River. < *Ann. Rec.*, 1875–76, p. 432.

850.

1875. BAIRD, SPENCER F. Animal Incrustation on the Great Eastern. < *Ann. Rec.*, 1875–76, p. 432.

851.

1875. BAIRD, SPENCER F. Physical Condition of the Herring-Fishery. < *Ann. Rec.*, 1875–76, p. 433.

852.

1875. BAIRD, SPENCER F. Food for Trout. < *Ann. Rec.*, 1875–76, p. 433.

853.

1875. BAIRD, SPENCER F. Electrical Fish-Bait. < *Ann. Rec.*, 1875–76, p. 434

854.

1875. BAIRD, SPENCER F. United States Salmon-Hatching Establishment. < *Ann. Rec.*, 1875–76, p. 434.

855.

1875. BAIRD, SPENCER F. New Fish Product. < *Ann. Rec.*, 1875–76, p. 435.

856.

1875. BAIRD, SPENCER F. Report of the Fish Commission of Virginia. < *Ann. Rec.*, 1875–76, p. 435.

857.

1875. BAIRD, SPENCER F.   Inspection of Fish in the Washington City Market.
< *Ann. Rec.*, 1875–76, p. 436.

858.

1875. BAIRD, SPENCER F.   Seventh Annual Report of the Fish Commissioners of New
York.   < *Ann. Rec.*, 1875–76, p. 437.

859.

1875. BAIRD, SPENCER F.   Gloucester Fisheries in 1875.   < *Ann. Rec.*, 1875–76, p. 439.

860.

1875. BAIRD, SPENCER F.   Fisheries of the Arctic Regions.   < *Ann. Rec.*, 1875–76, p.
439.

861.

1875. BAIRD, SPENCER F.   Failure in Introducing Salmon and Trout.   < *Ann. Rec.*,
1875–76, p. 439.

862.

1875. BAIRD, SPENCER F.   Yarmouth Aquarium.   < *Ann. Rec.*, 1875–76, p. 440.

863.

1875. BAIRD, SPENCER F.   Report of the American Museum of Natural History, New
York, for 1874.   < *Ann. Rec.*, 1875–76, p. 571.

864.

1875. BAIRD, SPENCER F.   Kirtland School of Natural Sciences.   < *Ann. Rec.*, 1875–
76, p. 571.

865.

1875. BAIRD, SPENCER F.   Normal School of Natural Sciences.   < *Ann. Rec.*, 1875–76,
p. 572.

866.

1875. BAIRD, SPENCER F.   First Annual Report of the Zoological Society of Cincinnati.   < *Ann. Rec.*, 1875–76, p. 572.

867.

1875. BAIRD, SPENCER F.   First Annual Report of the Geological and Agricultural
Survey of Texas.   < *Ann. Rec.*, 1875–76, p. 573.

868.

1875. BAIRD, SPENCER F.   Annual Report of the United States Geological and Geographical Survey of the Territories for 1873.   < *Ann. Rec.*, 1875–76, p. 573.

869.

1875. BAIRD, SPENCER F.   Arrangements for a Botanical Garden in Chicago.   < *Ann.
Rec.*, 1875–76, p. 574.

870.

1875. BAIRD, SPENCER F.   Report of the Icelandic Commission to Alaska.   < *Ann.
Rec.*, 1875–76, p. 576.

871.

1875. BAIRD, SPENCER F. Bequest to the Cincinnati Society of Natural History. <*Ann. Rec.*, 1875–76, p. 577.

872.

1875. BAIRD, SPENCER F. Index of Patents from 1790 to 1873. <*Ann. Rec.*, 1875–76, p. 579.

873.

1875. BAIRD, SPENCER F. Sums voted by the British Parliament for Scientific Instruction. <*Ann. Rec.*, 1875–76, p. 580.

874.

1875. BAIRD, SPENCER F. Annual Report of the Council of the Zoological Society of London. <*Ann. Rec.*, 1875–76, p. 581.

875.

1875. BAIRD, SPENCER F. Royal Society's Catalogue of Learned Societies and Scientific Papers. <*Ann. Rec.*, 1875–76, p. 584.

876.

1875. BAIRD, SPENCER F. Meeting of the American Fish-Culturists' Association. <*Ann. Rec.*, 1875–76, p. 585.

877.

1875. BAIRD, SPENCER F. National Park in the Island of Mackinaw. <*Ann. Rec.*, 1875–76, p. 586.

878.

1875. BAIRD, SPENCER F. Annual Report of the Librarian of Congress. <*Ann. Rec.*, 1875–76, p. 586.

879.

1875. BAIRD, SPENCER F. Additional Pay to the Survivors of the "Polaris." <*Ann. Rec.*, 1875–76, p. 587.

880.

1875. BAIRD, SPENCER F. Loan Exhibition of Scientific Apparatus. <*Ann. Rec.*, 1875–76, p. 588.

881.

1875. BAIRD, SPENCER F. Annual Report of the Peabody Museum of Archæology and Ethnology. <*Ann. Rec.*, 1876, p. 589.

882.

1876. BAIRD, SPENCER F. Appendix K. List of Birds collected by Charles S. McCarthy, Taxidermist.—Classified by Prof. Spencer F. Baird. <Simpson's Explorations across the Great Basin of Utah in 1859*, pp. 377–381. Bastard title : Explorations across the Great Basin of Utah. | = | Appendix K. | — | Ornithology. | A List of Birds, | by Prof. Spencer F. Baird, pp. [376–7].

List of 114 species, with localities, published as a separate, with cover and this title: Engineer's Department, U. S. Army. | Explorations across the Great Basin of Utah in 1859. | In charge of Capt. J. H. Simpson, Topographical Engineers. | = | Ornithology. | A | List of Birds. | Classified by | Prof. Spencer F. Baird. | — | Washington : | Government Printing Office. | 1876. 4to. pp. [10].

* Engineer Department, U. S. Army. | — | Report | of | Explorations | across the | Great Basin of the Territory of Utah | for a | direct Wagon-Route from Camp Floyd to Genoa, in Carson Valley, | in 1859. | By Captain J. H. Simpson, | Corps of Topographical Engineers, U. S. Army, | [now Colonel of Engineers, Bvt. Brig. Gen., U. S. A.] | made | by authority of the Secretary of War, and under instructions from Bvt. Brig. Gen. A. S. Johnston, | U. S. Army, commanding the Department of Utah. | — | Washington : | Government Printing Office. | 1876. 1 vol. 4to. pp. 518, maps and pll.

883.

1876. BAIRD, SPENCER F. Annual Record | of | Science and Industry | for | 1875. | Edited by | Spencer F. Baird, | with the assistance of eminent men of science. | [Cut.] | New York: | Harper & Brothers, Publishers, | Franklin Square. | 1876. (8vo. pp. ccxc, 656.)

884.

1876. BAIRD, SPENCER F. United States Fish Commission at Wood's Hole, Mass., 1875. < *Ann. Rec.*, 1875–76, p. cxxiv.

885.

1876. BAIRD, SPENCER F. Discoveries in the Biological History of Fishes. < *Ann. Rec.*, 1875–76, p. cxcviii.

886.

1876. BAIRD, SPENCER F. Pisciculture and the Fisheries. (General summary.) < *Ann. Rec.*, 1875–76, pp. ccxxiv–ccxxix.

887.

1876. BAIRD, SPENCER F. Necrology [of Science for 1875]. < *Ann. Rec.*, 1875–76, pp. 591–655.

888.

1876. BAIRD, SPENCER F. Fish in California. < *Rod and Gun*, vii, 1876, p. 326.

Quoting the *San Francisco Chronicle.*—Motive and method of fish protection in our waters and along our coast.

889.

1876. BAIRD, SPENCER F. United States Commission of Fish and Fisheries. | Part III. | Report | of | the Commissioner | for | 1873–4 and 1874–5. | — | A—Inquiry into the decrease of the food-fishes. | B—The propagation of food-fishes in the waters | of the United States. | — | Washington: | Government Printing Office. | 1876. 8vo. pp. (v) vi–xlvi. [U. S. F. C., 13.]

Report of Commissioner without supplemental papers.

890.

1876. BAIRD, SPENCER F. United States Commission of Fish and Fisheries. | Part III. | Report | of | the Commissioner | for | 1873–4 and 1874–5. | — | A.—Inquiry into the decrease of the food-fishes. | B.—The propagation of food-fishes in the waters | of the United States. | — | Washington: | Government Printing Office. | 1876. 8vo. pp. lii, 777. [U. S. F. C., 13.]

CONTENTS.

REPORT OF THE COMMISSIONER.

A.—Inquiry into the decrease of the food-fishes.

1876. BAIRD, SPENCER F.—Continued.

891.

1876. BAIRD, SPENCER F. "Report from Prof. Spencer F. Baird, Assistant Secretary, * * * (on) additions to the Museum and the various operations connected with it during the year 1875." < *Ann. Rep. Smithsonian Institution for the year 1875*, 1876, pp. 46–57, and 72–98.

892.

1876. BAIRD, SPENCER F. The U. S. Fish Commission. < *Forest and Stream*, vi, 1876, p. 147.

893.

1876. BAIRD, SPENCER F. "An account of the proposed plan of exhibition by the Smithsonian Institution at the International Centennial Exhibition, and the extent to which the work has been carried on." < *Ann. Rep. Smithsonian Institution for the year* 1875, 1876, pp. 58–71.

> Dated Washington, Jan. 3, 1876. Signed "Spencer F. Baird, Representative of the Smithsonian Institution and of the Department of Food-Fishes in Government Centennial Board."

894.

1876. BAIRD, SPENCER F. Connecticut River Shad for California. The shipment of a million Shad Fry from Holyoke, Mass., to the Sacramento River, Cal., under the care of F. N. Clark and T. H. Bean. < *Forest and Stream*, vii, p. 66.

895.

1876. BAIRD, SPENCER F. [Introductory to Dall's Classification of the Products of Sea and Shore.] < *Folio Circular, Centennial Series*, p. 1.

896.

1876. BAIRD, SPENCER F. Work accomplished by the Challenger. <*Ann. Rec.*, 1876–77, p. 240.*

897.

1876. BAIRD, SPENCER F. Explorations made under the direction of F. V. Hayden in 1876. <*Ann. Rec.*, 1876–77, p. 242.

898.

1876. BAIRD, SPENCER F. Exploration of the Rocky Mountain Region by J. W. Powell. <*Ann. Rec.*, 1876–77, p. 255.

899.

1876. BAIRD, SPENCER F. The Triassic Fauna in Illinois. <*Ann. Rec.*, 1876–77, p. 300.

900.

1876. BAIRD, SPENCER F. Remains of the Irish Elk. <*Ann. Rec.*, 1876–77, p. 300.

901.

1876. BAIRD, SPENCER F. Revision of the Glires. <*Ann. Rec.*, 1876–77, p. 301.

902.

1876. BAIRD, SPENCER F. Rapid Destruction of the Buffalo. <*Ann. Rec.*, 1876–77, p. 302.

903.

1876. BAIRD, SPENCER F. Geographical variations among North American Mammals, especially in respect to size. <*Ann. Rec.*, 1876–77, p. 302.

904.

1876. BAIRD, SPENCER F. A New Porpoise in New York Bay. <*Ann. Rec.*, 1876–77, p. 304.

* For full title of *Annual Record* see 1877.

905.

1876. BAIRD, SPENCER F.   Decrease of Birds in Massachusetts.   <*Ann. Rec.*, 1876–77, p. 309.

906.

1876. BAIRD, SPENCER F.   Catalogue of all the Birds Known up to This Day.   <*Ann. Rec.*, 1876–77, p. 310.

907.

1876. BAIRD, SPENCER F.   Domesticating the Prairie Chicken.   <*Ann. Rec.*, 1876–77, p. 310.   •

908.

1876. BAIRD, SPENCER F.   Additional Remains of the Moa.   <*Ann. Rec.*, 1876–77, p. 311.

909.

1876. BAIRD, SPENCER F.   The Migration of Birds.   <*Ann. Rec.*, 1876–77, p. 311.

910.

1876. BAIRD, SPENCER F.   Addition to North American Ornithology—Pyrrhophoæna.   <*Ann. Rec.*, 1876–77, p. 312.

911.

1876. BAIRD, SPENCER F.   The Habits of Birds.   <*Ann. Rec.*, 1876–77, p. 312.

912.

1876. BAIRD, SPENCER F.   New Fossil Giant Birds.   <*Ann. Rec.*, 1876–77, p. 313.

913.

1876. BAIRD, SPENCER F.   Reptiles of Costa Rica.   <*Ann. Rec.*, 1876–77, p. 315.

914.

1876. BAIRD, SPENCER F.   Snake-Eating Snakes.   <*Ann. Rec.*, 1876–77, p. 316.

915.

1876. BAIRD, SPENCER F.   Remarkable Habit of Frogs.   <*Ann. Rec.*, 1876–77, p. 316.

916.

1876. BAIRD, SPENCER F.   Reproduction in the Proteus.   <*Ann. Rec.*, 1876–78, p. 317.

917.

1876. BAIRD, SPENCER F.   Rafinesque's Fishes of Ohio.   <*Ann. Rec.*, 1876–77, p. 318.

918.

1876. BAIRD, SPENCER F.   The Pilot Fish.   <*Ann. Rec.*, 1876–77, p. 319.

919.

1876. BAIRD, SPENCER F.   New Work on European Fresh-Water Fishes.   <*Ann. Rec.*, 1876–77, p. 319.

920.

1876. BAIRD, SPENCER F.   Some Curious Australian Fishes.   <*Ann. Rec.*, 1876–77, p. 319.

921.

1876. BAIRD, SPENCER F.   Poey's Catalogue of Cuban Fishes.   < *Ann. Rec.*, 1876–77, p. 320.

922.

1876. BAIRD, SPENCER F.   Habits of the Salmon.   < *Ann. Rec.*, 1876–77, p. 320.

923.

1876. BAIRD, SPENCER F.   The Rainbow Fish.   < *Ann. Rec.*, 1876–77, p. 321

924.

1876. BAIRD, SPENCER F.   Incubation of *Chromis paterfamilias.*   < *Ann. Rec.*, 1876–77, p. 322.

925.

1876. BAIRD, SPENCER F.   Cause of the Black Spots on the Scales of Fish.   < *Ann. Rec.*, 1876–77, p. 323.

926.

1876. BAIRD, SPENCER F.   Remarkable Structure of Young Fishes.   < *Ann. Rec.*, 1876–77, p. 323.

927.

1876. BAIRD, SPENCER F.   Curious Habits of Fishes.   < *Ann. Rec.*, 1876–77, p. 324.

928.

1876. BAIRD, SPENCER F.   Eighth Report of the State Entomologist of Missouri. < *Ann. Rec..* 1876–77, p. 333.

929.

1876. BAIRD, SPENCER F.   Habits and Anatomy of a Nereid Worm.   < *Ann. Rec.*, 1876–77, p. 336.

930.

1876. BAIRD, SPENCER F.   Gathering of Euplectella.   < *Ann. Rec.*, 1876–77, p. 340.

931.

1876. BAIRD, SPENCER F.   Proposed Utilization of Fish Bones.   < *Ann. Rec.*, 1876–77, p. 372.

932.

1876. BAIRD, SPENCER F.   Report of the Maritime Fisheries of France.   < *Ann. Rec.*, 1876–77, p. 385.

933.

1876. BAIRD, SPENCER F.   Report of Bureau of Statistics.   < *Ann. Rec.*, 1876–77, p. 386.

934.

1876. BAIRD, SPENCER F.   Gloucester Fisheries for 1875.   < *Ann. Rec.*, 1876–77, p. 386.

935.

1876. BAIRD, SPENCER F.   Connection of Meteorology and Herring-Fisheries.   < *Ann. Rec.*, 1876–77, p. 387.

936.

1876. BAIRD, SPENCER F.   Potomac River Fisheries.   < *Ann. Rec.*, 1876–77, p. 388.

937.

1876. BAIRD, SPENCER F. Seal-Fisheries of 1876 on the Greenland Coast. $<$ *Ann. Rec.*, 1876–77, p. 389.

938.

1876. BAIRD, SPENCER F. Close time for Seals in the Northern Sea. $<$ *Ann. Rec.*, 1876–77, p. 389.

939.

1876. BAIRD, SPENCER F. Report on Alaska Seal Islands. $<$ *Ann. Rec.*, 1876–77, p. 389.

940.

1876. BAIRD, SPENCER F. Menhaden Fishery in 1875. $<$ *Ann. Rec.*, 1876–77, p. 390.

941.

1876. BAIRD, SPENCER F. New use for the Scrap of the Moss-Bunker. $<$ *Ann. Rec.*, 1876–77, p. 390.

942.

1876. BAIRD, SPENCER F. Utilizing the Offal of Codfish on the Gulf of St. Lawrence. $<$ *Ann. Rec.*, 1876–77, p. 391.

943.

1876. BAIRD, SPENCER F. Report of the Commissioner of Fisheries of Canada for 1875. $<$ *Ann. Rec.*, 1876–77, p. 391.

944.

1876. BAIRD, SPENCER F. Report of the Fish Commissioners of Maine. $<$ *Ann. Rec.*, 1876–77, p. 392.

945.

1876. BAIRD, SPENCER F. Report of the Fish Commissioners of New Hampshire. $<$ *Ann. Rec.*, 1876–77, p. 393.

946.

1876. BAIRD, SPENCER F. Tenth Report of the Massachusetts Fish Commissioners. $<$ *Ann. Rec.*, 1876–77, p. 393.

947.

1876. BAIRD, SPENCER F. Tenth Report of the Fish Commissioners of Connecticut. $<$ *Ann. Rec.*, 1876–77, p. 394.

948.

1876. BAIRD, SPENCER F. Eighth Report of the Fish Commissioners of New York. $<$ *Ann. Rec.*, 1876–77, p. 395.

949.

1876. BAIRD, SPENCER F. Fifth Annual Report of the Fish Commissioners of New Jersey. $<$ *Ann. Rec.*, 1876–77, p. 397.

950.

1876. BAIRD, SPENCER F. Sixth Annual Report of the Fish Commissioners of New Jersey. $<$ *Ann. Rec.*, 1876–77, p. 397.

951.

1876. BAIRD, SPENCER F. Action of the Kentucky Fish Commissioners. <*Ann. Rec.*, 1876–77, p. 398.

952.

1876. BAIRD, SPENCER F. Convention of the Western State Fish Commissioners. <*Ann. Rec.*, 1876–77, p. 399.

953.

1876. BAIRD, SPENCER F. First Report of the Iowa Fish Commissioners. <*Ann. Rec.*, 1876–77, p. 399.

954.

1876. BAIRD, SPENCER F. Second Report of the Commissioners of Fisheries of Wisconsin. <*Ann. Rec.*, 1876–77, p. 400.

955.

1876. BAIRD, SPENCER F. Second Report of the Fish Commissioners of Minnesota. <*Ann. Rec.*, 1876–77, p. 400.

956.

1876. BAIRD, SPENCER F. Arkansas Fish Commissioners. <*Ann Rec.*, 1876–77, p. ——.

957.

1876. BAIRD, SPENCER F. Biennial Report of the California Fish Commission. <*Ann. Rec.*, 1876–77, p. 401.

958.

1876. BAIRD, SPENCER F. Cultivation of Carp in California. <*Ann. Rec.*, 1876–77, p. 403.

959.

1876. BAIRD, SPENCER F. Capturing Eels in Cochin China. <*Ann Rec.*, 1876–77, p. 403.

960.

1876. BAIRD, SPENCER F. Hatching Whitefish in the Detroit River. <*Ann. Rec.*, 1876–77, p. 403.

961.

1876. BAIRD, SPENCER F. One Cause of the Death of Fishes. <*Ann. Rec.*, 1876–77, p. 405.

962.

1876. BAIRD, SPENCER F. Rapidity of Growth in certain Fishes. <*Ann. Rec.*, 1876–77, p. 405.

963.

1876. BAIRD, SPENCER F. Utilization of Warmed Waters in Fish-Culture. <*Ann. Rec.*, 1876–77, p. 406.

964.

1876. BAIRD, SPENCER F. Shad in the Mississippi. <*Ann. Rec.*, 1876–77, p. 406.

965.

1876. BAIRD, SPENCER F. Renewed attempt to send Salmon Eggs to New Zealand. <*Ann. Rec.*, 1876–77, p. 407.

966.

1876. BAIRD, SPENCER F.   Salmon in the Antipodes.   < *Ann. Rec.*, 1876–77, p. 408.

967.

1876. BAIRD, SPENCER F.   Salmon Eggs in South Africa.   <*Ann. Rec.*, 1876–77, p. 408.

968.

1876. BAIRD, SPENCER F.   Capture of Salmon in the Connecticut River.   <*Ann. Rec.*, 1876–77, p. 409.

969.

1876. BAIRD, SPENCER F.   Fattening of Oysters.   <*Ann. Rec. of Sci. and Ind.*, 1876–77, p. 410.

970.

1877. BAIRD, SPENCER F.   Annual Record | of | Science and Industry | for 1876. | Edited by | Spencer F. Baird, | with the assistance of Eminent Men of Science. | [Cut.] | New York : | Harper & Brothers, publishers, | Franklin Square. | 1877.   [8vo. pp. ccxxxvi, 609.]

"In compliance with a generally expressed wish, the names of the authors of the different portions of the Scientific Summary are given, for the first time in the present volume of the Record, in connection with their respective communications, all of them men occupying the front rank in America, as authors and investigators. Other collaborators not contributors to the first division of the volume are Prof. C. F. Hines, of Dickinson College, Carlisle, Pa.; Prof. F. W. Clarke, of the University of Cincinnati, Ohio; Prof. E. D. Cope, of Philadelphia; Prof. F. V. Hayden; Maj. J. W. Powell, Lieutenant Geo. M. Wheeler, U. S. A., and several others who prefer to remain unmentioned." (PREFACE.)

971.

1877. BAIRD, SPENCER F.   Necrology [of Science, 1876].   <*Ann. Rec. Sci. and Ind.*, 1876–77, p. 541.

972.

1877. BAIRD, SPENCER F.   "Many Fish are afflicted by crustaceous parasites, called *Argulus*, which adhere to their gills."   <*Ann. Rec.*, 1876–77, p. clx, gen. sum.

973.

1877. BAIRD, SPENCER F.   "A new form of Fishes discovered by Herr Bucholz in West Africa."   < *Ann. Rec.*, 1876–77, p. clxvii.

974.

1877. BAIRD, SPENCER F.   "A characteristic type of Fishes of the Northern Atlantic is exemplified in the species variously called 'Sea Wolf,' etc."   <*Ann. Rec.*, 1876–77, p. clxvii.

975.

1877. BAIRD, SPENCER F.   Genuine White Shad in the Ohio.   <*Forest and Stream*, viii, 1877, p. 280.

976.

1877. BAIRD, SPENCER F.   "Prof. Baird spoke of the Inception of the scheme to introduce California Salmon," &c.   Feb. 14, 1877.   <*Trans. American Fish Culturists' Association*, 1876–77 (*Sixth Annual Meeting*), 1877. p. 5.

977.

1877. BAIRD, SPENCER F.   "In regard to the introduction of Salmon" [and the work of the Commission].   <*Trans. American Fish Culturists' Association*, 1876–77 (*Sixth Annual Meeting*), 1877. pp. 64–70.

978.

1877. BAIRD, SPENCER F. "Extracts from Prof. Baird's Report, published in 1873" [concerning decrease of fish in New England]. *<Documents and Proceedings of the Halifax Commission, 1877, Appendix A (Case of Her Majesty's Government)*, p.—.

U. S. Reprint, 1878 [i], p. 98.

979.

1877. BAIRD, SPENCER F. "Extracts of a report on the condition of the sea-fisheries of the south coast of New England in 1871 and 1872, by Spencer F. Baird, Commissioner." *<Documents and Proceedings of the Halifax Commission, 1877, Appendix E,* xi, p. 34.

American Reprint, 1878 [i], pp. 229-231.

980.

1877. BAIRD, SPENCER F. "Extract from Eighth Report of the Commissioner of Fisheries of the State of Maine for the year 1874 (page 7)" [being the letter to E. M. Stilwell, Esq., already mentioned under No. 620]. *<Documents and Proceedings of the Halifax Commission, 1877, Appendix E,* xii, pp. 34-35.

American Reprint, 1878 [i], pp. 231-233.

981.

1877. BAIRD, SPENCER F. "No. 68. Prof. Spencer F. Baird, Assistant Secretary of the Smithsonian Institution, Washington, and United States Commissioner of Fish and Fisheries, called on behalf of the Government of the United States, sworn and examined." Thursday, Oct. 18, 1877. *<Documents and Proceedings of the Halifax Commission, 1877, Appendix L (United States evidence),* pp. 451–463.

U. S. Reprint, 1878 [iii], pp. 2795-2816.

982.

1877. BAIRD, SPENCER F. "The Conference met. The examination of Prof. Spencer F. Baird, called on behalf of the Government of the United States, resumed." Friday, Oct. 19, 1877. *<Documents and Proceedings of the Halifax Commission, 1877, Appendix L (United States evidence),* pp. 466–479.

American Reprint, 1878, [iii], pp. 2821-2849.

983.

1877. BAIRD, SPENCER F. [Various extracts.] *<Hind—The Effect of the Fishery Clauses of the Treaty of Washington on the Fisheries and Fishermen of British North America,* part i. Halifax, 1877. pp. vii, viii, xiii, xv, xvi, xix, 7, 10, 11, 12, 25, 37, 41, 47, 135, 136, 144.

984.

1877. BAIRD, SPENCER F. $(E)\frac{B}{3}$. | U. S. Commission of Fish and Fisheries. | = | Questions relative | to the | Food-fishes of the United States. foolscap, one sheet. 4 pp. Government Printing Office, 1877. [U. S. F. C., 15.]

A new edition of the circular bearing the same title, previously issued. [U. S. F. C., 3 = No. 217.]

985.

1877. BAIRD, SPENCER F. $(E)\frac{B}{6}$. | U. S. Commission of Fish and Fisheries. | — | Statistics of the Mackerel Fisheries, Etc. | — | Circular. | [1 page, foolscap. Washington, Government Printing Office, 1877.] [U. S. F. C., 16.] To accompany circular No. 984.

986.

1877. BAIRD, SPENCER F. (E)$\frac{B}{7}$. | U. S. Commission of Fish and Fisheries. | — | Statistics of the Cod Fisheries, Etc. | — | Circular. [1 page, foolscap. Washington, Government Printing Office, 1877.] [U. S. F. C., 17.] To accompany circular No. 984.

987.

1877. BAIRD, SPENCER F. (E)$\frac{B}{8}$. | U. S. Commission of Fish and Fisheries. | — | Statistics of the Mullet Fisheries, Etc. | — | Circular. [1 page, foolscap. Washington, Government Printing Office, 1877.] [U. S. F. C., 18.] To accompany circular No. 984.

988.

1877. BAIRD, SPENCER F. E$\frac{C}{3}$. | U. S. Commission of Fish and Fisheries. | = | Statistics of Coast and River Fisheries. | — | [4 pp., foolscap. Government Printign Office, 1877, Oct.] [U. S. F. C., 19.]

> Letter signed by Prof. Baird transmitting 55 questions prepared by G. Brown Goode.
> A. Fishing vessels..
> B. Shore and boat fishing.
> C. Pounds and weirs.
> D. Gill-nets.
> E. Seines.
> F. Fish-pots and eel-pots.
> G. Three-mile lines.
> H. Disposition of the fish.
> I. Estimates of annual yield.
> K. Fishermen.

989.

1877. BAIRD, SPENCER F. A request from the United States Commissioner of Fish and Fisheries. < *Forest and Stream*, 1877, x, p. 75.

990.

1877. BAIRD, SPENCER F. Salmon in the Hudson. < *Forest and Stream*, 1877, x, p. 154.

991.

1877. BAIRD, SPENCER F. Salmon in the Chesapeake. < *Forest and Stream*, 1877, x, p. 296.

992.

1877. BAIRD, SPENCER F. Report of Prof. Spencer F. Baird on the additions, &c., to the Museum in 1876. < *Ann. Rep. Smithsonian Institution for* 1876, 1877, pp. 38–63, and 84–115.

993.

1877. BAIRD, SPENCER F. Report of Professor Baird on the Centennial Exhibition of 1876. < *Ann. Rep. Smithsonian Institution for* 1876, 1877, pp. 64–83.

994.

1877. BAIRD, SPENCER F. Genuine White Shad in the Ohio. < *Forest and Stream*, viii, 1877, p. 280.

995.

1877. BAIRD, SPENCER F. Salmon in the Richelieu. < *Forest and Stream*, ix, 1877, p. 143.

996.

1877. BAIRD, SPENCER F. A New Fish. < *Forest and Stream*, ix, 1877, p. 381.
Quotations from letter concerning the discovery of *Chimaeia plumbea* by the Gloucester fishermen.

997.

1878. BAIRD, SPENCER F. The Delaware Salmon. < *Chicago Field*, ix, 1878, p. 165.

998.

1878. BAIRD, SPENCER F. Propagation of Eels. < *Sunbury (Pa.) American*, Aug. 30 or Sept. 6, 1878.
. Letter, dated Gloucester, Aug. 27, 1878, criticizing Eberhardt's article on propagation of eels, first published in the *Gartenlaube.*

999.

1878. BAIRD, SPENCER F. "The Herring Fishery of the Coast of Sweden." < *Cape Ann Advertiser*, Aug. 9, 1878.
Inclosing letter of Josua Lindahl regarding periodicity of occurrence of herring in Sweden.

1000.

1878. BAIRD, SPENCER F. The Periodicity of Herrings. < *Chicago Field*, x, 1878, p. 35.
Quoting *Cape Ann Advertiser.*

1001.

1878. BAIRD, SPENCER F. "The Fishery Statistics of the United States." < *Trans. American Fish-Cultural Association*, 1878, pp. 72–74.
An appeal to the fish dealers of New York to supply records of the amounts of fish handled by them.

1002.

1878. BAIRD, SPENCER F. United States Commission of Fish and Fisheries. | — | Part IV. | — | Report | of | the Commissioner | for | 1875–1876. | — | A.—Inquiry into the decrease of the food-fishes. | B.—The propagation of food-fishes in the waters of the United States. | — | Washington: | Government Printing Office. | 1878. 8vo. pp. ix, 50*. [U. S. F. C., 26.]
Report of Commissioner without supplementary papers.

1003.

1878. BAIRD, SPENCER F. United States Commission of Fish and Fisheries. | — | Part IV. | — | Report | of | the Commissioner | for | 1875–1876. | — | A.—Inquiry into the decrease of the food-fishes. | B.—The propagation of food-fishes in the waters of the United States. | — | Washington: | Government Printing Office. | 1878. 8vo. pp. ix, 50*, 1029, plates vi (Hist. of Whale Fishery). [U. S. F. C., 27.]

CONTENTS.

1.—REPORT OF THE COMMISSIONER.

A.—*General considerations.* Page.

1. Introductory remarks.. 1
Operations of previous years 1
Precaution and time required by the work............................. 1
Danger of hasty generalizations 1
Methods and direction of research 1
Utilization of work already performed by other departments of the government 2
Corresponding labors of other nations 2
Rapid increase in the work of the United States Fish Commissioner.............. 3

1004.

1878. BAIRD, SPENCER F. (*editor*). Annual Record | of | Science and Industry | for 1877. | Edited by | Spencer F. Baird, | with the assistance of eminent men of science. | [Cut.] | New York: | Harper & Brothers, Publishers. | Franklin Square. | 1878. 8vo. pp. xiv, 480.—Preface dated March 1, 1878.

"A modification of the original plan of the 'Annual Record' was commenced in the volume for 1877. Previous to that it consisted of two parts—first, a general summary of progress in the various branches of science; and, secondly, a series of abstracts of special papers, credited to the work in which they were published. These abstracts, although prepared by several specialists, were without indication of their authorship. The experience of several years showed that, in attempting to give abstracts of anything like the most important announcements of the year, more space was required than could be spared for the purpose; and it was therefore determined to enlarge the scope of the first division, and make it include a greater amount of detail, each summary to be prepared by some eminent specialist, and to be headed by his name."

1878. Baird, Spencer F.—Continued.

1005.

1878. Baird, Spencer F. (E)$\frac{B}{9}$. | U. S. Commission of Fish and Fisheries. | = | Questions relative | to the | Cod and the Cod Fisheries. | — | [Foolscap size. 4 pp. Washington, Government Printing Office, 1878.] [U. S. F. C., 28.]
Circular addressed to fishermen, transmitting 90 questions prepared by G. Brown Goode.

1006.

1878. Baird, Spencer F. $\frac{(E)}{101} \cdot \frac{B}{10}$. | U. S. Commission of Fish and Fisheries. | = | Questions relative | to the | Alewife and the Alewife Fisheries. | — | [Foolscap size. 4 pp. Washington, Government Printing Office, 1878.] [U. S. F. C., 29.]
Circular addressed to fishermen, transmitting 82 questions prepared by C. G. Atkins.

1007.

1878. Baird, Spencer F. $\frac{(E)}{102} \cdot \frac{B}{10}$. | U. S. Commission of Fish and Fisheries. | = | Questions relative | to the | Smelt and Smelt Fisheries. | — | [Foolscap size. 4 pp. Washington, Government Printing Office, 1878.] [U. S. F. C., 30.]
Circular addressed to fishermen, transmitting 69 questions prepared by C. G. Atkins.

1008.

1878. Baird, Spencer F. (E)$\frac{B}{12}$. | U. S. Commission of Fish and Fisheries. | = | Questions relative | to the | Mackerel and Mackerel Fisheries. | — | [Foolscap size. 4 pp. Washington, Government Printing Office, 1879.] [U. S. F. C., 32.]
Circular letter transmitting 78 questions prepared by G. Brown Goode.

1009.

1878. Baird, Spencer F. The Delaware Salmon. <*Chicago Field*, ix, 1878, p. 165.
Letter to Commissioner Anderson.

1010.

1878. Baird, Spencer F. "The 'Herald' Interviews Prof. Baird." <*Chicago Field*, x, 1878, p. 243.
Quoting *New York Herald*.
[Concerning the work of the U. S. Fish Commission at Gloucester, Mass.]

1011.

1878. Baird, Spencer F. A request from the United States Commissioner of Fish and Fisheries. <*Forest and Stream*, x, 1878, p. 75. Ed.
Asks for information concerning salmon, shad, &c., caught in western rivers where they have been introduced.

1012.

1878. Baird, Spencer F. Salmon in the Hudson. <*Forest and Stream*, x, 1878, p. 154.

1013.

1878. BAIRD, SPENCER F. Natural History of the Howgate Expedition. <*Forest and Stream*, ix, 1878, p. 413.
Memorandum given to Mr. L. Kumlien.

1014.

1878. BAIRD, SPENCER F. All about Eels. <*Forest and Stream*, xi, 1878, pp. 130, 131.
A short letter about intestinal worms and the young of eels.

1015.

1879. BAIRD, SPENCER F. Is it Herring Spawn? <*Forest and Stream*, xii, 1879, p. 5.

1016.

1879. BAIRD, SPENCER F. Fishes of the Deep Sea. <*Forest and Stream*, xii, 1879, p. 6.

1017.

1879. BAIRD, SPENCER F. Transportation of Alewife Eggs. <*Forest and Stream*, xii, 1879, p. 225.

1018.

1879. BAIRD, SPENCER F. The Hudson Salmon. <*Forest and Stream*, xii, 1879, p. 444.

1019.

1879. BAIRD, SPENCER F. Smithsonian Miscellaneous Collections. | —324— | Circular Relative to Scientific and Literary Exchanges. | 8vo. 2 pp. Dated Jan. 1, 1879.

1020.

1879. BAIRD, SPENCER F. (*editor*). Annual Record | of | Science and Industry | for 1878. | Edited by | Spencer F. Baird, | with the assistance of eminent men of science. | [Cut.] | New York: | Harper & Brothers, Publishers, | Franklin Square. | 1879. 8vo. pp. xvii (i), 715. Preface dated March 1, 1879.

TABLE OF CONTENTS.

1021.

BAIRD, SPENCER F. Circular [to observatories]. Letter size, 1 page. [August, 1879.]

1022.

BAIRD, SPENCER F. Prefatory Note [to Henry's Researches in Sound, with special reference to fog-signaling]. <*Report Smithsonian Institution*, 1878, pp. 455, 456.

1023.

BAIRD, SPENCER F. Advertisement [to Goode's Catalogue of Collections to illustrate the Animal Resources and the Fisheries of the United States]. <*Bulletin U. S. National Museum*, No. 14. Back of the title-page (ii). Dated April 3, 1879.

1024.

BAIRD, SPENCER F. Advertisement [to Eggers' "The Flora of St. Croix and the Virgin Islands"]. <*Bulletin U. S. National Museum*, No. 13; fly-leaf. Dated May, 1879.

1025.

BAIRD, SPENCER F. Advertisement [to Kumlien's Contributions to the Natural History of Arctic America]. <*Bulletin U. S. National Museum*, No. 14, p. 2. April 15, 1879.

1026.

BAIRD, SPENCER F. Advertisement [to Rhees' "The Smithsonian Institution: Journals of the Board of Regents, Reports of Committees, Statistics, etc."]. <*Smithsonian Miscellaneous Collections*, 329, p. iii. Dated December, 1879.

1027.

BAIRD, SPENCER F. Advertisement [to Rau's "The Palenque Tablet"]. <*Smithsonian Publications*, 331, p. iii. Dated November, 1879.

1028.

BAIRD, SPENCER F. Annual Report | of the | Board of Regents | of the | Smithsonian Institution | showing the | operations, expenditures, and condition of the | Institution for the year | 1878. | — | Washington: | Government Printing Office. | 1879. 8vo. pp. 575.

Report of Secretary for 1878 .... pp. 7–64
Government explorations and surveys in 1878..................................... 65–81
Additions to collections in the National Museum.................................... 83–112
Statistics of exchanges, &c.. 113–124
Acts and resolutions of Congress relative to the Smithsonian Institution 125
Report of the Executive Committee 126–132
Journal of Board of Regents 133–142

1029.

BAIRD, SPENCER F. Circular Relating to Fish Trade and Consumption of Fish. <*Chicago Field*, xii, 1879, p. 35.

1030.

BAIRD, SPENCER F. United States Commission of Fish and Fisheries. | — | Part V. | — | Report | of | The Commissioner | for | 1877. | — | A.—Inquiry into the decrease of food-fishes. | B.—The propagation of food-fishes in the | waters of the United States. | — | Washington: | Government Printing Office. | 1879. | 8vo. pp. 48. [U. S. F. C., 37.]

Report of Commissioner without supplementary papers.

1031.

1879. BAIRD, SPENCER F. United States Commission of Fish and Fisheries. | — | Part V. | — | Report | of | The Commissioner | for | 1877. | — | A.—Inquiry into the decrease of food-fishes. | B.—The propagation of food-fishes in the | waters of the United States. | — | Washington: | Government Printing Office. | 1879. | 8vo. pp. 48, 972. [U. S. F. C., 38.]

CONTENTS.

I.—REPORT OF THE COMMISSIONER.

A.—General considerations.

1032.

1879. BAIRD, SPENCER F. U. S. Commission of Fish and Fisheries | and | Census of 1880. | — | * * * | — | Circular inviting co-operation. | [Letter size, 1 p. —. July 1, 1879.] [U. S. F. C., 40.]

1033.

1879. BAIRD, SPENCER F. The National Fish Commission. <*Forest and Stream*, xiii, 1879, p. 725.

Summary of work at Gloucester, Mass.

1034.

1879. BAIRD, SPENCER F. The Farmer's Fish [Carp]. <*Forest and Stream*, xiii, 1879, p. 846.

1035.

1879. BAIRD, SPENCER F. Another New Fish on our Coast. <*Chicago Field*, xi, 1879, p. 117.

Letter to Eugene G. Blackford acknowledging receipt of the specimens of *Centropomus undecimalis*.

1036.

1879. BAIRD, SPENCER F. [A Letter Concerning California Salmon in Europe.] <*Chicago Field*, xi, 1879, p. 131.

1037.

1879. BAIRD, SPENCER F. Report of the U. S. Fish Commission. <*Chicago Field*, xii, 1879, p. 307.

Extracts from.

1038.

1879. BAIRD, SPENCER F. Advertisement [to Shakespeare's "The Nature of Reparatory Inflammation in Arteries after Ligatures, Acupressure, and Torsion"]. <*Smithsonian Publication*, 321. (Toner Lecture, vii,) p. iii. Dated April, 1879. (Title-page date, March, 1879.)

1039.

1879. BAIRD, SPENCER F. Advertisement [to Rhees' "The Smithsonian Institution: Documents relative to its origin and history"]. <*Smithsonian Miscellaneous Collections*, 328, p. iii. Dated April, 1879.

1040.

1879. BAIRD, SPENCER F. [7-031.] | — | U. S. Commission of Fish and Fisheries | and | Census of 1880. | — | * * * | — | Circular relating to Fish Trade and Consumption of Fish. | [Letter size, 1 p. Washington: Government Printing Office, 1879. July.] [U. S. F. C., 42.]

Letter transmitting Circular 43, prepared by G. Brown Goode.

1041.

1879. BAIRD, SPENCER F. Advertisement [to Elliott's "List of described species of Humming Birds]. <*Smithsonian Miscellaneous Collections*, 334, p. iii. Dated May, 1879.

1042.

1879. BAIRD, SPENCER F. Advertisement [to Rhees' "The Scientific Writings of James Smithson"]. <*Smithsonian Miscellaneous Collections*, 327, p. iii. Dated October, 1879.

1043.

1880. BAIRD, SPENCER F. Do Black Snakes Eat Fish. <*Forest and Stream*, xiii, 1880, p. 966.

1044.

1880. BAIRD, SPENCER F. Striped Bass and Shad in California. <*Forest and Stream*, xiv, 1880, p. 410.

1045.

1880. BAIRD, SPENCER F. Advertisement [to Wood's "Fever: a Study in Morbid and Normal Physiology]." <*Smithsonian Publication*, 357, p. iii. Dated October, 1880.

1046.

BAIRD, SPENCER F. Advertisement [to Goode's "Exhibit of the Fisheries and Fish-Culture of the United States of America made at Berlin in 1880. <*Bulletin. U. S. National Museum*, No. 18, p. (ii). Dated March 29, 1880.

1047.

BAIRD, SPENCER F. Introductory Note [to Catalogue of Exhibit of U. S. Fish Commission at the International Fishery Exhibition at Berlin]. < *Bulletin U. S. National Museum*, No. 18, pp. xii–xv. Dated March 29, 1880.

An account of the action of Congress and the organization of the Exhibition.

1048.

BAIRD, SPENCER F. California Salmon-Eggs. <*Forest and Stream*, xv, 1880, p. 107.

1049.

BAIRD, SPENCER F. Eggs for Distribution. <*Forest and Stream*, xv, 1880, p. 366.

1050.

BAIRD, SPENCER F. Advertisement [to Cope's "On the Zoological Position of Texas"]. <*Bulletin U. S. National Museum*, No. 17, p. 2. Dated May, 1880.

1051.

BAIRD, SPENCER F. United States Commission of Fish and Fisheries. | — | Part VI. | — | Report | of | the Commissioner | for | 1878. | — | A.—Inquiry into the decrease of food-fishes. | B.—The propagation of food-fishes in the waters of the United States. | — | Washington: | Government Printing Office. | 1880. 8vo. pp. lxiv.

Report of the Commissioner, without supplementary papers.

1052.

BAIRD, SPENCER F. United States Commission of Fish and Fisheries. | — | Part VI. | — | Report | of | the Commissioner | for | 1878. | — | A.—Inquiry into the decrease of food-fishes. | B.—The propagation of food-fishes in the waters of the United States. | — | Washington: | Government Printing Office. | 1880. 8vo. pp. lxiv (3), 988. pll. i–xvi, p. 32, i–xiii (with explanatory leaf preceding each); p. 462, i–vii (with explanatory leaves), p. 506.

CONTENTS.

I.—REPORT OF THE COMMISSIONER.

A.—GENERAL CONSIDERATIONS.

* The eggs of the California salmon were hatched out in 1878 but not distributed, for the most part, until 1879. The hatching and distribution were made by the State fishery commissions, except when otherwise stated. The imperfections of the returns will be remedied in the next report.

† Not hatched until 1879. The hatching and distribution made by the New Jersey Fish Commission.
The eggs were collected by F. N. Clark, of Northville, Mich.

---

* The species involved is the Quinnat salmon (*Salmo quinnat*).
    * Published in 1876.

1880. BAIRD, SPENCER F.—Continued.

1053.

1880. BAIRD, SPENCER F. Advertisement [to Waring's Suggestions for the Sanitary Drainage of New York City.] <*Smithsonian Publications*, 349, p. iii. Dated Washington, June, 1880.

1054.

1880. BAIRD, SPENCER F. Advertisement [to Smithsonian Miscellaneous Collections, Vol. XVI]. <*Smithsonian Miscellaneous Collections*, Vol. XVI, 1880, p. vii.

1055.

1880. BAIRD, SPENCER F. Advertisement [to Smithsonian Miscellaneous Collections, Vol. XVII]. <*Smithsonian Miscellaneous Collections*, Vol. XVII, 1880, p. vii.

1056.

BAIRD, SPENCER F. 46th Congress, } Senate. { Mis. Doc., | = | Annual Report | of the | Board of Regents | of the | Smithsonian Institution, | showing | the operations, expenditures, and condition of | the Institution | for | the year 1879. | — | Washington: | Government Printing Office. | 1879. 8vo. pp. (7) 8-631 (1). Report of the Secretary, pp. 11-116. Dated January, 1880. Report of Executive Committee, pp. 117-142. James Smithson and his Bequest, pp. 143-210. General Appendix, pp. 211-589. List of Illustrations, pp. 591-595. Index, pp. 597-631.

REPORT OF PROFESSOR BAIRD.

1057.

BAIRD, SPENCER F. [Copy of Circular Issued to State Commissioners Regarding California Salmon Eggs.] < *Chicago Field*, xiv, 1880, p. 91.

1058.

BAIRD, SPENCER F. Destruction of Fish. <*Chicago Field*, xiv, 1880, p. 236.

1059.

BAIRD, SPENCER F. Salmon Eggs. <*Chicago Field*, xiv, 1880, p. 284.

1060.

BAIRD, SPENCER F. A Good Fish for Barren Waters [Carp]. <*Chicago Field*, xii, 1880, p. 323, 324.
Quoting Newark *Daily Advertiser*.

ADDENDA.

49½.

BAIRD, SPENCER F.   On the | Serpents of New York; | with a notice of a species not hitherto included | in the fauna of the State. | By Spencer F. Baird. | — | Albany: | C. Van Benthuysen, Printer. | 1854. | 8vo, pp. (2)-28.

74½.

BAIRD, SPENCER F.   Description of New Genera and Species of North American Lizards in the Museum of the Smithsonian Institution.. <*Proc. Acad Nat. Sci. Phila.*, Dec., 1858, pp. 253-256.

Family IGUANIDÆ.

1858. BAIRD, SPENCER F.—Continued.

Sceloporus longpipes, Baird, n. s..
    Fort Tejon, Cal.
Sceloporus couchii, Baird, n. s ....
    Santa Caterina, N. Léon.
Anolis cooperi, Baird, n. s ....
    California.
                    Family GECKOTIDÆ.
Sphoeriodactylus notatus, Baird, n. s ....
    Key West, Fla.
Stenodactylus variegatus, Baird, n. s ....
    Rio Grande and Gila Valleys.
                    Family XANTUSIDÆ.
Xantusia, Baird, n. g ....
Xantusia vigilis, Baird, n. s..
    Fort Tejon, Cal.
                    Family LACERTIDÆ.
Cnemidophorus inornatus, Baird, n. s..
    New Leon.
Cnemidophorus octolineatus, Baird, n. s ....
    New Leon.
                    Family ZONURIDÆ.
Gerrhonotus webbii, Baird, n. s.....:..
    Near San Diego, Cal.
Gerrhonotus infernalis, Baird, n. s...
    Devil's River, Tex.
Gerrhonotus olivaceus, Baird, n. s ....
    Near San Diego.
Lepidosternon floridanum, Baird, n. s..
    Micanopy, Fla.
                    Family SCINCIDÆ.
Plestiodon leptogrammus, Baird, n. s ....
    Platte River Valley.
Plestiodon inornatus, Baird, n. s...
    Sand Hills of Platte.
Plestiodon tetragrammus, Baird, n. s...
    Lower Rio Grande.
Plestiodon egregius, Baird, n. s ....
    Indian Key, Fla.
Plestiodon septentrionalis, Baird, n. s...
    Minnesota and Nebraska.

632⅓.

1874. BAIRD, SPENCER F., T. M. BREWER, and R. RIDGWAY. A | History | of | North American Birds | By | S. F. Baird, T. M. Brewer, and R. Ridgway | Land Birds | Illustrated by 64 Colored Plates and 593 Woodcuts. | Volume [1-2-3] | [Cut] | Boston | Little, Brown, and Company | 1874. Three volumes, 4to, pp.: Vol. I, XXVIII, 596, VI; Vol. II, 590, VI; Vol. III, 560, XXVIII.

The present work is designed to meet the want, which has long been felt, of a descriptive account of the Birds of North America, with notices of their geographical distribution, habits, methods of nesting, character of eggs, their popular nomenclature, and other points connected with their life history.

For many years past the only systematic treatises bearing upon this subject have been "The American Ornithology" of Alexander Wilson, finished by that author in 1814, and brought down to the date of 1827 by George Ord; the "Ornithological Biography" of Audubon, bearing the date of 1838, with a second edition "Birds of America," embracing a little more of detail, and completed in 1844; and "A Manual of the Ornithology of the United States and Canada," by Nuttall, of which a first edition was published in 1832 and a second in 1840. Since then no work relating to American Ornithology, of a biographical nature, has been presented to the public, with the exception of some of limited extent, such as those of Girard, on the "Birds of Long Island," in 1844; DeKay's "Birds of N. Y.," 1844; Samuel's "Ornithology and Oölogy of New England," 1868, and a few others; together with quite a number of minor papers on the birds of particular localities of greater or less moment, chiefly

74. BAIRD, SPENCER F., T. M. BREWER, *and* R. RIDGWAY—Continued.

, published in periodicals and the Proceedings of Societies. The reports of many of the government exploring parties also contain valuable data, especially those of Dr. Newberry, Dr. Heermann, Dr. J. G. Cooper, Dr. Suckley, Dr. Kennerly, and others.

More recently (in 1870) Professor Whitney, Chief of the Geological Survey of California, has published a very important volume on the ornithology of the entire west coast of North America, written by Dr. J. G. Cooper, and containing much original detail in reference to the habits of the western species. This is by far the most valuable contribution to the biography of North American birds that has appeared since the time of Audubon, and, with its typographical beauty and numerous and excellent illustrations, all on wood and many of them colored, constitutes one of the most noteworthy publications in American Zoölogy.

Up to the time of the appearance of the work of Audubon, nearly all that was known of the great region of the United States west of the Missouri River was the result of the journey of Lewis & Clark up the Missouri and across to the Pacific Coast, and that of John K. Townsend and Mr. Nuttall, both of whom made some collections and brought back notices of the country, which, however, they were unable to explore to any great extent. The entire region of Texas, New Mexico, Colorado, Arizona, Nevada, and California was unvisited, as also a great portion of territory north of the United States boundary, including British Columbia and Alaska.

A work by Sir John Richardson, forming a volume in his series of "Fauna Boreali-Americana," in reference to the ornithology of the region covered by the Hudson Bay Company's operations, was published in 1831, and has been much used by Mr. Audubon, but embraces little or nothing of the great breeding grounds of water birds in the neighborhood of the Great Slave and Bear Lakes, the Upper Yukon, and the shores of the Arctic coast.

It will thus be seen that a third of a century has elapsed since any attempt has been made to present a systematic history of the birds of North America.

The object of the present work is to give, in as concise a form as possible, an account of what is known of the birds, not only of the United States, but of the whole region of North America north of the boundary-line of Mexico, including Greenland on the one side, and Alaska with its islands on the other. The published materials for such a history are so copious that it is a matter of surprise that they have not been sooner utilized, consisting, as they do, of numerous scattered biographies and reports of many government expeditions and private explorations. But the most productive source has been the great amount of manuscript contained in the archives of the Smithsonian Institution, in the form of correspondence, elaborate reports, and field-notes of collectors and travelers, the use of which, for the present work, has been liberally allowed by Professor Henry. By far the most important of these consist of notes made by the late Robert Kennicott in British America, and received from him and other gentlemen in the Hudson Bay Territory, who were brought into intimate relationship with the Smithsonian Institution through Mr. Kennicott's efforts. Among them may be mentioned more especially Mr. R. MacFarlane, Mr. B. R. Ross, Mr. James Lockhart, Mr. Lawrence Clark, Mr. Strachan Jones, and others, whose names will appear in the course of the work. The especial value of the communications received from these gentlemen lies in the fact that they resided for a long time in a region to which a large proportion of the rapacious and water birds of North America resort during the summer for incubation, and which until recently has been sealed to explorers.

Equally serviceable has been the information received from the Yukon River and Alaska generally, including the Aleutian Islands, as supplied by Messrs. Robert Kennicott, William H. Dall, Henry M. Bannister, Henry W. Elliott, and others.

It should be understood that the remarks as to the absence of general works on American Ornithology, since the time of Audubon, apply only to the life history of the species, as, in 1858, one of the authors of the present work published a systematic account of the birds of North America, constituting volume IX of the series of Pacific Railroad Reports; while from the pen of Dr. Elliott Coues, a well-known and eminent ornithologist, appeared in 1872 a comprehensive volume, entitled "A Key to North American Birds," containing descriptions of the species and higher groups.

The technical, or descriptive, matter of the present work has been prepared by Messrs. Baird and Ridgway, that relating to the *Raptores* entirely by Mr. Ridgway; and all the accounts of the habits of the species are from the pen of Dr. Brewer. In addition to the matter supplied by these gentlemen, Professor Theodore N. Gill has furnished that portion of the introduction defining the class of birds as compared with other vertebrates; while to Dr. Coues is to be given the entire credit for the pages embracing the tables of the Orders and Families, as well as for the Glossary beginning on page 535 of Vol. III.

Nearly all the drawings of the full-length figures of birds contained in the work were made directly on the wood, by Mr. Edwin L. Sheppard, of Philadelphia, from original sketches taken from nature; while the heads were executed for the most part by Mr. Henry W. Elliott and Mr. Ridgway. Both series have been engraved by Mr. Hobart H. Nichols, of Washing-

1874. BAIRD, SPENCER F., T. M. BREWER, *and* R. RIDGWAY—Continued.

ton. The generic outlines were drawn by Anton H. Schönborn, and engraved by the peculiar process of Jewett, Chandler & Co., of Buffalo. All of these, it is believed, speak for themselves, and require no other commendation.

A considerable portion of the illustrations were prepared, by the persons mentioned above, for the Reports of the Geological Survey of California, and published in the volume on Ornithology. To Professor Whitney, Chief of the Survey, acknowledgments are due for the privilege of including many of them in the present History of North American Birds, and also for the Explanation of Terms on page 526 of Vol. III.

A few cuts, drawn by Wolf and engraved by Whymper, first published in "British Birds in their Haunts," and credited in their proper places, were kindly furnished by the London Society for the Diffusion of Christian Knowledge; and some others prepared for an unpublished volume by Dr. Blasius, on the Birds of Germany, were obtained from Messrs. Vieweg and Son, of Braunschweig.

The volume on the Water Birds is in an advanced state of preparation, and will be published with the least possible delay.

SPENCER F. BAIRD.

SMITHSONIAN INSTITUTION,
 Washington, January 8, 1874.
 < *Baird, Brewer, and Ridgway.—Birds of North America, Vol. I, Preface.*

1874. BAIRD, SPENCER F., T. M. BREWER, *and* R. RIDGWAY—Continued.

1874. BAIRD, SPENCER F., T. M. BREWER, *and* R. RIDGWAY—Continued.

APPENDIX. Vol. III

I.—ADDITIONS AND CORRECTIONS.

II. SYSTEMATIC CATALOGUE.

NOTE.—In the Systematic Catalogue which follows, Professor Baird's writings are classified with reference to the topics of which they treat. It should be noted that the articles in the *Annual Record of Science and Industry*, to which frequent reference is made, are in large part critical notices and reviews of the work of other writers.

ANALYSIS.

MAMMALS.

22.—1851. On the ruminating animals of North America and their susceptibility of domestication. *<Report of Commissioner of Patents for 1851*, pp. 104–128, 8 plates.

29.—1852. Mammals of the Valley of the Great Salt Lake. *<Stansbury's Exploration and Survey of the Valley of the Great Salt Lake of Utah, etc.* Philadelphia, 1852. [App. C.] pp. 309–313.

30.—1852. Mammals Collected by Lieutenant Abert in New Mexico. *<Stansbury's Exploration and Survey of the Valley of the Great Salt Lake of Utah, etc.* Philadelphia, 1852. [App. C.] p. 313.

34.—1852. [Note in reference to Vulpes Utah, Aud. and Bach.] *<Proc. Acad. Nat. Sci. Phila.*, vi, 1852–3 (1854), p. 124.

59.—1854. Characteristics of some New Species of Mammalia, collected by the U. S. and Mexican Boundary Survey. Part I. *<Proc. Acad. Nat. Sci. Phila.*, vii. pp. 331–3.

60.—1854. Characteristics of some New Species of North American Mammalia, collected chiefly in connection with the U. S. Surveys of a Railroad Route to the Pacific. Part I. *<Proc. Acad. Nat. Sci. Phila.*, vii, pp. 333–6.

64.—1855. Mammals of Chili. *<Gillis, Naval Astron. Exp.*, ii, pp. 153–162.*

65.—1855. List of Mammalia found in Chile. *<Gillis, Naval Astron. Exp.*, ii. pp. 163–171.*

72.—1857. [Name *Tamias pallasii* proposed instead of *Sciurus striatus*, Pallas, nec Linn.] *<Eleventh Ann. Rep. Smithsonian Institution*, p. 55.

75.—1857. Catalogue of North American Mammals, chiefly in the Museum of the Smithsonian Institution, Washington. Smithsonian Institution, July 1857. 4to. pp. 21.

76.—1857. Explorations and Surveys for a Railroad Route from the Mississippi River to the Pacific Ocean: War Department. Mammals. By Spencer F. Baird, Assistant Secretary of the Smithsonian Institution, Washington, D. C., 1857. <Reports of Explorations and Surveys to ascertain the most practicable and economical Route for a Railroad from the Mississippi River to the Pacific Ocean. Volume viii, 1857. 4to. pp. xlviii, 757, pl. xviii—lx.

•82·—1858. Description of a Phyllostome Bat from California, in the Museum of the Smithsonian Institution. <*Proc. Acad. Nat. Sci. Phila.*, x, 1858, pp. 116, 117.

86.—1859. United States and Mexican Boundary Survey, under the order of Lieut. Col. W. H. Emory, Major First Cavalry, and United States Commissioner. Mammals of the Boundary, by Spencer F. Baird, Assistant Secretary of the Smithsonian Institution. With notes by the naturalists of the Survey. <Report on the United States and Mexican Boundary Survey. Volume II, 1859, 4to, 2d part Zoology, pp. (2)+3–62, pll. xxvii (Mammals); (2)×3–1.

93.—1859. Report upon Mammals collected on the Survey. <Explorations and Surveys for a Railroad Route from the Mississippi River to the Pacific Ocean. War Department. Report of Lieut. E. G. Beckwith, Third Artillery, upon Explorations for a Railroad Route, near the 38th and 39th parallels of north latitude, by Captain J. W. Gunnison, Corps of Topographical Engineers, and near the forty-first parallel of north latitude, by Lieut. E. G. Beckwith, Third Artillery. 1854. Zoological Report, xx, 1857, in Report P. R. R. Surv., vol. x, 1859.

97.—1859. Report on Mammals collected on the Survey. <Explorations and Surveys for a Railroad Route from the Mississippi River to the Pacific Ocean. War Department. Routes in California, with the routes near the Thirty-fifth and Thirty-second Parallels, explored by Lieut. R. S. Williamson, Corps of Top. Eng., in 1853. Zoological Report. Washington, D. C. 1859.

101.—1859. Mammals of North America; the descriptions of species based chiefly on the collections in the Museum of the Smithsonian Institution. Philadelphia: J. B. Lippincott & Co. 1859. 4to. pp. xxxiv, 1–735+1–55+ 737–764, pll. i–lxxxvi.

172.—1871. Immunity of the Pig from injury by serpent bite. <*Ann. Rec.*, 1871–72, pp. 255, 256.

196.—1871. American Tapirs. <*Ann. Rec.*, 1871–72, p. 278.

261.—1872. Characteristics of higher groups of Mammals. <*Ann. Rec.*, 1872–73, p. 238.

262.—1872. Fur-bearing Animals of New Jersey. <*Ann. Rec.*, 1872–72, p. 240.

263.—1872. Capture of Bassaris in Ohio. <*Ann. Rec.*, 1872–73, p. 240.

264.—1872. Law in regard to Killing Buffalo. <*Ann. Rec.* 1872–73, p. 241.

265.—1872. New American Mastodon. <*Ann. Rec.*, 1872–73, p. 242.

296.—1872. The Proboscidians of the American Eocene. <*Ann. Rec.*, 1872–73, p. 307.

297.—1872. The Armed Metalophodon. <*Ann. Rec.*, 1872–73, p. 308.

304.—1872. Scammon on the West Coast Cetaceans. <*Ann. Rec.*, 1872–73, p. 312.

306.—1871. Origin of the Domestic Dog. <*Ann. Rec.*, 1872–73, p. 317.

307.—1872. The Mammals of Thibet. <*Ann. Rec.*, 1872–73, p. 318.

313.—1872. A fossil Lemuroid from the Eocene of Wyoming. <*Ann. Rec.*, 1872–73, p. 326.

317.—1872. Peculiar Mound Crania. <*Ann. Rec.*, 1872–73, p. 321.

318.—1872. Migrations of the California Gray Whale. *<Ann. Rec.*, 1872–73, p. 332.

320.—1872. Fossil Elephant in Alaska. *<Ann. Rec.*, 1872–73, p. 333.

324.—1872. Fossil Mammals from the West. *<Ann. Rec.*, 1872–73, p. 337.

328.—1872. Report of the United States Agricultural Department of Cattle Diseases. *<Ann. Rec.*, 1872–73, p. 381.

383.—1872. Catch of Fur Seals in 1872. *<Ann. Rec.*, 1872–73, p. 435.

384.—1872. Oil Works on Unalaschka. *<Ann. Rec.*, 1872–73, p. 436.

387.—1872. American Whale Fishery in 1871. *<Ann. Rec.*, 1872–73, p. 437.

402.—1872. Rat Catching. *<Ann. Rec.*, 1872–73, p. 458.

479.—1873. New Fossil Carnivora. *<Ann. Rec.*, 1873–75, p. 284.

480.—1873. Orohippus agilis. *<Ann. Rec.*, 1873–75, p. 286.

481.—1873. Maynard on the Mammals of Florida. *<Ann. Rec.*, 1873–75, p. 286.

488.—1873. Number of Glyptodonts, or Extinct Giant Armadillos. *<Ann. Rec.*, 1873–75, p. 302.

489.—1873. International Exhibition of Horns. *<Ann. Rec.*, 1873–75, p. 303.

495.—1873. New Vertebrate Fossils. *<Ann. Rec.*, 1873–75, p. 308.

496.—1873. Aboriginal Monkey. *<Ann. Rec.*, 1873–75, p. 310.

500.—1873. The Relationship of the Coyote to the Pointer Dog. *<Ann. Rec.*, 1873–75, p. 314.

505.—1873. Relations of the Megatheriidæ. *<Ann. Rec.*, 1873–75, p. 318.

535.—1873. Extermination of Field-Mice. *<Ann. Rec.*, 1873–75, p. 412.

541.—1873. Is Seal Oil Fish Oil? *<Ann. Rec.*, 1873–75, p. 430.

648.—1874. Paucity of Mammals in Cuba. *<Ann. Rec.*, 1874–75, p. 300.

649.—1874. Embryology of the Lemurs. *<Ann. Rec.*, 1874–75, p. 301.

650.—1874. Genesis of the Horse. *<Ann. Rec.*, 1874–75, p. 301.

651.—1874. Extermination of Buffaloes. *<Ann. Rec.*, 1874–75, p. 303.

674.—1874. The Species of American Squirrels. *<Ann. Rec.*, 1874–75, p. 332.

675.—1874. Relationship of American Deer to their British analogues. *<Ann. Rec.*, 1874–75, p. 335.

676.—1874. The Characters and Relations of the Hyopotamidæ. *<Ann. Rec.*, 1874–75, p. 336.

678.—1874. History of the Pacific Coast Marine Mammals. *<Ann. Rec.*, 1874–75, p. 338.

679.—1874. Decrease in the European Bison. *<Ann. Rec.*, 1874–75, p. 339.

681.—1874. Discovery of Putorius nigripes. *<Ann. Rec.*, 1874–75, p. 339.

682.—1874. The Fossil Hog of America. *<Ann. Rec.*, 1874–75, p. 340.

698.—1874. Taming the Zebra. *<Ann. Rec.*, 1874–75, p. 352.

704.—1874. Destructiveness of Rodents in California. *<Ann. Rec.*, 1874–75, p. 397.

822.—1875. Fisheries and Seal-Hunting in the White Sea and Northern Ocean. *<Ann. Rec.*, 1875–76, p. 413.

823.—1875. Close time for the Capture of Seals. *<Ann. Rec.*, 1875–76, p. 414.

824.—1875. Bad Condition of the Hair-Seal Fisheries. *<Ann. Rec.*, 1875–76, p. 414.

900.—1876. Remains of the Irish Elk. *<Ann. Rec.*, 1876–77, p. 300.

901.—1876. Revision of the Glires. *<Ann. Rec.*, 1876–77, p. 301.

902.—1876. Rapid Destruction of the Buffalo. *<Ann. Rec.*, 1876–77, p. 302.

903.—1876. Geographical variations among North American Mammals, especially in respect to size. *<Ann. Rec.*, 1876–77, p. 302.

904.—1876. A New Porpoise in New York Bay. *<Ann. Rec.*, 1876–77, p. 304.

937.—1876. Seal-Fisheries of 1876 on the Greenland Coast. *<Ann. Rec.*, 1876–77, p. 389.

938.—1876. Close time for Seals in the Northern Sea. *<Ann. Rec.*, 1876–77, p. 389.

939.—1876. Report on Alaska Seal Islands. *<Ann. Rec.*, 1876–77, p. 389.

BIRDS.

1.—1843. Descriptions of two Species, supposed to be new, of the Genus Tyrannula, Swainson, found in Cumberland County, Pennsylvania. <*Proc. Acad. Nat. Sci. Phila.*, i, pp. 283–285, 1843.

2.—1844. List of Birds found in the vicinity of Carlisle, Cumberland County, Penn., about Lat. 40° 12′ W., Lon. 77° 11′ W. <*Amer. Journ. Sci. and Arts*, xlvi, 1844, pp. 261–273.

3.—1844. Descriptions of two species, supposed to be new, of the genus Tyrannula (Swainson), found in Cumberland Co., Penn. <*Amer. Journ. Sci. and Arts*, xlvi, 1846, pp. 273–276.

6.—1845. Catalogue of birds found in the neighborhood of Carlisle, Cumberland Co., Pa. <*Lit. Rec. and Journ. Linnæan Assoc. Pennsylvania College*, i, 1845, pp. 249–257.

9.—1847. Dr. Leidy read a letter from Prof. Spencer F. Baird, of Carlisle, Pa., describing a Hybrid between the Canvass back Duck and the Common Duck. <*Proc. Acad. Nat. Sci. Phila.*, iii, 1846 and 1847, p. 209.

31.—1852. Birds [of the Valley of the Great Salt Lake]. <*Stansbury's Exploration and Survey of the Valley of the Great Salt Lake of Utah, etc.* Philadelphia, 1852. [App. C.] pp. 314–325.

31.—1852. Birds collected in New Mexico by Lieut. Abert. <*Stansbury's Exploration and Survey of the Valley of the Great Salt Lake of Utah, etc.* Philadelphia, 1852. [App. C.] pp. 325, 326.

32.—1852. List of birds inhabiting America, west of the Mississippi, not described in Audubon's Ornithology. <*Stansbury's Exploration and Survey of the Valley of the Great Salt Lake, etc.* Philadelphia, 1852. [App. C.] pp. 327–335.

40.—1852. Description of a new Species of Sylvicola. <*Ann. Lyc. Nat. Hist. N. Y.*, v, 1852, pp. 217, 218, pl. vi.

56.—1854. Descriptions of New Birds collected between Albuquerque, N. M., and San Francisco, California, during the winter of 1853–54, by Dr. C. B. R. Kennerly and H. B. Möllhausen, Naturalists attached to the Surveys of the Pacific R. R. Route, under Lt. A. W. Whipple. <*Proc. Acad. Nat. Sci. Phila.*, vii, pp. 118–20.

74.—1857. American Oology. <*Edinb. New Philos. Journal*, new ser., v, 1857, p. 374.

78.—1858. Explorations and Surveys for a Railroad Route from the Mississippi River to the Pacific Ocean. War Department. Birds. By Spencer F. Baird, Assistant Secretary Smithsonian Institution, with the co-operation of John Cassin and George N. Lawrence. Washington, D. C., 1858, pp. i–lvi 1–1005. (No illustrations.) Dated Washington, Oct. 20, 1853.

79.—1858. Birds found at Fort Bridger, Utah. <*Pacific Railroad Report*, ix, 1858, App. B, pp. 926, 927.

80.—1858. Catalogue of North American Birds, chiefly in the Museum of the Smithsonian Institution. Washington: Smithsonian Institution. October, 1858, 4to. pp. xv–lvi.

83.—1858. "Description of a new Sparrow collected by Mr. Samuels in California." <*Proc. Bost. Soc. Nat. Hist.*, vi, pp. 379, 380, Aug., 1858. Read June 2.

87.—1859. United States and Mexican Boundary Survey, under the order of Lieut. Col. W. H. Emory, Major First Cavalry, and United States Commissioner. Birds of the Boundary, by Spencer F. Baird, Assistant Secretary of the Smithsonian Institution. With notes by the Naturalists of the Survey. Report on the United States and Mexican

87·—1859. United States and Mexican Boundary—Continued.

Boundary Survey, made under the direction of the Secretary of the Interior, by William H. Emory, Major First Cavalry, and United States Commissioner. Volume ii, Washington: Cornelius Wendell, Printer. 1859. 4to. pp. (2) 3-33 (1) pll. xxv.

90.—1859. Smithsonian Miscellaneous Collections. Catalogue of North American Birds, chiefly in the Museum of the Smithsonian Institution. [First octavo edition.] Washington: Smithsonian Institution. 1859. 8vo. 2 p. ll., pp. 19 + 2.

94.—1859. No. 2. Report on Birds collected on the Survey. <Explorations and Surveys for a Railroad Route from the Mississippi River to the Pacific Ocean. War Department. Report of Lieut. E. G. Beckwith, Third Artillery, upon Explorations for a Railroad Route, near the 38th and 39th parallels of north latitude, by Captain J. W. Gunnison, Corps of Topographical Engineers, and near the forty-first parallel of north latitude, by Lieut. E. G. Beckwith, Third Artillery. 1854. pp. 11-16, pll. xii, xiii, xiv, xv, xvii, xxxii, xxxv, in Rep. P. R. R. Surv., vol. x. Third article.

98.—1859. "[Résumé of Ornithological Field Operations in progress in America, etc.] <Ibis, i," 1859, pp. 334, 335.

99.—1859. Notes on a collection of Birds made by Mr. John Xantus, at Cape St. Lucas, Lower California, and now in the Museum of the Smithsonian Institution. <Proc. Acad. Nat. Sci. Phila., xi, 1859 (1860), pp. 299-306.

103 (104).—1860. The Birds of North America; the descriptions of Species based chiefly on the Collections in the Museum of the Smithsonian Institution. By Spencer F. Baird, Assistant Secretary of the Smithsonian Institution, with the co-operation of John Cassin, of the Academy of Natural Sciences of Philadelphia, and George N. Lawrence, of the Lyceum of Natural History of New York. With an Atlas of One Hundred Plates. Text. Philadelphia: J. B. Lippincott & Co. 1860. 4to. pp. 1vi, 1005.

106.—1861. Smithsonian Miscellaneous Collections; instructions in reference to collecting Nests and Eggs of North American Birds. 8vo. pp. 82. (No title-page.)

108.—1861. Report upon the Colorado River of the West, Explored in 1857 and 1858 by Lieutenant Joseph C. Ives, Corps of Topographical Engineers, under the direction of the Office of Explorations and Surveys, A. A. Humphreys, Captain Topographical Engineers, in charge. By order of the Secretary of War. Washington: Government Printing Office. 1861. 1 vol., 4to.

List of birds. pp. 5-6. (This part dated 1860.)

111.—1863. [Notice of R. Kennicott's and J. Xantus's Movements in North America.] <Ibis, v, 1863, pp. 238, 239.

112.—1863. [Letter on J. Xantus's Collections at Colima, Mexico.] <Ibis, v, 1863, p. 476.

114.—1863. Notes on the Birds of Jamaica. By W. T. March, with remarks by S. F. Baird. <Proc. Acad. Nat. Sci. Phila., 1863, i, pp. 150-154 (May); ii, pp. 283-304 (Nov.); iii, 1864, pp. 62-72.

115.—1864-66. Smithsonian Miscellaneous Collections. Review of American Birds, in the Museum of the Smithsonian Institution. By S. F. Baird. Part I. North and Middle America. Washington: Smithsonian Institution. [No date on title: June, 1864, to p. 33; July, 1864, to p. 81; Aug., 1864, to p. 129; Sept., 1864, to p. 145; Oct., 1864, to p. 161; Nov., 1864, to p. 177; Apr., 1865, to p. 241; May, 1865, to p. 321; May, 1865, to p. 417; June, 1865, to end.] 1 vol. 8vo. pp. iv, 450.

118, 119, 120.—1866. The Distribution and Migrations of North American Birds. By Spencer F. Baird, Assistant Secretary Smithsonian Institution. (Abstract of a memoir presented to the National Academy of Sciences, Jan., 1865.) $<$*Amer. Jour. Sci. and Arts* (2), xli, pp. 78–90, 184–92, 337–47.

121.—1866. Smithsonian Miscellaneous Collections—210.—Arrangement of Families of Birds. [Adopted provisionally by the Smithsonian Institution.] 8vo. pp. 8.

122.—1866. The origin of the domestic Turkey. $<$*Report of the Commissioner of Agriculture for* 1866, pp. 288–290.

123.—1866. Die Verbreitung und Wanderungen der Vögel Nord-Amerika's. $<$*Journal für Ornithologie*, xiv, 1866, pp. 244–269, 338–352.

124.—1867. The Distribution and Migrations of North American Birds. $<$*Ibis*, 1867, 2d ser., iii, pp. 257–293.

125.—1867. Note on the Pelican in "The Pelican in Cayuga County, N. Y., by W. J. Beal." $<$*American Naturalist*, i, pp. 323–4, Aug., 1867.

129.—1869. On additions to the Bird-Fauna of North América, made by the Scientific Corps of the Russo-American Telegraph Expedition. $<$*Trans. Chicago Acad.*, i, pt. ii, 1869, pp. 311–325, pll. xxvii–xxxiv.

130.—1870. Fossil Birds of the United States. $<$*Harper's New Monthly Mag.*, xl, 1870, pp. 467, 469, and 470.

131.—1871. New link between Reptiles and Birds. $<$*Harper's New Monthly Mag.*, xl, 1870, p. 628. · See, also, p. 469.

147.—1871. Variation of color in Birds with the locality $<$*Ann. Rec.*, 1871–72, pp. 190, 191.

173.—1871. Peculiarities of the Florida Wild Turkey. $<$*Ann. Rec.*, 1871–72, p. 257.

174.—1871. Existing specimens of the Great Auk. $<$*Ann. Rec.*, 1871–72, pp. 258, 259.

191.—1871. Allen on the Birds of East Florida. $<$*Ann. Rec.*, 1871–72, p. 2.

206.—1871. Oil from Birds. $<$*Ann. Rec.*, 1871–72, pp. 466, 467.

266.—1872. American Birds in Europe. $<$*Ann. Rec.*, 1872–73, p. 247.

301.—1872. Coues's Work on American Birds. $>$*Ann. Rec.*, 1872–73, p. 310.

02.—1872. New Ornithological Periodical. $<$*Ann. Rec.*, 1872–73, p. 311.

309.—1872. Maynard on the Birds of Florida. $<$*Ann. Rec.*, 1872–73, p. 320.

310.—1872. Allen on the Birds of Kansas, etc. $<$*Ann. Rec.*, 1872–73, p. 321.

311.—1872. Filaria in the Brain of the Water-Turkey. $<$*Ann. Rec.*, 1872–73, p. 324.

312.—1872. Use of the Bill of the Huia Bird. $<$*Ann. Rec.*, 1872–73, p. 324.

322.—1872. Carpal and Tarsal Bones of Birds. $<$*Ann. Rec.*, 1872–73, p. 336.

323.—1872. Coues on the Birds of the United States. $<$*Ann. Rec.*, 1872–73, p. 336.

482.—1873. A New Fossil Bird. $<$*Ann. Rec.*, 1873–75, p. 288.

515.—1873. Bird Collections in London. $<$*Ann. Rec.*, 1873–75, p. 336.

523.—1873. Number of American Birds. $<$*Ann. Rec.*, 1873–75, p. 340.

585.—1873. Oil from Birds. $<$*Ann. Rec.*, 1873–75, p. 566.

632½.—1874. A History of North American Birds. By S. F. Baird, T. M. Brewer, and R. Ridgway. Land Birds. Illustrated by 64 Colored Plates and 593 Woodcuts. Volume [1] [Cut] Boston. Little, Brown, and Company, 1874. Three volumes. 4to. pp.: Vol. I, xxviii, 596, vi; Vol. II, 590, iv; Vol. III, 560, xxviii.

652.—1874. The new Fossil Bird of the Sheppey Clay. $<$*Ann. Rec.* 1874–75, p. 305.

683.—1874. Dall on the Birds of Alaska. $<$*Ann. Rec.*, 1874–75, p. 340.

634.—1874. Lawrence's Birds of Northwestern Mexico. $<$*Ann. Rec.*, 1874–75, p. 341.

685.—1874. Collection of Birds of Paradise. $<$*Ann. Rec.*, 1874–75, p. 341.

686.—1874. Geographical Distribution of Asiatic Birds. $<$*Ann. Rec.*, 1874–75, p. 341.

687.—1874. Suggested Introduction of the Rook into the United States. $<$*Ann. Rec.*, 1874–75, p. 342.

663.—1874. Dr. Coues' Manual of Field Ornithology. <*Ann. Rec.*, 1874–75, p. 343.

689.—1874. Catalogue of American Birds. <*Ann. Rec.*, 1874–75, p. 343.

692.—1874. Dall's Catalogue of the Shells of Behring's Strait. <*Ann. Rec.*, 1874–75, p. 345.

703.—1874. Introduction of Prairie Chickens into the Eastern States. <*Ann. Rec.*, 1874–75, p. 391.

786.—1875. Is Sex Distinguishable in Egg-Shells ? <*Ann. Rec.*, 1875–76, p. 320.

790.—1875. Discovery in Newfoundland of the Great Auk. <*Ann. Rec.*, 1875–76, p. 339.

791.—1875. Habits of Kingfishers. <*Ann. Rec.*, 1875–76, p. 339.

792.—1875. Professor Alfred Newton on the Migration of Birds. <*Ann. Rec.*, 1875–76, p. 340.

808.—1875. Introduction of the American Turkey. <*Ann. Rec.*, 1875–76, p. 354.

882.—1876. Appendix K. List of Birds collected by Charles S. McCarthy, Taxidermist. Classified by Professor Spencer F. Baird. <*Simpson's Explorations across the Great Basin of Utah in* 1859, pp. 377–381.

905.—1876. Decrease of Birds in Massachusetts. <*Ann. Rec.*, 1875–77, p. 309.

906.—1876. Catalogues of all the Birds Known up to This Day. <*Ann. Rec.*, 1876–77, p. 310.

907.—1876. Domesticating the Prairie Chicken. <*Ann. Rec.*, 1876–77, p. 310.

908.—1876. Additional Remains of the Moa. <*Ann. Rec.*, 1876–77, p. 311.

909.—1876. The Migration of Birds. <*Ann. Rec.*, 1876–77, p. 311.

910.—1876. Addition to North American Ornithology—Pyrrhophæna. <*Ann. Rec.*, 1876–77, p. 312.

911.—1876. The Habits of Birds. <*Ann. Rec.*, 1876–77, p. 312.

912.—1876. New Fossil Giant Birds. <*Ann. Rec.* 1876–77, p. 313.

REPTILES AND BATRACHIANS.

10.—1849. Revision of the North American Tailed-Batrachia, with descriptions of new genera and Species. <*Journ. Acad. Nat. Sci. Phila.*, 2d ser., i, pp. 281–294, Oct., 1849.

11.—1849. Descriptions of four new species of North American Salamanders, and one new species of Scink. <*Jour. Acad. Nat. Sci. Phila.*, 2d ser., i, Oct., 1849, pp. 282–294.

12.—1850. Descriptions of four new species of North American Salamanders, and one new species of Scink. <*Amer. Journ. Sci. and Arts*, ix, 2d ser., Jan., 1850, pp. 137–9.

14.—1850· On the Urodelian Batrachians. <*Proc. Amer. Assoc. Adv. Sci.*, ii, 1850, p. 402.

21.—1851. Article "Reptiles" in "Outlines of General Zoology". Reprinted from the Iconographic Encyclopædia of Science, Literature, and Art. (New York: Rudolph Garrigue, publisher,) 1851. [8vo. pp. xxii, 502, xvi].

27.—1852. (With CHARLES GIRARD.) Characteristics of some New Reptiles in the Museum of the Smithsonian Institution. (First Part.) <*Proc. Acad. Nat. Sci. Phila.*, vi, 1852–3 (1854). pp. 68–70.

33.—1852. (With CHARLES GIRARD.) Reptiles [of the Valley of the Great Salt Lake]. <*Stansbury's Exploration and Survey of the Valley of the Great Salt Lake.* Philadelphia, 1852.

35.—1852. (With CHARLES GIRARD.) Characteristics of some New Reptiles in the Museum of the Smithsonian Institution. Second Part. Containing the species of the Saurian order, collected by John H. Clark, under Col. J. D. Graham, head of the Scientific Corps, U. S. and Mexican Boundary Commission, and a few others from the same adjoining territories, obtained from other sources, and mentioned under their special headings. <*Proc. Acad. Nat. Sci. Phila.*, vi, pp. 125–9.

36.—1852. (With CHARLES GIRARD.) Characteristics of some New Reptiles in the Museum of the Smithsonian Institution. By Spencer F. Baird and Charles Girard. Third Part. Containing the Batrachians in the collection made by J. H. Clark, esq., under Col. J. D. Graham, on the United States and Mexican Boundary. <*Proc. Acad. Nat. Sci. Phila.*, vi, p. 173.

37.—1852. (With CHARLES GIRARD.) Descriptions of New Species of Reptiles, collected by the U. S. Exploring Expedition under the command of Capt. Charles Wilkes, U. S. N. First Part.—Including the species from the Western Coast of America. <*Proc. Acad. Nat. Sci. Phila.*, vi, pp. 174–177.

38.—1853. (With CHARLES GIRARD.) List of Reptiles collected in California by Dr. John L. Leconte, with description of New Species. <*Proc. Acad. Nat. Sci. Phila.*, vi, pp. 300–302.

39.—1853. (With CHARLES GIRARD.) Catalogue of North American Reptiles in the Museum of the Smithsonian Institution. Part I.—Serpents. By S. F. Baird and C. Girard. Washington: Smithsonian Institution. January, 1853. *January* 1, 1853. 8vo. pp. xvi, 172.

44.—1853. (With CHARLES GIRARD.) "A communication * * * upon a species of frog, and another of toad * * * recently described from specimens in the Herpetological Collections of the U. S. Exploring Expedition." <*Proc. Acad. Nat. Sci. Phila.*, vi, 1853, pp. 378–379.

49.—1853. (With CHARLES GIRARD.) Reptiles [of the Red River Region]. <*Marcy and McClellan's Exploration of the Red River of Louisiana in the year* 1852. Washington, 1853. (Appendix F.) 8vo. pp. 217–244.

49½.—1854. On the Serpents of New York; with a notice of a species not hitherto included in the fauna of the State. By Spencer F. Baird. Albany: C. Van Benthuysen, Printer. 1854. 8vo. pp. (2)–28.

55.—1854. Descriptions of New Genera and Species of North American Frogs. <*Proc. Acad. Nat. Sci. Phila.*, vii, pp. 59–62.

74½.—1858. Description of New Genera and Species of North American Lizards in the Museum of the Smithsonian Institution. <*Proc. Acad. Nat. Sci. Phila.*, Dec., 1858, pp. 253–6.

88.—1859. Reptiles of the Boundary, by Spencer F. Baird, Assistant Secretary of the Smithsonian Institution, with notes by the Naturalists of the Survey. <*United States and Mexican Boundary Survey, Executive Documents*, 1st Session 34th Congress, vol. 14, part 3, 1855–'56, Doc. No. 135.

92.—1859. Exploration and Surveys for a Railroad Route from the Mississippi River to the Pacific Ocean. War Department. Reptiles: By Spencer F. Baird, Assistant Secretary of the Smithsonian Institution, Washington, D. C., 1859. Reports of explorations and surveys to ascertain the most practicable and economical route for a railroad from the Mississippi River to the Pacific Ocean. Volume X. 1859. Parts iii and iv. 4to. First article. No text, pll. xxiv–xxvi.

95.—1859. Report on Reptiles collected on the Survey. No. 3. <Exploration and Surveys for a Railroad Route from the Mississippi River to the Pacific Ocean. War Department. Report of Lieut. E. G. Beckwith, Third Artillery, upon Explorations for a Railroad Route, near the 38th and 39th parallels of north latitude, by Captain J. W. Gunnison, Corps of Topographical Engineers, and near the forty-first parallel of north latitude, by Lieut. E. G. Beckwith, Third Artillery. 1854. Pll. xvii, xviii, xxiii, xxiv, in Rep. P. R. R. Surv., vol. x.

96.—1859. Report upon the Reptiles of the Route. No. 4. <Explorations and Surveys for a Railroad Route from the Mississippi River to the Pacific Ocean. War Department. Route near the Thirty-fifth Parallel, ex-

96.—1859. Report upon the Reptiles of the Route—Continued.
plored by Lieutenant A. W. Whipple, Topographical Engineers, in
1853 and 1854. Zoological Report. Washington, D. C., 1859. Pll.
xxv, xxvi, xxvii. [In vol. x, Rep. P. R. R. Surv., part vi, No. 4, pp.
37–45.]

100.—1859. Description of New Genera and Species of North American Lizards in
the Museum of the Smithsonian Institution. <*Proc. Acad. Nat. Sci.
Phila.*, x, 1858 (1859), pp. 253–256.

172.—1871. Immunity of Pig from Injury by Serpent-bite. <*Ann. Rec.*, 1871–72.
pp. 255–256.

175.—1871. Relation of weight to length in Crocodiles and Alligators. <*Ann. Rec.*
1871–72, p. 259.

208.—1871.—Fayrer on Snake Bites. <*Ann. Rec.*, 1871–72, p. 577.

267.—1872.—Antagonism of Harmless Serpents to Poisonous ones. <*Ann. Rec.*,
1872–73, p. 255·

268.—1872. Blood from the Eye of the Horned Toad. <*Ann. Rec.*, 1872–73, p. 256.

269.—1872. Horned Frogs Viviparous. <*Ann. Rec.*, 1872–73, p. 258.

291.—1872. A Sea-Serpent in Highland Loch. <*Ann. Rec.*, 1872–73, p. 296.

321.—1872. Enumerations of American Serpents. <*Ann. Rec.*, 1872–73, p. 334.

483.—1873. Development of a Guadeloupe Frog. <*Ann. Rec.*, 1873–75, p. 290.

503.—1873. Pterodactyl in the Cambridge Museum. <*Ann. Rec.*, 1873–75, p. 316.

642.—1874.—New Experiments on the Venom of East Indian Serpents. <*Ann.
Rec.*, 1874–75, p. 287.

653.—1871. Affinities of Heloderma horridum. <*Ann. Rec.*, 1874–75, p. 308.

654.—1874. Occurrence of a Cuban Crocodile in Florida. <*Ann. Rec.*, 1874–75,
p. 308.

696.—1874. Respiration in the Amphibia. <*Ann. Rec.*, 1874–75, p. 350·

793.—1875. The Batrachia and Reptilia of North America. <*Ann. Rec.*, 1875–76,
p. 343.

913.—1876. Reptiles of Costa Rica. <*Ann. Rec.*, 1876–77, p. 315.

914.—1876. Snake-Eating Snakes. <*Ann. Rec.*, 1876–77, p. 216.

915.—1876. Remarkable Habit of Frogs. <*Ann. Rec.*, 1876–77, p. 316.

916.—1876. Reproduction in the Proteus. <*Ann. Rec.*, 1876–78, p. 317.

1043.—1880. Do Black Snakes Eat Fish. <*Forest and Stream*, xiii, 1880, p. 966.

FISHES, FISHERIES, AND FISH CULTURE.

7·—1846. The Sea-Serpent in Norway. <*Lit. Rec. and Journ. Linnæan Assoc.
Pennsylvania College*, ii, 1846, pp. 106–107.

21.—1851. Article "Fishes" in "Outlines of General Zoology". Reprinted from
the Iconographic Encyclopædia of Science, Literature and Art,
(New York: Rudolph Garrigue, publisher). 1851. [8vo. pp. xxii,
502, xvi.]

43.—1853. (With CHARLES GIRARD.) Descriptions of some new fishes from the
River Zuni. <*Proc. Acad. Nat. Sci. Phila.*, vi, 1853, pp. 368, 339.

45.—1853. (With CHARLES GIRARD.) Descriptions of New Species of Fishes col-
lected by Mr. John H. Clark, on the U. S. and Mexican Boundary
Survey, under Lt. Col. Jas. D. Graham. <*Proc. Acad. Nat. Sci.
Phila.*, vi, pp. 387–390.

46.—1853. (With CHARLES GIRARD.) Description of New Species of Fishes collected by Captains R. B. Marcy and Geo. B. McClellan in Arkansas.
<*Proc. Acad. Nat. Sci. Phila.*, vi, pp. 390–392.

47.—1853. (With CHARLES GIRARD.) Fishes [of the Zuñi River]. <*Sitgreaves'
Report of an Expedition down the Zuñi and Colorado Rivers*. Washington, 1853. pp. 148–152.

50.—1853. (With CHARLES GIRARD.) Fishes [of the Red River Region]. <*Marcy and McClellan's Exploration of the Red River of Louisiana in the year 1852. Washington, 1853. [Appendix F.] pp. 245–252.

51.—1854. (With CHARLES GIRARD.) Descriptions of new species of Fishes collected in Texas, New Mexico, and Sonora, by Mr. John H. Clark, on the U. S. and Mexican Boundary Survey, and in Texas by Capt. Stewart Van Vliet, U. S. A. Second Part. <*Proc. Acad. Nat. Sci. Phila., vii, 1854, pp. 24–29.

52.—1854. (With CHARLES GIRARD.) Notice of a new genus of Cyprinidæ. <*Proc. Acad. Nat. Sci. Phila., vii, 1854, p. 158.

54.—1854. (With CHARLES GIRARD.) Descriptions of new species of Fishes collected in Texas, New Mexico, and Sonora, by Mr. John H. Clark, on the U. S. and Mexican Boundary Survey, and in Texas by Capt. Stewart Van Vliet, U. S. A. Second Part. <*Proc. Acad. Nat. Sci. Phila., vii, 1854–55, pp. 24–29.

57.—1854. (With CHARLES GIRARD.) [Cyprinidæ of Heerman's Collection.] <*Girard's Descriptions of New Fishes collected by Dr. A. S. Heerman. (Proc. Acad. Nat. Sci. Phila., 1854, pp. 129–156)=(pp. 135–138.)

58.—1854. (With CHARLES GIRARD.) Notice of a new genus of Cyprinidæ. <*Proc. Acad. Nat. Sci. Phila., vii, 1854, p. 158.

63.—1855. Report on the Fishes observed on the coasts of New Jersey and Long Island during the summer of 1854, by Spencer F. Baird, Assistant Secretary of the Smithsonian Institution. <*Ninth Annual Report of the Smithsonian Institution (for 1854), 1855, pp. 317 325+337.

67.—1856. List of Fishes inhabiting the State of New York, sent to the New York State Cabinet of Natural History by the Smithsonian Institution in May, 1855. <*Ninth Annual Report of the Regents of the University of the State of New York on the Condition of the State Cabinet of Natural History, and the Historian and Antiquarian Collection annexed thereto. 1856, pp. 22–29.

70.—1856. [A description of the genus Ceratichthys.] <*Proc. Acad. Nat. Sci. Phila., 1856, viii, p. 212.

77.—1857. Catalogue of Fishes copied from a "Report on the Fishes observed on the coasts of New Jersey and Long Island during the summer of 1854, by Spencer F. Baird, Assistant Secretary of the Smithsonian Institution." From the Ninth Annual Report of the Smithsonian Institution for 1854. <*Catalogue of Zoological and Botanical Productions of the County of Cape May, in Geology of the County of Cape May, State of New Jersey, 1857, pp. 146–149.

127.—1868. The Basking Shark—The "Great Sea Monster." <*American Agriculturist, xxvii, 1868, p. 130.

132.—1871. Spawning of the Goose Fish (Lophius Americanus.) <*American Naturalist, v, 1871, p. 785.

133.—1871. See below under special list of publications issued as Commissioner of Fisheries.

148.—1871. Catalogue of Fishes in the British Museum. <*Ann. Rec., 1871–72, p. 206.

149.—1871. British Museum Fishes. <*Ann. Rec., 1871–72, p. 206.

150.—1871. Spawning of Herring. <*Ann. Rec., 1871–72, p. 207.

151.—1871. The Food of the Sea-Herring. <*Ann. Rec., 1871–72, pp. 208–210.

152.—1871. Fishes of Cuba. <*Ann. Rec., 1871–72, p. 212.

153.—1871. Phosphorescence of Dead Fish. <*Ann. Rec., 1871–72, p. 211.

154.—1871. Fresh-Water Fishes of Algeria. <*Ann. Rec., 1871–72, p. 211.

155.—1871. Confusion of names of Fishes. <*Ann. Rec., 1871–72, p. 207.

156.—1871. Tame Codfish. <*Ann. Rec., 1871–72, p. 212.

157.—1871. Teeth of the Sturgeon. <*Ann. Rec., 1871–72, p. 213.

Development of the Lamprey. <*Ann. Rec.*, 1871–72, p. 213.
Lütken on Ganoid Fishes. <*Ann. Rec.*, 1871–72, p. 214.
Gourami Fish. <*Ann. Rec.*, 1871–72, p. 214.
A new Lophioid Fish. <*Ann. Rec.*, 1871–72, p. 214.
Peculiarities of Salmon Kelts. <*Ann. Rec.*, 1871–72, p. 215
Salmon-Fishing in Loch Tay. <*Ann. Rec.*, 1871–72, p. 215.
"Landlocked Salmon." <*Ann. Rec.*, 1871–72, p. 216.
Food for Young Trout. <*Ann. Rec.*, 1871–72, p. 217.
Tailless Trout in Scotland. <*Ann. Rec.*, 1871–72, p. 217.
Fossil Fishes of Wyoming. <*Ann. Rec.*, 1871–72, p. 248.
Cephalaspis in America. <*Ann. Rec.*, 1871–72, p. 248.
Cod-fisheries of Alaska. <*Ann. Rec.*, 1871–72, p. 259.
Occurrence of the Pompano (and Spanish Mackerel) northward. <*Ann. Rec.*, 1871–72, p. 260.
Increase of Salmon in the British Provinces. <*Ann. Rec.*, 1871–72, p. 260.
Use of the pectoral fins of fish. ∧*Ann Rec.*, 1871–72, p. 261.
Relations of Ganoids to Plagiostomes. <*Ann. Rec.*, 1871–72, pp. 261–263.
Theory of the Salmon Fly. <*Ann. Rec.*, 1871–72, p. 263.
Capture of Horse Mackerel in Buzzard's Bay. <*Ann. Rec.*, 1871–72, p. 363.
Did Hendrik Hudson find salmon in the Hudson River? <*Ann. Rec.*, 1871–72, p. 264.
Black Bass in the Potomac. <*Ann. Rec.*, 1871–72, p. 264.
Cause of Death of Fresh-water Fish in Salt Water. <*Ann. Rec.*, 1871–72, p. 265.
Proper Fish for Stocking Rivers. <*Ann. Rec.*, 1871–72, p. 265.
Living Eyeless Fish. <*Ann. Rec.*, 1871–72, p. 266.
Stocking waters of New York with Fish. <*Ann. Rec.*, 1871–72, p. 266.
Killing Fish with Torpedoes in Florida. <*Ann. Rec.*, 1871–72, p. 267.
Fungus growth on Fish and their Eggs. <*Ann. Rec.*, 1871–72, p. 267.
Schools of Young Bluefish. <*Ann. Rec.*, 1871–72, p. 278.
Fish-Guano Flour from Loffodon. <*Ann. Rec.*, 1871–72, pp. 342, 343.
Report of the Connecticut Fish Commission. <*Ann. Rec.*, 1871–72, p. 348. Review.
Nutrition of Young Fish in Hatching Establishments. <*Ann. Rec.*, 1871–72, p. 350.
Irish Oyster Fisheries. <*Ann. Rec.*, 1871–72, pp. 352, 353.
Artificial Ice Packing Fish. <*Ann. Rec.*, 1871–1872, p. 355.
Preservation of Dead Salmon for an indefinite time. <*Ann. Rec.*, 1871–72, p. 356.
Importance of Killing freshly-captured Fish. <*Ann. Rec.*, 1871–72. p. 387.
Pegging Lobster Claws. <*Ann. Rec.*, 1871–72, pp. 552, 553.
Commissioner of Fisheries. <*Ann. Rec.*, 1871–72, pp. 605, 606.
Fishing Steamer. <*Ann. Rec.*, 1871–72, p. 606. Notice of a steamer devised in England for sea-fishing.
See below, in List of publications issued as Commissioner of Fisheries.
See below, in List or publications issued as Commissioner of Fisheries.
See below, in List of publications issued as Commissioner of Fisheries.
Professor Agassiz's Prophecies. <*Ann. Rec.* 1872–73, p. 215.
Cope on the Fossil Fish of the Kansas Cretaceous. <*Ann. Rec.*, 1872–73, p. 258.
Nest-building Fish. <*Ann. Rec.*, 1872–73, p. 259.
Another Pelagic Fish-Nest. <*Ann. Rec.*, 1872–73, p. 260.

BD

273.—1872. Respiration in Fish. <*Ann. Rec.*, 1872–73, p. 261.

274.—1872. Genesis of Hippocampus. <*Ann. Rec.*, 1872–73, p. 261.

275.—1872. Chinese Cyprinidæ. <*Ann. Rec.*, 1872–73, p. 262.

276.—1872. Death of an Aged Carp. <*Ann. Rec.*, 1872–73, p. 262.

277.—1872. Salmon Fly-fishing on the Northwest Coast of America. <*Ann. Rec.*, 1872–73, p. 262.

278.—1872. Venomous Fish in the Mauritius. <*Ann. Rec.*, 1872–73, p. 263.

279.—1872. Teeth in Young Sturgeons. <*Ann. Rec.*, 1872–73, p. 264.

280.—1872. Monster Cod. <*Ann. Rec.*, 1872–73, p. 264.

281.—1872. Stones in the Stomachs of Codfish. <*Ann. Rec.*, 1872–73, p. 264.

282.—1872. Bluefish on the Southern Coast. <*Ann. Rec.*, 1872–73, p. 265.

288.—1872. Parasites and Commensals of Fish. <*Ann. Rec.*, 1872–73, p. 272.

289.—1872. Worms in the Trout of Yellowstone Lake. <*Ann. Rec.*, 1872–73, pp. 284, 275.

291.—1872. A Sea-Serpent in a Highland Loch. <*Ann. Rec.*, 1872–73, p. 296.

292.—1872. Generation of Eels. <*Ann. Rec.*, 1872–73, p. 299.

293.—1872. Alleged Gigantic Pike. <*Ann. Rec.*, 1872–73, p. 301.

294.—1872. Nature of the Blue Coloring Matter in Fishes. <*Ann. Rec.*, 1872–73, p. 302.

308.—1872. Change of Color in Fishes. <*Ann. Rec.*, 1872–73, p. 319.

329.—1872. R. D. Cutts on Sea Fisheries. <*Ann. Rec.*, 1872–73, p. 396.

330.—1872. French Fisheries for 1870. <*Ann. Rec.*, 1872–73, p. 397.

331.—1872. Comparison of American and French Fisheries. <*Ann. Rec.*, 1872–73, p. 398.

332.—1872. German Fishery Association. <*Ann. Rec.*, 1872–73, p. 399.

333.—1872. Fisheries of the Gulf of Naples. <*Ann. Rec.*, 1872–73, p. 399.

334.—1872. French Fish-Breeding Establishment. <*Ann. Rec.*, 1872–73, p. 399.

335.—1872. Fishery Exposition at Gothenburg (Sweden). <*Ann. Rec.*, 1872–73, p. 400.

336.—1872. Fish-Culturists' Association at Albany. <*Ann. Rec.*, 1872–73, p. 400.

337.—1872. U. S. Appropriation for the Propagation of Fish. <*Ann. Rec.*, 1872–73, p. 401.

338.—1872. Report of Maine Fish Commissioners for 1871. <*Ann. Rec.*, 1872–73, p. 403.

339.—1872. Fish Culture in New Hampshire. <*Ann. Rec.*, 1872–73, p. 405.

340.—1872. Prizes of the Massachusetts Agricultural Society for Fish-Culture. <*Ann. Rec.*, 1872–73, p. 406.

341.—1872. Alabama Fish Commissioners. <*Ann. Rec.*, 1872–73, p. 406.

342.—1872. Fish-Culture in California. <*Ann. Rec.*, 1872–73, p. 407.

343.—1872. Report of California Fish Commissioners. <*Ann. Rec.*, 1872–73, p. 408.

344.—1872. Stocking California Waters with Trout. <*Ann. Rec.*, 1872–73, p. 409.

345.—1872. Transporting Black Bass to California. <*Ann. Rec.*, 1872–73, p. 409.

346.—1872. Report of Connecticut Fish Commissioners for 1771. <*Ann. Rec.*, 1872–73, p. 409.

347.—1872. Planting of Shad in the Valley of the Mississippi and the Lakes. <*Ann. Rec.*, 1872–73, p. 410.

348.—1872. Report of the New York Fish Commissioners for 1871. <*Ann. Rec.* 1872–72, p. 412.

349.—1872. Second Report of New Jersey Fish Commissioners. <*Ann. Rec.*, 1872–73, p. 413.

350.—1872. Transportation of Black Bass to England. <*Ann. Rec.*, 1872–73, p. 414.

351.—1872. Fisheries of the North Carolina Coast. <*Ann. Rec.*, 1872–73, p. 414.

352.—1872. Consumption of Bluefish in New York. <*Ann. Rec.*, 1872–72, p. 414.

353.—1872. Breeding Salmon and Trout in Inclosures. <*Ann. Rec.*, 1872–73, p. 415.

354.—1872. Capture of Rhine Salmon in Holland. <*Ann. Rec.*, 1872–73, p. 116.

355.—1872. Cost of Salmon Eggs in Europe. $<$*Ann. Rec.*, 1872–73, p. 418.

356.—1872. Artificial Breeding of Salmon. $<$*Ann. Rec.*, 1872–73, p. 418.

357.—1872. Do Salmon need to Reside in Salt Water? $<$*Ann. Rec.*, 1872–73, p. 418.

358.—1872. Profitable Result of Salmon Planting in Germany. $<$*Ann. Rec.*, 1872–73, p. 419.

359.—1872. Trout-Breeding in France. $<$*Ann. Rec.*, 1872–73, p. 420.

360.—1872. Renewal of Salmon-Planting in the Delaware. $<$*Ann. Rec.*, 1882–73, p. 421.

361.—1872. Salmon and Trout in Australia. $<$*Ann. Rec.*, 1872–73, p. 422.

362.—1872. Spawning of Herring. $<$*Ann. Rec.*, 1872–73, p. 422.

363.—1872. Breeding of Smelt in Europe. $<$*Ann. Rec.*, 1872–73, p. 423.

364.—1872. Raising Otsego Bass. $<$*Ann. Rec.*, 1872–73, p. 423.

365.—1872. Fisheries on the Coast of Norway. $<$*Ann. Rec.*, 1872–73, p. 423.

366.—1872. Loffoden Codfishery. $<$*Ann. Rec.*, 1872–73, p. 424.

367.—1872. Spawning of Menhaden. $<$*Ann. Rec.*, 1872–73, p. 425.

368.—1872. Herring-Fishery in Great Britain. $<$*Ann. Rec.*, 1872–73, p. 425.

369.—1872. Reappearance of a Peculiar Herring on the Norway Coast. $<$*Ann. Rec.*, 1872–73, p. 426.

370.—1872. Use of Fishes as Manure in England. $<$*Ann. Rec.*, 1872–73, p. 426.

371.—1872. Food of Shad. $<$*Ann. Rec.*, 1872–73, p. 426.

372.—1872. Bryan on the Decrease of Shad. $<$*Ann. Rec.*, 1872–73, p. 427.

373.—1872. Shad in Alabama. $<$*Ann. Rec.*, 1872–73, p. 427.

374.—1872. Shad in Red River, Arkansas. $<$*Ann. Rec.*, 1872–73, p. 428.

375.—1872. Shad Hatching in the Hudson River. $<$*Ann. Rec.*, 1872–73, p. 428.

376.—1872. Planting of Shad in the Genesee River. $<$*Ann. Rec.*, 1872–73, p. 429.

377.—1872. Planting of Shad in Lake Champlain. $<$*Ann. Rec.*, 1872–73, p. 429.

378.—1872. Stocking California with Shad. $<$*Ann. Rec.*; 1872–73, p. 430.

379.—1872. Transferring Shad to the Sacramento River. $<$*Ann. Rec.*, 1872–73, p. 430.

380.—1872. *Cyprinus orfus* as an Ornamental and Food Fish. $<$*Ann. Rec.*, 1872–73, p. 431.

381.—1872. Prize Essay on the Reproduction of Eels. $<$*Ann. Rec.*, 1872–73, p. 434.

382.—1872. Marking Whitefish. $<$*Ann. Rec.*, 1872–73, p. 435.

383.—1872. Catch of Fur Seals in 1872. $<$*Ann. Rec.*, 1872–73, p. 435.

384.—1872. Oil Works on Unalaschka. $<$*Ann. Rec.*, 1872–73, p. 436.

385.—1872. Spawning of Codfish in Alaska. $<$ *Ann. Rec.*, 1872–73, p. 436.

386.—1872. Codfishing in the Shumagin Islands. $<$*Ann. Rec.*, 1872–73, p. 436.

387.—1872. American Whale Fishing in 1871. $<$*Ann. Rec.*, 1872–73, p. 473.

388.—1872. Salmon Fisheries in the Columbia River. $<$*Ann. Rec.*, 1872–73, p. 440.

389.—1872. Capture of Sacramento Salmon with the Hook. $<$*Ann. Rec.*, 1872–73, p 441.

390.—1872. Breeding of Leeches. $<$*Ann. Rec.*, 1872–73, p. 441.

391.—1872. Spawning of the Sterlet. $<$*Ann. Rec.*, 1872–73, p. 442.

392.—1872. Tunny Fisheries on the South Shore of the Mediterranean. $<$*Ann. Rec.*, 1872–73, p. 442.

393.—1872. Fishing Statistics of Great Britain for 1859. $<$*Ann. Rec.*, 1872–73, p. 443.

394.—1872. Fisheries of the Shumagin Islands. $<$*Ann. Rec.*, 1872–73, p. 444.

395.—1872. Utilization of Refuse Fish. $<$*Ann. Rec.*, 1872–73, p. 444.

396.—1872. Peculiarities of Reproduction of California Salmon. $<$*Ann. Rec.*, 1872–73, p. 445.

397.—1872. Winter-Quarters of Nova Scotia Salmon. $<$*Ann. Rec.*, 1872–73, p. 446.

398.—1872. Growth of Salmon. $<$*Ann. Rec.*, 1872–73, p. 446.

399.—1872. Best kind of Water for Salmon Hatching. $<$*Ann. Rec.*, 1872–73, p. 446.

400.—1872. Alleged Discovery of Young Shad in the Sacramento River. $<$*Ann. Rec.*, 1872–73, p. 447.

401.—1872. Report of Fish Commissioners of Vermont (for 1871-72). <*Ann. Rec.,* 1872-73, p. 447.

418.—1873. A bill to regulate the use of Stationary Apparatus in the Capture of Fish fol. bill form. 6 pp.

419.—1873. Statistics of the Menhaden Fisheries, etc. [Questions addressed to fishermen, etc.—Washington, December 20, 1873.] 4to, letter form, 2 l. [U. S. F. C., 5.]

419.—1873. See below, in special list of publications issued as Commission of Fisheries.

421.—1873. See below, in special list of publications issued as Commission of Fisheries.

422.—1873. See below, in special list of publications issued as Commission of Fisheries.

423.—1873. Memoranda of inquiry relative to the Food-Fishes of the United States. <*Rep. U. S. Comm. Fish and Fisheries,* part i, 1873, pp. 1-3.

424.—1873. Questions relative to the Food-Fishes of the United States. <*Rep. U. S. Comm. Fish and Fisheries,* part i, 1873, pp. 1-3.

425.—1873. Testimony in regard to the present condition of the Fisheries, taken in 1871. <*Rep. U. S. Comm. Fish and Fisheries,* part i, 1873, pp 7-72.

426.—1873. Report of conference held by the U. S. Commissioner of Fish and Fisheries at Boston, October 5, 1871, with the Fishery Commissioners of Massachusetts and Rhode Island. <*Rep. U. S. Comm. Fish and Fisheries,* 1873, part i, pp. 125-131.

427.—1873. Draught of law proposed for the consideration of and enactment by the Legislatures of Massachusetts, Rhode Island, and Connecticut. <*Rep. U. S. Comm. Fish and Fisheries,* part i, 1873, pp. 132-134.

428.—1873. Notices in regard to the Abundance of Fish on the New England Coast in former times. <*Rep. U. S. Comm. Fish and Fisheries,* part i, 1873, pp. 148-172.

429.—1873. Statistics of Fish and Fisheries on the South Shore of New England. <*Rep. U. S. Comm. Fish and Fisheries,* part i, 1873, pp. 172-181.

430.—1873. Supplementary Testimony and Information relative to the Condition of the Fisheries of the South Side of New England, taken in 1872. <*Rep. U. S. Comm. Fish and Fisheries,* part i, 1873, pp. 182-195.

431.—1873. Natural History of some of the more important Food-fishes of the South Shore of New England.—I. The Scup, *Stenotomus argyrops* (Linn.), Gill. <*Rep. U. S. Comm. Fish and Fisheries,* part i, 1873, pp. 228-235.

432.—1873. [Natural History of some of the more important Food-fishes of the South Shore of New England.]—II. Blue-Fish, *Pomatomus saltatrix* (Linn.), Gill. <*Rep. U. S. Comm. Fish and Fisheries,* part i, 1873, pp. 235-252.

433.—1873. Description of Apparatus used in capturing Fish on the Sea-coast and Lakes of the United States. <*Rep. U. S. Comm. Fish and Fisheries,* part i, 1873, pp. 253-274.

434.—1873. List of Patents granted by the United States to the end of 1872, for inventions connected with the Capture, Utilization, or Cultivation of Fishes and Marine Invertebrates. <*Rep. U. S. Comm. Fish and Fisheries,* part i, 1873, pp. 275-280.

435.—1873. List of Fishes collected at Wood's Hole. <*Rep. U. S. Comm. Fish and Fisheries,* part i, 1873, pp. 823-827.

436.—1873. Salmon in the Hudson. <*Forest and Stream,* i, 1873, p. 233.

437.—1873. Planting California Salmon at Fort Edward. <*Forest and Stream,* i, 1873, p. 298.

438.—1873. The New England Fisheries. <*American Sportsman,* iii, 1873, p. 23.

439.—1873. Signal Telegraphy and the Herring Fishery. <*Ann. Rec.,* 1873-75, p. 73.*

472.—1873. Opening of the "Anderson School of Natural History." <*Ann. Rec.*, 1873–75, p. 266.

473.—1873. The Brighton Aquarium. <*Ann. Rec.*, 1873–75, p. 267.

484.—1873. Geographical Distribution of Percoid Fishes. <*Ann. Rec.*, 1873–75, p. 291.

487.—1873. Blood Corpuscles of the Salmonidæ. <*Ann. Rec.*, 1873–75, p. 299.

490.—1873. Respiration in Fishes at Different Ages. <*Ann. Rec.*, 1873–75, p. 304.

491.—1873. Absence of Fish above the Yosemite Falls. <*Ann. Rec.*, 1873–75, p. 305.

492.—1873. Reproduction of the Eel. <*Ann. Rec.*, 1873–75, p. 306.

504.—1873. A Large Fish. <*Ann. Rec.*, 1873–75, p. 317.

508.—1873. Habits of the Black Bass. <*Ann. Rec.*, 1873–75, p. 322.

509.—1873. Food of the Basking Shark. <*Ann. Rec.*, 1873–75, p. 328.

516.—1873. Brighton Aquarium. <*Ann. Rec.*, 1873–75, p. 336.

517.—1873. New Scaphirhynchus in Turkestan. <*Ann. Rec.*, 1873–75, p. 336.

518.—1873. An Aquarium for Central Park. <*Ann. Rec.*, 1873–75, p. 337.

525.—1873. Curious Fish. <*Ann. Rec.*, 1873–75, p. 342.

529.—1873. Alleged Shower of Fish-Scales. <*Ann. Rec.*, 1873–75, p. 350.

533.—1873. Fish Guano. <*Ann. Rec.*, 1873–75, p. 387.

534.—1873. Value of Sea-Weed Manure. <*Ann. Rec.*, 1873–75, p. 395.

537.—1873. Statistics of Canada Fisheries for 1859. <*Ann. Rec.*, 1873–75, p. 427.

538.—1873. British Exhibition of Fishing Products at Vienna. •<*Ann. Rec.*, 1873–75, p. 427.

539.—1873. Exhibition of Fishery Products at Vienna. <*Ann. Rec.*, 1873–75, p. 429.

540.—1873. Fishery Models at the late Scandinavian Exhibition. <*Ann. Rec.*, 1873–75, p. 429.

541.—1873. Is Seal Oil Fish Oil? <*Ann. Rec.*, 1873–75, p. 430.

542.—1873. Gloucester Winter Herring Fishery. <*Ann. Rec.*, 1873–75, p. 431.

543.—1873. Emden Herring-Fishery for 1872. <*Ann. Rec.*, 1873–75, p. 431.

544.—1873. Trade in Frozen Herring. <*Ann. Rec.*, 1873–75, p. 432.

545.—1873. Improvement in Value of the British Salmon Fisheries. <*Ann. Rec.*, 1873–75, p. 433.

546.—1873. Fishery Laws in Germany. <*Ann. Rec.*, 1873–75, p. 433.

547.—1873. Shipments Eastward of California Salmon. <*Ann. Rec.*, 1873–75, p. 433.

548.—1873. Meeting of the American Fish-Culturists' Association. <*Ann. Rec.*, 1873–75, p. 434.

549.—1873. Culture of Sea-Fish in Fresh-Water. <*Ann. Rec.*, 1873–75, p. 435.

550.—1873. Sixth Report of the Maine Commissioners of Fisheries for 1872. <*Ann. Rec.*, 1873–75, p. 436.

551.—1873. Report of the Fish Commission of Rhode Island for 1872. <*Ann. Rec.*, 1873–75, p. 437.

552.—1873. Report of the Fish Commission of New York for 1872. <*Ann. Rec.*, 1873–75, p. 438.

553.—1873. Ohio Fish Commission. <*Ann. Rec.*, 1873–75, p. 441.

554.—1873. Michigan Fishery Bill. <*Ann. Rec.*, 1873–75, p. 441.

555.—1873. Hybrids of Salmon and Trout. <*Ann. Rec.*, 1873–75, p. 442.

556.—1873. Cultivation of Fish in Ditches and Ponds. <*Ann. Rec.*, 1873–75, p. 443.

557.—1873. United States Salmon-Breeding Establishment at Bucksport, Maine. <*Ann. Rec.*, 1873–75, p. 443.

558.—1873. Marked Salmon on the American coast. <*Ann. Rec.*, 1873–75, p. 444.

559.—1873. Transporting Salmon Eggs to New Zealand. <*Ann. Rec.*, 1873–75, p. 445.

560.—1873. Naturalization of Trout in New Zealand. <*Ann. Rec.*, 1873–75, p. 447.

561.—1873. Food for Diminutive Trout. <*Ann. Rec.*, 1873–75, p. 447.

562.—1873. Alleged occurrence of Shad in the Mississippi. <*Ann. Rec.*, 1873–75, p. 448.

563.—1873. Increase in the Growth of Trout. <*Ann. Rec.*, 1873–75, p. 448.

564.—1873. Shad in the Sacramento River. <*Ann. Rec.*, 1873–75, p. 449.

565.—1873. Shad in California Waters. <*Ann. Rec.*, 1873–75, p. 449.

566.—1873. Hatching Striped Bass Artificially. <*Ann. Rec.*, 1873–75, p. 450.

567.—1873. Shad in the Altahama River. <*Ann. Rec.*, 1873–75, p, 450.

568.—1873. Treatment of Fish-Ponds. <*Ann. Rec.*, 1873–75, p. 452.

569.—1873. Culture of the Sterlet. <*Ann. Rec.*, 1873–75, p. 452.

570.—1873. Maritime Fisheries of France for 1871. <*Ann. Rec.*, 1873–75, p. 453.

571.—1873. Laws regulating the Fewfoundland Fisheries. <*Ann. Rec.*, 1873–75, p. 454.

572.—1873. Fish Inspection Law of Canada. <*Ann. Rec.*, 1873–75, p. 455.

573.—1873. Recent Fishery and Game Laws of the Ohio Legislature. <*Ann. Rec.*, 1873–75, p. 457.

574.—1873. Pacific Cod-Fisheries of 1873. <*Ann. Rec.*, 1873–75, p. 458.

575.—1873. German Report of United States Fisheries and Fish-Culture. <*Ann. Rec.*, 1873–75, p. 458.

576.—1873. The Fish of the Caspian Sea. <*Ann. Rec.*, 1873–75, p. 459.

577.—1873. Prices of American Fish-Eggs and Fry in England. <*Ann. Rec.*, 1873–75, p. 459.

578.—1873. Gloucester Halibut Fishery. <*Ann. Rec.*, 1873–75, p. 460.

579.—1873. Statistics of Egyptian Fisheries. <*Ann. Rec.*, 1873–75, p. 460.

580.—1873. Arrival of Salmon Eggs in New Zealand. <*Ann. Rec.*, 1873–75, p. 462.

581.—1873. Shad in the Alleghany River. <*Ann. Rec.*, 1873–75, p. 462.

582.—1873. Second Annual Meeting of the American Fish-Culturists' Association. <*Ann. Rec.*, 1873–75, p. 463.

583.—1873. Taking California Salmon with a Hook. <*Ann. Rec.*, 1873–75, p. 464.

584.—1873. The Fresh-Water Fisheries of India. <*Ann. Rec.*, 1873–75, p. 465.

586.—1873. Influence of External Pressure in the Life of Fishes. <*Ann. Rec.*, 1873–75, p. 467.

587.—1873. Utilization of Old Fish Pickle. <*Ann. Rec.*, 1873–75, p. 502.

588.—1873. The Sponge Trade. <*Ann Rec.*, 1873–75, p. 569.

589.—1873. Action of Cod-Liver Oil. <*Ann. Rec.*, 1873–75, p. 623.

614.—1874. Food-Fishes of the United States. <*Forest and Stream*, i, 1874, p. 330.

615.—1874. [The introduction of Young Salmon into New York waters.] <*Forest and Stream*, i, 1874, p. 347.

616.—1874. The Garfish. <*Forest and Stream*, i, 1874, p. 475.

617.—1874. Pisciculture and the Fisheries. <*Forest and Stream*, ii, 1874, p. 52.

818.—1874. Fly-fishing for Shad. <*Forest and Stream*, ii, 1874, p. 155.

619.—1874. Introduction of California Salmon into Australia. <*Forest and Stream*, ii, 1874, p. 229.

620.—1874. Letters concerning Fishways. *Forest and Stream*, ii, 1874, p. 340.

621.—1874. The Tarpum. <*Forest and Stream*, ii, 1874, p. 389.

622.—1874. The Blue-Black Trout. <*Forest and Stream*, iii, 1874, p. 277.

623.—1874. Prof. Baird's Report. Extracts. <*Forest and Stream*, iii, 1874, pp. 276, 292, 308, 324, 340, 356, 388.

624.—1874. [Extracts from Prof. Baird's Report.] <*American Sportsman*, v. 1874, pp. 148, (?) 162, 178.

625.—1874. "Opinion as to the probable cause of the rapid diminution of the supply of food fishes on the coast of New England, and especially of Maine." Letter to E. M. Stilwell, Esq., Bangor, Me. <*Rep. Comm. Inl. Fish. Mass.*, 1874, pp. 42–46.

627.—1874. See below under Papers published as Commissioner of Fisheries.

628.—1874. See below under Papers published as Commissioner of Fisheries.

630.—1874. Reports of Special Conferences of the U. S. Commissioner of Fisheries with the American Fish-Culturists' Association and State Commissioners of Fisheries. <*Rep. U. S. Comm. Fish and Fisheries*, part ii, 1874, pp. 757-773.

631.—1874. See below in list of publications issued as Commissioner of Fisheries.

643.—1874. Composition of the Cartilage of the Shark. <*Ann. Rec.*, 1874-75, p. 289.

644.—1874. Composition of the Body Fluids of Fish and Invertebrates. <*Ann. Rec.*, 1874-75, p. 291.

645.—1874. Malformation of Fish Embryos. <*Ann. Rec.*, 1874-75, p. 294.

646.—1874. The Sea-Serpent on the Scotch Coast. <*Ann. Rec.*, 1874-75, p. 296.

655.—1874. Food of the Shad. <*Ann. Rec.*, 1874-75, p. 310.

656.—1874. The Structure of the Lancelet. <*Ann. Rec.*, 1874-75, p. 310.

657.—1874. Fish Living in Dried Mud. <*Ann. Rec.*, 1874-75, p. 311.

658.—1874. The "Nerfling" Fish. <*Ann. Rec.*, 1874-75, p. 311.

659.—1874. Longevity of Fishes. <*Ann. Rec.*, 1874-75, p. 312.

660.—1874. Spawning of Whiting-Pout. <*Ann. Rec.*, 1874-75, page 312.

661.—1874. Sensibility of Fish to Poisons. <*Ann. Rec.*, 1874-75, p. 313.

662.—1874. Structure of the Embryonic Cellule in the Eggs of Bony Fishes. <*Ann. Rec.*, 1874-75, p. 314.

664.—1874. The Food of the Oyster, and a New Parasite. <*Ann. Rec.*, 1874-75, p. 315.

677.—1874. Shad in the Gulf of Mexico. <*Ann. Rec.*, 1874-75, p. 338.

680.—1874. Eggs of the Siluridæ. <*Ann. Rec.*, 1874-75, p. 339.

694.—1874. Change of Volume of Fish in Swimming. <*Ann. Rec.*, 1874-75, p. 348.

695.—1874. The Development of Sharks and Rays. <*Ann. Rec.*, 1874-75, p. 349.

697.—1874. The Basking Shark. <*Ann. Rec.*, 1874-75, p. 351.

705.—1874. Oil from Sharks' Livers. <*Ann. Rec.*, 1874-75, p. 419.

706.—1874. Fisheries and Sea Temperatures. <*Ann. Rec.*, 1874-75, p. 419.

707.—1874. Marine Fisheries of Maine in 1873. <*Ann. Rec.*, 1874-75, p. 420.

708.—1874. Consumption of Marine Products in Washington. <*Ann. Rec.*, 1874-75, p. 420.

709.—1874. The French Fisheries. <*Ann. Rec.*, 1874-75, p. 422.

710.—1874. The Seal and Herring Fisheries of Newfoundland. <*Ann. Rec.*, 1874-75, p. 424.

711.—1874. Alaska Cod-Fisheries in 1873. <*Ann. Rec.*, 1874-75, p. 424.

712.—1874. Fish-Culture in Castalia Springs. <*Ann. Rec.*, 1874-75, p. 425.

713.—1874. Restocking Otsego Lake, N. Y., with Fish. <*Ann. Rec.*, 1874-75, p. 426.

714.—1874. Introduction of British Fish into India. <*Ann. Rec.*, 1874-75, p. 426.

715.—1874. Transporting Living Trout. <*Ann. Rec.*, 1874-75, p. 427.

716.—1874. Caution in Planting Young Salmon. <*Ann. Rec.*, 1874-75, p. 427.

717.—1874. Destruction of Fish on the Oregon Coast by Nitro-Glycerine. <*Ann. Rec.*, 1874-75, p. 428.

718.—1874. Sterlet from St. Petersburg at the Brighton Aquarium. <*Ann. Rec.*, 1874-75, p. 428.

719.—1874. Stocking a Pond in Utah with Eels. <*Ann. Rec.*, 1874-75, p. 428.

720.—1874. Spinal Column of the Sturgeon as an Article of Food. <*Ann. Rec.*, 1874-75, p. 447.

759.—1875. Fishway at Holyoke. <*Ann. Rec.*, 1873-75, p. cx.

760.—1875. Pisciculture and the Fisheries. (A general summary of progress.) <*Ann. Rec.*, 1873-75, pp. cx, cxviii.

765.—1875. Mr. Balfour on the Embryology of Sharks. <*Ann. Rec.*, 1874-75, p. civ, general summary.

766.—1875. Fisheries (and Pisciculture) general summary. <*Ann. Rec.*, 1874-75, p. clxix.

767.—1875. Prof. Baird's Report. (Editorial, quoting.) <*Forest and Stream*, iii, 1875, p. 340.

768.—1875. Prof. Baird's Report. Comparative Value of Anadromous and other Fishes. <*Forest and Stream*, iii, 1875, p. 356.

769.—1875. Prof. Baird's Report. Different Methods of Multiplying Fish. (Editorial. quoting.) <*Forest and Stream*, iii, 1875, p. 388.

770.—1875. See below in list of papers published as Commissioner of Fisheries.

772.—1875. Conclusions as to Decrease of Cod-Fisheries on the New England Coast (Report of U. S. Commissioner of Fisheries). <*Rep. Comm. Inland Fisheries Massachusetts*, 1875, pp. 38–41.

773.—1875. Soles and Turbot for American Waters. <*Rod and Gun*, vii, 1875, p. 150. Letter to Frank Buckland.

774.—1875. Protection of Salmon. <*Forest and Stream*, v, 1875, p. 166.

775.—1875. Fish Culture in Kentucky. <*Forest and Stream*, v, 1875, p. 243.

776.—1875. "Prof. Baird made a brief statement as to the action of the United States Fish Commission." Abstract. <*Proc. American Fish-Culturists' Association*, 4th ann. meeting, 1875, pp. 8, 9.

794.—1875. Report of the Occurrence of Large Codfish off Mazatlan. <*Ann. Rec.*, 1875–76, p. 344.*

795.—1875. Grayling in the Au Sable River, Mich. <*Ann Rec.*, 1875–76, p. 344.

797.—1875. Monograph on the Anguilliform Fish. <*Ann. Rec.*, 1875–76, p. 345.

798.—1875. Largest Pike ever taken in England. <*Ann. Rec.*, 1875–76, p. 346.

799.—1875. Habits of Eels. <*Ann. Rec.*, 1875–76, p. 346.

800.—1875. Fossil Lepidosteus. <*Ann. Rec.*, 1875–76, p. 347.

801.—1875. Productive Season of the Cod on the Faroe Islands. <*Ann. Rec.*, 1875–76, p. 347.

802.—1875. Softness of Bones in Old Congers. <*Ann. Rec.*, 1875–76, p. 347.

803.—1875. Leptocephali are Larval Forms of Congers, etc. <*Ann Rec.*, 1875–76, p. 348.

805.—1875. Giant Cuttle-fish Found on the Grand Bank December, 1874. *Ann. Rec.*, 1875–76.

809.—1875. Report of the Fish Commission of Canada for 1874. <*Ann. Rec.*, 1875–76, p. 405.

810.—1875. Ninth Annual Report of the Massachusetts Commissioners of Fisheries. <*Ann. Rec.*, 1875–76, p. 406.

811.—1875. Ninth Report of the Fish Commissioners of Connecticut. <*Ann. Rec.*, 1875–76, p. 407.

812.—1875. First Report of the Commissioners of Fisheries of Michigan. <*Ann. Rec.*, 1875–76, p. 407.

813.—1875. First Annual Report of the Fish Commissioners of Minnesota. <*Ann. Rec.*, 1875–76, p. 408.

814.—1875. Fifth Report of the Fish Commissioners of Rhode Island. <*Ann. Rec.*, 1875–76, p. 408.

815.—1875. Report of the Fish Commissioners of Pennsylvania for 1874. <*Ann. Rec.*, 1875–76, p. 409.

816.—1875.—Report of the Fish Commissioners of New Hampshire for 1874. <*Ann. Rec.*, 1875–76, p. 410.

817.—1875.—Second Report of the Fish Commissioners of Vermont. <*Ann. Rec.*, 1875–76, p. 411.

818.—1875.—First Report of the Fish Commissioners of Wisconsin. <*Ann. Rec.*, 1875–76, p. 411.

819.—1875.—Third Annual Report of the American Fish-Culturists' Association. <*Ann. Rec.*, 1875–76, p. 412.

820.—1875.—Meeting of the American Fish-Culturists' Association. <*Ann. Rec.*, 1875–76, p. 412.

821.—1875.—Objection to the Use of Submerged Net-Weirs. <*Ann. Rec.*, 1875–76, p. 413.

886.—1876. Pisciculture and the Fisheries. (General summary.) $<$*Ann. Rec.,* 1875-76, pp. ccxxiv-ccxxix.

888.—1876. Fish in California. $<$*Rod and Gun,* vii, 1876, p. 326.

889.—1876. See below, under Papers published as Commissioners of Fisheries.

890.—1876. See below, under Papers published as Commissioners of Fisheries.

892.—1876. The United States Fish Commission. $<$*Forest and Stream,* vi, 1876, p. 147.

894.—1876. Connecticut River Shad for California. The shipment of a million Shad Fry from Holyoke, Mass., to the Sacramento River, Cal., under the care of F. N. Clark and T. H. Bean. $<$*Forest and Stream,* vii, p. 66.

917.—1876. Rafinesque's Fishes of Ohio. $<$*Ann. Rec.* 1876-77, p. 318.

918.—1876. The Pilot Fish. $<$*Ann. Rec.,* 1876-77, p. 319.

919.—1876. New Work on European Fresh-Water Fishes. $<$*Ann. Rec.,* 1876-77, p. 319

920.—1876. Some Curious Australian Fishes. $<$*Ann. Rec.,* 1876-77, p. 319.

921.—1876. Poey's Catalogue of Cuban Fishes. $<$*Ann. Rec.,* 1876-77, p. 320.

922.—1876. Habits of the Salmon. $<$*Ann. Rec.,* 1876-77, p. 320.

923.—1876. The Rainbow Fish. $<$*Ann. Rec.,* 1876-77, p. 321.

924.—1876. Incubation of *Chromis paterfamilias.* $<$*Ann. Rec.,* 1876-77, p. 322.

925.—1876. Cause of the Black Spots on the Scales of Fish. $<$*Ann. Rec.,* 1876-77. p. 323.

926.—1876. Remarkable Structure of Young Fishes. $<$*Ann. Rec.,* 1876-77' p, 323.

927.—1876. Curious Habits of Fishes. $<$*Ann. Rec.,* 1876-77, p. 324.

930.—1876. Gathering of Euplectella. $<$*Ann. Rec.,* 1876-77, p. 340.

931.—1876. Proposed Utilization of Fish Bones. $<$*Ann. Rec.,* 1876-77, p. 372.

932.—1876. Report of the Maritime Fisheries of France. $<$*Ann. Rec.,* 1876-77, p. 385.

934.—1876. Gloucester Fisheries for 1875. $<$*Ann. Rec.,* 1876-77, p. 386.

935.—1876. Connection of Meteorology and Herring-Fisheries. $<$*Ann. Rec.,* 1876-77, p. 387.

936.—1876. Potomac River Fisheries. $<$*Ann. Rec.,* 1876-77, p. 388.

937.—1876. Seal-Fisheries of 1876 on the Greenland Coast. $<$*Ann. Rec.,* 1876-77, p. 389.

938.—1876. Close time for Seals in the Northern Sea. $<$*Ann. Rec.,* 1876-77, p. 389.

939.—1876. Report on Alaska Seal Islands. $<$*Ann. Rec.,* 1876-77, p. 389.

940.—1876. Menhaden Fishery in 1875. $<$*Ann. Rec.,* 1876-77, p. 390.

941.—1876. New use for the Scrap of the Moss-Bunker. $<$*Ann. Rec.,* 1876-77, p. 390.

942.—1876. Utilizing the Offal of Codfish on the Gulf of St. Lawrence. $<$*Ann. Rec.,* 1876-77, p. 391.

943.—1876. Report of the Commissioner of Fisheries of Canada for 1875. $<$*Ann. Rec.,* 1876-77, p. 391.

944.—1876. Report of the Fish Commissioners of Maine. $<$*Ann. Rec.,* 1876-77, p. 392.

945.—1876. Report of the Fish Commissioners of New Hampshire. $<$*Ann. Rec.,* 1876-77, p. 393.

946.—1876. Tenth Report of the Massachusetts Fish Commissioners. $<$*Ann. Rec.,* 1876-77, p. 393.

947.—1876. Tenth Report of the Fish Commissioners of Connecticut. $<$*Ann. Rec.,* 1876-77, p. 394.

948.—1876. Eighth Report of the Fish Commissioners of New York. $<$*Ann. Rec.,* 1876-77, p. 395.

949.—1876. Fifth Annual Report of the Fish Commissioners of New Jersey. $<$*Ann. Rec.,* 1876-77, p. 397.

950.—1876. Sixth Annual Report of the Fish Commissioners of New Jersey. $<$*Ann. Rec.,* 1876-77, p. 397.

951.—1876. Action of the Kentucky Fish Commissioners. $<$*Ann. Rec.,* 1876-77, p. 398.

952.—1876. Convention of the Western State Fish Commissioners. $<$*Ann. Rec.,* 1876-77, p. 399.

953.—1876. First Report of the Iowa Fish Commissioners. <*Ann. Rec.*, 8176–77, p. 399.

954.—1876. Second Report of the Commissioners of Fisheries of Wisconsin. <*Ann. Rec.*, 1876–77, p. 400.

955.—1876. Second Report of the Fish Commissioners of Minnesota. <*Ann. Rec.*, 1876–77, p. 400.

956.—1876. Arkansas Fish Commissioners. <*Ann. Rec.*, 1876–77, p. —.

957.—1876. Biennial Report of the California Fish Commission. <*Ann. Rec.*, 1876–77, p. 401.

958.—1876. Cultivation of Carp in California. <*Ann. Rec.*, 1876–77, p. 403.

959.—1876. Capturing Eels in Cochin China. <*Ann. Rec.*, 1876–77, p. 403.

960.—1876. Hatching Whitefish in the Detroit River. <*Ann. Rec.*, 1876–77, p. 403.

961.—1876. One Cause of the Death of Fishes. <*Ann. Rec.*, 1876–77, p. 405.

962.—1876. Rapidity of Growth in certain Fishes. <*Ann. Rec.*, 1876–77, p. 405.

963.—1876. Utilization of Warmed Waters in Fish-Culture. <*Ann. Rec.*, 1876–77, p. 406.

964.—1876. Shad in the Mississippi. <*Ann. Rec.*, 1876–77, p. 406.

965.—1876. Renewed attempt to send Salmon Eggs to New Zealand. <*Ann. Rec.* 1876–77, p. 407.

966.—1876. Salmon in the Antipodes. <*Ann. Rec.*, 1876–77, p. 408.

967.—1876. Salmon Eggs in South Africa. <*Ann. Rec.*, 1876–77, p. 408.

968.—1876. Capture of Salmon in the Connecticut River. <*Ann. Rec.*, 1876–77, p. 409.

969.—1876. Fattening of Oysters. <*Ann. Rec. of Sci. and Ind.*, 1876–77, p. 410.

972.—1876. Many Fish are afflicted by crustaceous parasites, called *Argulus*, which adhere to their gills. <*Ann. Rec.*, 1876–77, p. clx., gen. sum.

973.—1876. A new form of Fishes discovered by Herr Bucholz in West Africa. <*Ann. Rec.*, 1876–77, p. clxvii.

974.—1877. A characteristic type of Fishes of the Northern Atlantic as exemplified in the species of variously called "Sea Wolf," etc. <*Ann. Rec.*, 1876–77, p. clxvii.

975.—1877. Genuine White Shad in the Ohio. <*Forest and Stream*, viii, 1877, p. 280.

976.—1877. "Prof. Baird spoke of the Inception of the scheme to introduce California Salmon," &c. Feb. 14, 1877. <*Trans. American Fish-Culturists' Association*, 1876–77 (*Sixth Annual Meeting*), 1877, p. 5.

977.—1877. "In regard to the introduction of Salmon" and the work of the United States Fish Commission. <*Trans. American Fish-Culturist's Association*, 1876–77, (*Sixth Annual Meeting*), 1877, pp. 64–70.

978.—1877. "Extracts from Prof. Baird's Report, published in 1873" [concerning decrease of fish in New England]. <*Documents and Proceedings of the Halifax Commission*, 1877, *Appendix A* (*Case of Her Majesty's Government*), p. —. U. S. Reprint, 1878 [i], p. 98.

979.—1877. "Extracts of a report on the condition of the sea-fisheries of the south coast of New England in 1871 and 1872, by Spencer F. Baird, Commissioner." <*Documents and Proceedings of the Halifax Commission*, 1877, *Appendix E*, xi, p. 34.—. U. S. Reprint, 1878 [i], pp. 229–231.

980.—1877. "Extracts from Eighth Report of the Commissioner of Fisheries of the State of Maine for the year 1874 (page 7)" [being the letter to E. M. Stilwell, Esq., already mentioned under No. 620]. <*Documents and Proceedings of the Halifax Commission*, 1877, *Appendix E*, xii, pp. 34, 35. U. S. Reprint, 1878 [i], pp. 231–233.

981.—1877. "No. 68. Prof. Spencer F. Baird, Assistant Secretary of the Smithsonian Institution, Washington, and United States Commissioner of Fish and Fisheries, called on behalf of the Government of the United States,

981.—1877. No. 68—Continued.
sworn and examined." Thursday, October 18, 1877. <*Documents and Proceedings of the Halifax Commission, 1877, Appendix L (United States evidence*), pp. 451–463. U. S. Reprint, 1878 [iii], pp. 2795–2816.

982.—1877. "The Conference met. The examination of Prof. Spencer F. Baird, called on behalf of the Government of the United States, resumed." Friday, October 19, 1877. <*Documents and Proceedings of the Halifax Commission, 1877, Appendix L (United States evidence*), pp. 466–479. American Reprint, 1878 [iii], pp. 2821–2849.

983.—1877. [Various extracts.] · <*Hind—The Effect of the Fishery Clauses of the Treaty of Washington on the Fisheries and Fishermen of British North America,* part i, Halifax, 1877, pp. vii, viii, xiii, xv, xvi, xix, 7, 10, 11, 12, 25, 37, 41, 47, 135, 136, 144.

984.—1877. See below, in List of Papers published as Commissioner of Fisheries.

985.—1877. See below, in List of Papers published as Commissioner of Fisheries.

986.—1877. See below, in List of Papers published as Commissioner of Fisheries.

987.—1877. See below, in List of Papers published as Commissioner of Fisheries.

988.—1877. See below, in List of Papers published as Commissioner of Fisheries.

989.—1877. A request for the United States Commissioner of Fish and Fisheries. <*Forest and Stream,* 1877, x, p. 75.

990.—1877. Salmon in the Hudson. <*Forest and Stream,* 1877, x, p. 154.

991.—1877. Salmon in the Chesapeake. <*Forest and Stream,* 1877, x, p. 296.

994.—1877. Genuine White Shad in the Ohio. <*Forest and Stream,* viii, 1877, p. 280.

995.—1877. Salmon in the Richelieu. <*Forest and Stream,* ix, 1877, p. 143.

996.—1877. A New Fish. <*Forest and Stream,* ix, 1877, p. 381.

997.—1878. The Delaware Salmon. <*Chicago Field,* ix, 1878, p. 165.

998.—1878. Propagation of Eels. <*Sunbury (Pa.) American,* Aug. 30 or Sept. 6, 1878.

999.—1878. "The Herring Fishery of the Coast of Sweden. <*Cape Ann Advertiser,* Aug. 9, 1878.

1000.—1878. The Periodicity of Herrings. <*Chicago Field,* x, 1878, p. 35.

1001.—1878. "The Fishery Statistics of the United States." <*Trans. American Fish-Culturalists Association,* 1878, pp. 72–74.

1002.—1878. See below, under Papers published as Commissioner of Fisheries.

1003.—1878. See below, under Papers published as Commissioner of Fisheries.

1005.—1878. See below, under Papers published as Commissioner of Fisheries.

1006.—1878. See below, under Papers published as Commissioner of Fisheries.

1007.—1878. See below, under Papers published as Commissioner of Fisheries.

1008.—1878. See below, under Papers published as Commissioner of Fisheries.

1009.—1878. The Delaware Salmon. <*Chicago Field,* ix, 1878, p. 165.

1010.—1878. The "Herald" Interviews Prof. Baird. <*Chicago Field,* x, 1878, p. 243.

1011.—1878. A request from the United States Commissioner of Fish and Fisheries. <*Forest and Stream,* x, 1878, p. 75. Ed.

1012.—1878. Salmon in the Hudson. <*Forest and Stream,* x, 1878, p. 154.

1014.—1878. All about Eels. <*Forest and Stream,* xi, 1878, pp. 130, 131.

1015.—1879. Is it Herring Spawn? <*Forest and Stream,* xii, 1879, p. 5.

1016.—1879. Fishes of the Deep Sea. <*Forest and Stream,* xii, 1879, p. 6.

1017.—1879. Transportation of Alewife Eggs. <*Forest and Stream,* xii, 1879, p. 225.

1018.—1879. The Hudson Salmon. <*Forest and Stream,* xii, 1879, p. 444.

1029.—1879. Circular Relating to Fish Trade and Consumption of Fish. <*Chicago Field,* xii, 1879, p. 35.

1030.—1879. See below, under Papers published as Commissioner of Fisheries.

1031.—1878. See below, under Papers published as Commissioner of Fisheries.

1032.—1879. See below, in List of Papers published as Commissioner of Fisheries.

1033.—1879. The National Fish Commission. <*Forest and Stream,* xiii, 1879, p. 725.

1034.—1879. The Farmer's Fish [Carp]. <*Forest and Stream*, xiii, 1879, p. 846.

1035.—1879. Another New Fish on our Coast. <*Chicago Field*, xi, 1879, p. 117.

1036.—1879. A Letter Concerning California Salmon in Europe. <*Chicago Field*, xi, 1879, p. 131.

1037.—1879. Report of the U. S. Fish Commission. <*Chicago Field*, xii, 1879, p. 307.

1040.—1879. See below, in List of Papers published as Commissioner of Fisheries.

1043.—1880. Do Black Snakes Eat Fish? <*Forest and Stream*, xiii, 1880, p. 966.

1044.—1880. Striped Bass and Shad in California. <*Forest and Stream*, xiv, 1880, p. 410.

1048.—1880. California Salmon-Eggs. <*Forest and Stream*, xv, 1880, p. 107.

1049.—1880. Eggs for Distribution. <*Forest and Stream*, xv, 1880, p. 366.

1050.—1880. See below, in List of Papers published as Commissioner of Fisheries.

1051.—1880. See below, in List of Papers published as Commissioner of Fisheries.

1058.—1880. Destruction of Fish. <*Chicago Field*, xiv, 1880, p. 236.

1059.—1880. Salmon Eggs. <*Chicago Field*, xiv, 1880, p. 284.

1060.—1880. A Good Fish for Barren Waters [Carp]. <*Chicago Field*, xii, 1880, p. 323, 324.

INVERTEBRATES.

139.—1871. Explorations in Vineyard Sound by Prof. Verrill. <*Ann. Rec.*, 1871–72, p. 140.

167.—1871. Hermit Crabs Climbing Trees. <*Ann. Rec.*, 1871–72, p. 229.

192.—1871. Zoological Stations in the Gulf of Naples. <*Ann. Rec.*, 2871–72, pp. 274, 275.

193.—1871. Verrill's Exploration in New Jersey. <*Ann. Rec.*, 1871–72, p. 276.

194.—1871. Dr. Stimpson's Exploration in Florida. <*Ann. Rec.* 1871–72, p. 377.

200.—1871. Irish Oyster Fisheries. <*Ann. Rec.* 1871–72, pp. 352, 353.

207.—1871. Pegging Lobster Claws. <*Ann. Rec.*, 1871–72, pp. 552, 553.

213.—1871. Return of Mr. Gwyn Jeffreys to England. <*Ann. Rec.*, 1871–72, p. 608.

233.—1872. Explorations of Dr. Stimpson. <*Ann. Rec.*, 1872–73, p. 154.

236.—1872. Explorations of William H. Dall. <*Ann. Rec.*, 1872–73, p. 175.

243.—1872. Marine Zoology of the Bay of Fundy. <*Ann. Rec.*, 1872–73, p. 201.

244.—1872. The Voyage of the Hassler. <*Ann. Rec.*, 1872–73, p. 204.

249.—1872. Fulfillment of the Predictions of Professor Agassiz. <*Ann. Rec.*, 1872–73, p. 214.

250.—1872. Professor Agassiz's Prophecies. <*Ann. Rec.*, 1872–73, p. 215.

283.—1872. Edward's Work on North American Butterflies. <*Ann. Rec.*, 1872–73, p. 267.

284.—1872. Have Trilobites Legs? <*Ann. Rec.*, 1872–73, p. 268.

285.—1872. Early stages of the American Lobster. <*Ann. Rec.*, 1872–73, p. 269.

286.—1872. Professor Gill's Arrangement of Mollusks. <*Ann. Rec.*, 1872–73, p. 269.

287.—1872. Embryology of Terebratulina and Ascidia, and Protective Coloration of Mollusca. <*Ann. Rec.*, 1872–73, p. 271.

288.—1872. Parasites and Commensals of Fish. <*Ann. Rec.*, 1872–73, p. 272.

289.—1872.—Worms in the Trout of Yellowstone Lake. <*Ann. Rec.*, 1872–73, pp. 274, 275.

298.—1872. Fossil Fishes and Insects from the Nevada Shales. <*Ann. Rec.*, 1872–73, p. 308.

303.—1872. Edwards on North American Butterflies. <*Ann. Rec.*, 1872–73, p. 311. Review.

311.—1872. Filaria in the Brain of the Water-Turkey. <*Ann. Rec.*, 1872–73, p. 324.

327.—1872. Report of C. V. Riley, State Entomologist of Missouri, for 1871. *<Ann. Rec.*, 1872–73, p. 378.

390.—1872. Breeding of Leeches. *<Ann. Rec.*, 1872–73, p. 441.

413.—1872. American Journal of Conchology. *<Ann. Rec.*, 1872–73, p. 611.

447.—1873. Explorations in the Gulf of St. Lawrence in 1872. *<Ann. Rec.*, 1873–75, p. 216.

448.—1873. Fauna of the St. George's Bank and adjacent waters. *<Ann. Rec.*, 1873–75, p. 218.

459.—1873. Explorations of W. H. Dall in the Aleutian Islands. *<Ann. Rec.*, 1873–75, p. 246.

471.—1873. "Revision of the Echini," by Alexander Agassiz. *<Ann. Rec.*, 1873–75, p. 265.

485.—1873. Allman on Tubularian Hydroids. *<Ann. Rec.*, 1873–75, p. 296.

486.—1873. Haeckel on the Calcareous Sponges. *<Ann. Rec.*, 1873–75, p. 298.

493.—1873. Influence of External Conditions on the Structure of Insects. *<Ann. Rec.*, 1873–75, p. 307.

507.—1873. Binney on the Geographical Distribution of Mollusks. *<Ann. Rec.*, 1873–75, p. 320.

510.—1873. Determining Sex in Butterflies. *<Ann. Rec.*, 1873–75, p. 329.

511.—1873. Distribution of California Moths. *<Ann. Rec.*, 1873–75, p. 330.

512.—1873. Pavonaria Blakei, a New Alcyonoid Polyp. *<Ann. Rec.*, 1873–75, p. 332.

513.—1873. Terrestrial Mollusca in the Bahamas. *<Ann. Rec.*, 1873–75, p. 333.

521.—1873. Catalogue of Rhode Island Mollusca. *<Ann. Rec.*, 1873–75, p. 339.

530.—1873. Habits of the Craw-Fish. *<Ann. Rec.*, 1873–75, p. 351.

588.—1873. The Sponge Trade. *<Ann. Rec.*, 1873–75, p. 569.

644.—1873. Composition of the Body Fluids of Fish and Invertebrates. *<Ann. Rec.*, 1874–75, p. 291.

663.—1874. The Embryology of Terebratulina. *<Ann. Rec.*, 1874–75, p. 314.

664.—1874. The Food of the Oyster, and a New Parasite. *<Ann. Rec.*, 1874–75, p. 315.

665.—1874. Explanation of the alleged occurrence of the King-Crab in Holland. *<Ann. Rec.*, 1874–75, p. 316.

691.—1874. An "Army Worm." *<Ann. Rec.*, 1874–75, p. 344.

722.—1874. Cambridge Entomological Club. *<Ann. Rec.*, 1874–75. p. 574.

757.—1875. Further Contributions to the Minute Anatomy of the Tæniæ, which prey on Fish. *<Ann. Rec.*, 1873–75, p. xcvii.

758.—1875. Further Observation on the Cercariæ in the Intestines of Fish. *<Ann. Rec.*, 1873–75, p. xcvii.

796.—1875. Respiration of the Loach. *<Ann. Rec.*, 1875–76, p. 345.

804.—1875. Have Jelly-Fishes a Nervous System? *<Ann. Rec.*, 1875–76, p. 348.

805.—1875. Giant Cuttle-Fish found on the Grand Bank, December, 1874. *<Ann. Rec.*, 1875–76, p. 351.

807.—1875. Giant Cuttle-Fish found on the Grand Bank, December, 1874. *<Ann. Rec.*, 1875–76, p. 351.

850.—1875. Animal Incrustation on the Great Eastern. *<Ann. Rec.*, 1875–76, p. 432.

895.—1876. Introductory to Dall's Classification of the Products of Sea and Shore. *<Folio Circular, Centennial Series*, p. 1.

928.—1876. Eighth Report of the State Entomologist of Missouri. *<Ann. Rec.* 1876–67, p. 333.

929.—1876. Habits and Anatomy of a Nereid Worm. *<Ann. Rec.*, 1876–77, p. 336.

930.—1876. Gathering of Euplectella. *<Ann. Rec.*, 1876–77, p. 340.

969.—1876. Fattening of oysters. *<Ann. Rec.*, 1876–77, page 410.

972.—1877. Many Fish are afflicted by crustaceous parasites, called *Argulus*, which adhere to their gills. *<Ann. Rec.*, 1876–77, p. clx, gen. sum.

PLANTS.

5.—1845. Contributions toward a catalogue of the trees and shrubs of Cumberland County, Pa. <*Lit. Rec. and Journ. Linnæan Assoc. Pennsylvania College,* i, No. 4, Feb., 1845, pp. 57–63.

105.—1860. (*Translator.*) On the principal plants used as food by man. Sketch of the plants chiefly used as food by man in different parts of the world and at various periods. By Dr. F. Unger. (Translated from the German for this Report.) <*Report of the Commissioner of Patents for the year* 1859. *Agriculture,* 1860, pp. 299–362.

190.—1881. Fungus growth on Fish and their Eggs. <*Ann. Rec.,* 1871–72, p. 267.

323.—1872. Algæ of Rhode Island. <*Ann. Rec.,* 1872–73, p. 342.

326.—1872. Relation of Recent North American Flora to Ancient. <*Ann. Rec.* 1872–73, p. 352.

403.—1872. Department Report on the Preparation of Timber. <*Ann. Rec.,* 1872–73. p. 485.

446.—1873. Lesquereux on the Fossil Plants of the Northern Hemisphere. <*Ann. Rec.,* 1873–75, p. 210.

531.—1873. The Sequoias of California, and their History. <*Ann. Rec.,* 1873–75, p. 363.

532.—1873. Forest Growth in the Wabash Valley. <*Ann. Rec.,* 1873–75, p. 367.

534.—1873. Tenth Annual Report of the Massachusetts Agricultural College. <*Ann. Rec.,* 1873–75, p. 418.

700.—1874. Sea-Weeds of the Bay of Fundy. <*Ann. Rec.,* 1874–75, p. 356.

701.—1874. New Yearly Report of the Progress of Botany. <*Ann. Rec.,* 1874–75, p. 361.

702.—1874. Recent Publications in Systematic Botany. <*Ann. Rec.,* 1874–75, p. 363.

744.—1874. Botanical Conservatory of the Maryland Academy of Sciences. <*Ann. Rec.,* 1874–75, p. 590.

747.—1874. Additions to the National Herbarium in 1874. <*Ann. Rec.,* 1874–75, p. 593.

GEOGRAPHICAL DISTRIBUTION; FAUNAS; ACCLIMATATION.

2.—1844. List of Birds found in the vicinity of Carlisle, Cumberland County, Penn., about Lat. 40° 12' W., Lon. 77° 11' W. By William M. & Spencer F. Baird. <*Amer. Journ. Sci. and Arts,* xlvi, 1844, No. 2, Jan.–Mar., art. vi, pp. 261–273.

5.—1845. Contributions towards a catalogue of the trees and shrubs of Cumberland County, Pa. <*Lit. Rec. and Journ. Linnæan Assoc. Pennsylvania College,* i, No. 4, Feb., 1845, pp. 57–63.

6.—1845. Catalogue of birds found in the neighborhood of Carlisle, Cumberland Co., Pa. <*Lit. Rec. and Journ. Linnæan Assoc. Pennsylvania College,* i, No. 12, Oct., 1845, pp. 249–257.

28.—1852. Zoology of the Valley of the Great Salt Lake of Utah. <*Stansbury's Exploration and Survey of the Valley of the Great Salt Lake of Utah, etc.,* Philadelphia, 1852, App. C, pp. 305–378, pll. i–x.

31.—1852. Birds [of the Valley of the Great Salt Lake]. <*Stansbury's Exploration and Survey of the Valley of the Great Salt Lake of Utah, etc.* Philadelphia, 1852. [App. C.] pp. 314–325 [+ 325–335 extraneous bird matter.]

33.—1852. (CHARLES GIRARD.) Reptiles of the Valley of the Great Salt Lake. *<Stansbury's Exploration and Survey of the Valley of the Great Salt Lake.* Philadelphia, 1852. [App. C.] pp. 336-353.

39.—1853. (CHARLES GIRARD.) Catalogue of North American Reptiles in the Museum of the Smithsonian Institution. Part I.—Serpents. By S. F. Baird and C. Girard. Washington: Smithsonian Institution. January, 1853. 8vo. pp. xvi, 172.

63.—1855. Report on the fishes observed on the coasts of New Jersey and Long Island during the summer of 1854, by Spencer F. Baird, Assistant Secretary of the Smithsonian Institution. *<Ninth Annual Report of the Smithsonian Institution [for* 1854], 1855, pp. 317-325 + *337.

65.—1855. List of Mammalia found in Chili. *<Gillis, Naval Astron. Exp.,* ii, pp. 163-171.*

67.—1856. List of Fishes inhabiting the State of New York: sent to the New York State Cabinet of Natural History by the Smithsonian Institution in May, 1855 (by Professor S. F. Baird). *<Ninth Annual Report of the Regents of the University of the State of New York on the Condition of the State Cabinet of Natural History and the Historical and Antiquarian Collection annexed thereto.* * * " 1856, pp. 22-29.

75.—1857. Catalogue of North American Mammals, chiefly in the Museum of the Smithsonian Institution. By Spencer F. Baird, Assistant Secretary of the Smithsonian Institution. Washington: Smithsonian Institution, July, 1857. 4to. pp. 21.

76.—1857. Explorations and Surveys for a Railroad Route from the Mississippi River to the Pacific Ocean. War Department. Mammals: By Spencer F. Baird, Assistant Secretary of the Smithsonian Institution. Washington, D. C., 1857. Reports of explorations and surveys, to ascertain the most practicable and economical route for a railroad from the Mississippi River to the Pacific Ocean. Volume VIII. 1857. 4to. pp. xlviii, 757, pl. xviii-lx.

77.—1857. Catalogues of Fishes, copied from a "Report on the Fishes observed on the Coasts of New Jersey and Long Island during the summer of 1854. *<Catalogue of Zoological and Botanical Productions of the County of Cape May, in Geology of the County of Cape May, State of New Jersey,* 1857. pp. 146-148.

78.—1858. Explorations and Surveys for a Railroad Route from the Mississippi River to the Pacific Ocean. War Department. Birds: By Spencer F. Baird, Assistant Secretary Smithsonian Institution, with the co-operation of John Cassin and George N. Lawrence. Washington, D. C. 1858, pp. i-lvi, 1-1005. (No illustrations.) Dated Washington, Oct. 20, 1853. Reports of Explorations and Surveys, to ascertain the most practicable and economical route for a railroad from the Mississippi River to the Pacific Ocean. Volume IX. 1858. 4to. pp. lvi, 1005.

79.—1858. Birds found at Fort Bridger, Utah. *<Pacific Railroad Report,* ix, 1858, App. B, pp. 926, 927.

80.—1858. Catalogue of North American Birds, chiefly in the Museum of the Smithsonian Institution. By Spencer F. Baird, Assistant Secretary of the Smithsonian Institution. Washington: Smithsonian Institution. October, 1858. 4to. pp. xv-lvi.

87.—1859. Birds of the Boundary, by Spencer F. Baird, Assistant Secretary of the Smithsonian Institution. With notes by the Naturalists of the Survey. United States and Mexican Boundary Survey, under the order of Lieut.-Col. W. H. Emory, Major First Cavalry, and United States Commissioner. *<*pp. (2) 3-35 (1) pll. xli.

88.—1859. United States and Mexican Boundary Survey, under the order of Lieut. Col. W. H. Emory, Major First Cavalry, and United States Commissioner. Reptiles of the Boundary, by Spencer F. Baird, Assistant Secretary of the Smithsonian Institution. With notes by the Naturalists of the Survey. pp. (2) 3–35, pll. xli (for title of volume see under 80 and 81). <*Ibid.*

91.—1859. Smithsonian Miscellaneous Collections. Catalogue of North American Birds, chiefly in the Museum of the Smithsonian Institution. By Spencer F. Baird. Washington: Smithsonian Institution. 1859. 8vo. 2 p. ll., pp. 19 + 2.

93.—1859. Report upon Mammals collected on the Survey. <Explorations and Surveys for a Railroad Route from the Mississippi River to the Pacific Ocean. Report of Lieut. E. G. Beckwith, Third Artillery, upon Explorations for a Railroad Route, near the 38th and 39th parallels of north latitude, by Captain J. W. Gunnison, Corps of Topographical Engineers, and near the forty-first parallel of north latitude, by Lieut. E. G. Beckwith, Third Artillery. 1854. Zoölogical Report, xx, 1857, in Report P. R. R. Surv., vol. x, 1859. Third Article, pp. (1) 7–9 (1), pll. v–x.

94.—1859. Report on Birds collected on the Survey. <Explorations and Surveys for a Railroad Route from the Mississippi River to the Pacific Ocean. War Department. Report of Lieut. E. G. Beckwith, Third Artillery, upon Explorations for a Railroad Route, near the 38th and 39th parallels of north latitude, by Captain J. W. Gunnison, Corps of Topographical Engineers, and near the forty-first parallel of north latitude, by Lieut. E. G. Beckwith, Third Artillery. 1854. pp. 27, pll. xii, xiii, xiv, xv, xvii, xxxii, xxxv, in Rep. P. R. R. Surv., vol. x.

95.—1859. No. 3. Report on Reptiles collected on the Survey. <Explorations and Surveys for a Railroad Route from the Mississippi River to the Pacific Ocean. War Department. Report of Lieut. E. G. Beckwith, Third Artillery, upon Explorations for a Railroad Route, near the 38th and 39th parallels of north latitude, by Captain J. W. Gunnion, Corps of Topographical Engineers, and near the forty-first parallel of north latitude, by Lieut E. G. Beckwith, Third Artillery. 1854. pp. 17–20 [in Rep. P. R. R. Surv., vol. x, third article], pll. xvii, xviii, xxiv.

96.—1859. No. 4. Report upon the Reptiles of the Route. <Explorations and Surveys for a Railroad Route from the Mississippi River to the Pacific Ocean. War Department. Route near the Thirty-fifth Parallel, explored by Lieutenant A. W. Whipple, Topographical Engineers, in 1853 and 1854. Zoological Report. Washington, D. C. 1859. [In vol. x, Report P. R. R. Surv., part vi, No. 4.] pp. 37–45, pll. xxv, xxvi, xxvii.

97.—1859. Report on Mammals collected on the Survey. <Explorations and Surveys for a Railroad Route from the Mississippi River to the Pacific Ocean. War Department. Routes in California, to connect with the routes near the Thirty-fifth and Thirty-second Parallels, explored by Lieut. R. S. Williamson, Corps of Top. Eng., in 1853. Zoological Report. Washington, D. C. 1859. [In vol. x, Report P. R. R. Surv., part iv, art. 3.] pp. 81, 82.

99.—1859. Notes on a collection of Birds made by Mr. John Xantus, at Cape St. Lucas, Lower California, and now in the Museum of the Smithsonian Institution. <*Proc. Acad. Nat. Sci. Phila.*, xi, 1859 (1860), pp. 299–306.

103.—1860. The Birds of North America; the descriptions of species based chiefly on the collections in the Museum of the Smithsonian Institution. By Spencer F. Baird, Assistant Secretary of the Smithsonian Institution, with the co-operation of John Cassin, of the Academy of Natural Sciences of Philadelphia, and George N. Lawrence, of the Lyceum of Natural History of New York. With an Atlas of One Hundred Plates. Text. Philadelphia: J. B. Lippincott & Co. 1860. 4to. pp. lvi, 1005.

115.—[1864–66.] Smithsonian Miscellaneous Collections. Review of American Birds, in the Museum of the Smithsonian Institution. By S. F. Baird. Part I. North and Middle America. [Medallion.] Washington; Smithsonian Institution. [No date on title: June, 1864, to p. 33; July 1864, to p. 81; Aug., 1864, to p. 129; Sept., 1864, to p. 145; Oct., 1864, to p. 161; Nov., 1864, to p. 177; Apr., 1865, to p. 241; May, 1865, to p. 321; May, 1866, to p. 417; June, 1866, to end.] 1 vol. 8vo. Originally issued in sheets as successively printed, at above dates. pp. iv, 450.

118.—1866. The Distribution and Migrations of North American Birds; by Spencer F. Baird, Assistant Secretary Smithsonian Institution. (Abstract of a memoir presented to the National Academy of Sciences, Jan., 1865.) $<$ Amer. Jour. Sci. and Arts, (2), xli, pp. 78–90, 184–92, 337–47.

123.—1866. Prof. Spencer F. Baird. Die Verbreitung und Wanderungen der Vögel Nord-Amerika's. $<$ Journal für Ornithologie, xiv, 1866, pp. 244–269, 338–352.

    Translated from American Journal of Science and Arts, vol. xli, 1866.

124.—1867. The Distribution and Migrations of North American Birds. $<$ Ibis, 1867, 2d ser., iii, pp. 257–293.

    Reprinted from Am. Journ. Sci. and Arts, xli, Jan., Mar., May, 1866.

129.—1869. On additions to the Bird-Fauna of North America, made by the Scientific Corps of the Russo-American Telegraph Expedition. $<$ Trans. Chicago Acad. Sciences, i, pt. ii, 1869, pp. 311–325, pll. xxvii–xxxiv.

143.—1871. Faunal peculiarities of the Azores. $<$ Ann. Rec., 1871–72, pp. 149, 150.

144.—1871. Faunal provinces of the west coast of America. $<$ Ann. Rec., 1871–72, p. 152.

243.—1872. Marine Zoology of the Bay of Fundy. $<$ Ann. Rec., 1872–73, p. 201.

247.—1872. Grandidier on the Zoology of Madagascar. $<$ Ann. Rec., 1872–73, p. 213.

345.—1872. Transporting Black Bass to California. $<$ Ann. Rec., 1872–73, p. 409.

347.—1872. Planting of Shad in the Valley of the Mississippi and the Lakes. $<$ Ann. Rec., 1872–73, p. 410.

350.—1872. Transportation of Black Bass to England. $<$ Ann. Rec., 1872–73, p. 414.

361.—1872. Salmon and Trout in Australia. $<$ Ann. Rec., 1872–73, p. 422.

369.—1872. Reappearance of a Peculiar Herring on the Norway Coast. $<$ Ann. Rec., 1872–73, p. 426.

373.—1872. Shad in Alabama. $<$ Ann. Rec., 1872–73, p. 427.

374.—1872. Shad in Red River, Arkansas. $<$ Ann. Rec., 1872–73, p. 428.

376.—1872. Planting of Shad in the Genesee River. $<$ Ann. Rec., 1872–73, p. 429.

377.—1872. Planting of Shad in Lake Champlain. $<$ Ann. Rec., 1872–73, p. 429.

378.—1872. Stocking California with Shad. $<$ Ann. Rec., 1872–73, p. 430.

379.—1872. Transferring Shad to the Sacramento River. $<$ Ann. Rec., 1872–73, p. 430.

400.—1872. Alleged Discovery of Young Shad in the Sacramento River. $<$ Ann. Rec., 1872–73, p. 447.

468.—1873. Effects of Seasons on the Distribution of Animals and Plants. $<$ Ann. Rec., 1873–75, p. 263.

481.—1873. Maynard on the Mammals of Florida. <*Ann. Rec.*, 1873–75, p. 286.

484.—1873. Geographical Distribution of Percoid Fishes. <*Ann. Rec.*, 1873–75, p. 291.

491.—1873. Absence of Fish Above the Yosemite Falls and in the headwaters of the Hudson. <*Ann. Rec.*, 1873–75, p. 305.

507.—1873. Binney on Geographical Distribution of Mollusks. <*Ann. Rec.*, 1873–75, p. 320.

511.—1873. Distribution of California Moths. <*Ann. Rec.*, 1873–75, p. 330.

521.—1873. Catalogue of Rhode Island Mollusca. [By H. F. Carpenter.] <*Ann. Rec.*, 1873–75, p. 339.

523.—1873. Number of American Birds. <*Ann. Rec.*, 1873–75, p. 340.

547.—1873. Shipments Eastward of California Salmon. <*Ann. Rec.*, 1873–75, p. 433.

559.—1873. Transporting Salmon Eggs to New Zealand. <*Ann. Rec.*, 1873–75, p. 445.

560.—1873. Naturalization of Trout in New Zealand. <*Ann. Rec.*, 1873–75, p. 447.

564.—1873. Shad in the Sacramento River. <*Ann. Rec.*, 1873–75, p. 449.

565.—1873. Shad in California Waters. <*Ann. Rec.*, 1873–75, p. 449.

580.—1873. Arrival of Salmon Eggs in New Zealand. <*Ann. Rec.*, 1873–75, p. 462.

641.—1874. Natural History of the Bermudas. <*Ann. Rec.*, 1874–75, p. 283.

648.—1874. Paucity of Mammals in Cuba. <*Ann. Rec.*, 1874–75, p. 300.

651.—1874. Extermination of Buffaloes. <*Ann. Rec.*, 1874–75, p. 303.

654.—1874. Occurrence of a Cuban Crocodile in Florida. <*Ann. Rec.*, 1874–75, p. 308.

686.—1874. Geographical Distribution of Asiatic Birds. <*Ann. Rec.*, 1874–75, p. 341. Review.

690.—1874. American King-Crab on the European Coast. <*Ann. Rec.*, 1874–75, p. 344.

703.—1874. Introduction of Prairie Chickens into the Eastern States. <*Ann. Rec.*, 1874–75, p. 391.

714.—1874. Introduction of British Fish into India. <*Ann. Rec.*, 1874–75, p. 426.

719.—1874. Stocking a Pond in Utah with Eels. <*Ann. Rec.*, 1874–75, p. 428.

785.—1875. Fauna of the Mammoth Cave. <*Ann. Rec.*, 1875–76, p. 31.

792.—1875. Professor Alfred Newton on the Migration of Birds. <*Ann. Rec.*, 1875–76, p. 340.

793.—1875. The Batrachia and Reptilia of North America. <*Ann. Rec.*, 1875–76, p. 343.

806.—1875. Fauna of the Caspian. <*Ann. Rec.*, 1875–76, p. 351.

808.—1875. Introduction of the American Turkey. <*Ann. Rec.*, 1875–76, p. 354.

830.—1875. Increase of English Fishes in Tasmania. <*Ann. Rec.*, 1875–76, p. 419.

832.—1875. Distribution of Trout Eggs from Tasmania to the Neighboring Colonies. <*Ann. Rec.*, 1875–76, p. 420.

833.—1875. Importation of Gourami into Paris. <*Ann. Rec.*, 1875–76, p. 420.

882.—1876. List of Birds collected by Charles S. McCarthy, Taxidermist.—Classified by Prof. Spencer F. Baird. <Simpson's Explorations across the Great Basin of Utah in 1859, pp. 377–381. Bastard title: Explorations across the Great Basin of Utah. Appendix K. Ornithology. A List of Birds, by Prof. Spencer F. Baird, pp. [376–7.]

894.—1876. Connecticut River Shad for California. The shipment of a million Shad Fry from Holyoke, Mass., to the Sacramento River, Cal., under the care of F. N. Clark and T. H. Bean. <*Forest and Stream*, vii, p. 66.

899.—1876. The Triassic Fauna in Illinois. <*Ann. Rec.*, 1876–77, p. 300.

909.—1876. The Migration of Birds. <*Ann. Rec.*, 1876–77, p. 311.

990.—1877. Salmon in the Hudson. <*Forest and Stream*, 1877, x, p. 154.

991.—1877. Salmon in the Chesapeake. <*Forest and Stream*, 1877, x, p. 296.

994.—1877. Genuine White Shad in the Ohio. <*Forest and Stream*, viii, 1877, p. 280.

997.—1878. The Delaware Salmon. <*Chicago Field*, ix, 1878, p. 165.

1009.—1878. The Delaware Salmon. <*Chicago Field*, ix, 1878, p. 165.

1018.—1879. The Hudson Salmon. <*Forest and Stream*, xii, 1879, p. 444.

GEOLOGY; MINERALOGY; AND PALEONTOLOGY.

13.—1850. On the Bone Caves of Pennsylvania. <*Proc. Amer. Assoc. Adv. Sci.*, ii, 1850, pp. 352-355. (Cambridge Meeting, Aug., 1849.) Read August 20, 1849.

130.—1870. Fossil Birds of the United States. <*Harper's New Monthly Mag.*, xl 1870, pp. 467, 469, and 470.

169.—1871. Fossil Fishes of Wyoming. <*Ann. Rec.*, 1871–72, p. 248.

170.—1871. Cephalaspis in America. <*Ann. Rec.*, 1871–72, p. 248.

171.—1871. Port Kennedy Bone-Cave. <*Ann. Rec.*, 1871–72, p. 249.

174.—1871. Existing specimens of the Great Auk. <*Ann. Rec.*, 1871–72, pp. 258–259.

224.—1872. Second Report of the Geological Survey of Indiana for 1870. <*Ann. Rec.*, 1872–73, p. 125.

225.—1872. Report of the Geological Survey of Ohio for 1870. <*Ann. Rec.*, 1872–73, p. 125.

227.—1872. Report of Mr. Clarence King—Vol. 5. <*Ann. Rec.*, 1872–73, p. 127.

228.—1872. Progress of the Geological Survey of California. <*Ann. Rec.*, 1872–73, p. 128.

229.—1872. Geology of the Bermudas. <*Ann. Rec.*, 1872–73, p. 137.

232.—1872. Professor Marsh's Explorations in 1871. <*Ann. Rec.*, 1872–73, p. 153.

234.—1872. Exploration of Prof. Hartt in Brazil. <*Ann. Rec.*, 1872–73, p. 157.

284.—1872. Have Trilobites Legs? <*Ann. Rec.*, 1872–73, p. 268.

295.—1872. New American Fossil Vertebrates. <*Ann. Rec.*, 1872–73, p. 305.

296.—1872. Proboscidians of the American Eocene. <*Ann. Rec.*, 1872–73, p. 307.

297.—1872. The Armed Metalophodon. <*Ann. Rec.*, 1872–73, p. 308.

298.—1872. Fossil Fishes and Insects from the Nevada Shales. <*Ann. Rec.*, 1872–73, p. 308.

313.—1872. A Fossil Lemuroid from the Eocene of Wyoming. <*Ann. Rec.*, 1872–73, p. 326.

316.—1872. Prehistoric Remains in Wyoming. <*Ann. Rec.*, 1872–73, p. 330.

320.—1872. Fossil Elephant in Alaska. <*Ann. Rec.*, 1872–73, p. 333.

326.—1872. Relation of Recent North American Flora to Ancient. <*Ann. Rec.*, 1872–73, p. 352.

440.—1873. Third and Fourth Annual Report of the Geological Survey of Indiana for 1871, 1872. <*Ann. Rec.*, 1873–75, p. 202.

441.—1873. Report for 1872 on the Geology of New Jersey. <*Ann. Rec.*, 1873–75, p. 203.

442.—1873. Geological Survey of Canada for 1871–72. <*Ann. Rec.*, 1873–75, p. 202.

443.—1873. Fourth Annual Report of Mining Statistics for 1872. <*Ann. Rec.*, 1873–75, p. 203.

444.—1873. Geological Survey of Ohio. <*Ann. Rec.*, 1873–75, p. 204.

445.—1873. Final Report of the Geological Survey of Ohio. <*Ann. Rec.*, 1873–75, p. 205.

446.—1873. Lesquereux on the Fossil Plants of the Northern Hemisphere. <*Ann. Rec.*, 1873–75, p. 210.

479.—1873. New Fossil Carnivora. <*Ann. Rec.*, 1873–75, p. 284.

480.—1873. Orophippus agilis. <*Ann. Rec.*, 1873–75, p. 286.

488.—1873. Number of Glyptodonts, or Extinct Giant Armadillos. <*Ann. Rec.*, 1873–75, p. 302.

495.—1873. New Vertebrate Fossils. <*Ann. Rec.*, 1873–75, p. 308.

496.—1873. Aboriginal Monkey. <*Ann. Rec.*, 1873–75, p. 310.

501.—1873. The Fossils Discovered by Professor Cope. <*Ann. Rec.*, 1873–75, p. 315.

505.—1873. Relations of the Megatheriidæ. <*Ann. Rec.*, 1873–75, p. 318.

652.—1874. The new Fossil Bird of the Sheppey Clay. <*Ann. Rec.*, 1874–75, p. 305

668.—1874. Fossil Vertebrates in Ohio. <*Ann. Rec.*, 1874–75, p. 322.

682.—1874. The Fossil Hog of América. <*Ann. Rec.*, 1874–75, p. 340.

784.—1875. Discovery of Animal Remains in the Lignite Beds of the Saskatchewan District. <*Ann. Rec.*, 1875–76. p. 311.

790.—1875. Discovery in Newfoundland of the Great Auk. <*Ann. Rec.*, 1875–76, p. 339.

868.—1875. Annual Report of the United States Geological and Geographical Survey of the Territories. <*Ann. Rec.*, 1875–76, p. 573.

899.—1876. The Triassic Fauna in Illinois. <*Ann. Rec.*, 1876–77, p. 300.

900.—1876. Remains of the Irish Elk. <*Ann. Rec.*, 1876–77, p. 300.

908.—1876. Additional Remains of the Moa. <*Ann. Rec.*, 1876–77, p. 311.

912.—1876. New Fossil of Giant Birds. <*Ann. Rec.*, 1876–77, p. 313.

ANTHROPOLOGY.

13.—1859. On the Bone Caves of Pennsylvania. <*Proc. Amer. Assoc. Adv. Sci.*, ii, 1850, pp. 352–355. (Cambridge Meeting, Aug., 1849.) Read Aug. 20, 1849.

145.—1871. Darwin on The Descent of Man. <*Ann. Rec.*, 1871–72, p. 156.

146.—1871. Shell-heaps in New Brunswick. <*Ann. Rec.*, 1871–72, p. 182.

168.—1871. Ancient City in New Mexico. <*Ann. Rec.*, 1871–72, p. 241.

171.—1871. Port Kennedy Bone-Cave. <*Ann. Rec.*, 1871–72, p. 249.

253.—1872. Prehistoric Beads. <*Ann. Rec.*, 1872–73, p. 221.

254.—1872. The Tanis Stone. <*Ann. Rec.*, 1872–73, p. 230.

255.—1872. German Central Museum of Ethnology. <*Ann. Rec.*, 1872–73, p. 232.

256.—1872. Stranding of a Japanese Junk on the Aleutian Islands. <*Ann. Rec.*, 1872–73, p. 232.

257.—1872. Use of the Boomerang by American Indians. <*Ann. Rec.*, 1872–73, p. 235.

258.—1872. Dwarfed Human Head. <*Ann. Rec.*, 1872–73, p. 235.

259.—1872. Shell Mound near Newburyport. <*Ann. Rec.*, 1872–73, p. 236.

260.—1872. Journal of the Anthropological Institute of New York. <*Ann. Rec.*, 1872–73, p. 236.

290.—1872. Prehistoric (?) Man in America. <*Ann. Rec.*, 1872–73, p. 295.

305.—1872. People using the Boomerang. <*Ann. Rec.*, 1872–73, p. 316.

314.—1872. Prehistoric Remains in Unalashka. <*Ann. Rec.*, 1872–73, p. 327.

315.—1872. Archæology in America. <*Ann. Rec.*, 1872–73, p. 329.

316.—1872. Prehistoric Remains in Wyoming. <*Ann. Rec.*, 1872–73, p. 330.

317.—1872. Peculiar Mound Crania. <*Ann. Rec.*, 1872–73, p. 321.

319.—1872. Objects from Florida Mounds. <*Ann. Rec.*, 1872–73, p. 332.

459.—1873. Explorations of W. H. Dall in the Aleutian Islands. <*Ann. Rec.*, 1873–75, p. 246.

476.—1873. Preservation of British Prehistoric Monuments. <*Ann. Rec.*, 1873–75, p. 276.

477.—1873. Prehistoric Cannibalism in Florida. <*Ann. Rec.*, 1873–75, p. 281.

478.—1873. Working of Mica Mines in North Carolina in Prehistoric Times. <*Ann. Rec.*, 1873–75, p. 282.

497.—1873. The Prehistoric Races of America. <*Ann. Rec.*, 1873–75, p. 311.

498.—1873. The Cesnola Collection. <*Ann. Rec.*, 1873–75, p. 312.

499.—1873. The Canstadt Race of Mankind. <*Ann. Rec.*, 1873–75, p. 312.

522.—1873. The Mummied Heads of the Peruvian Indians. <*Ann. Rec.*, 1873–75, p. 339.

524.—1873. Ethnology of the Peat Bogs. <*Ann. Rec.*, 1873–75, p. 341.

527.—1873. The Fossils Discovered by Professor Cope. <*Ann. Rec.*, 1873–75, p. 348.

528.—1873. Antiquities of the Southern Indians. <*Ann. Rec.*, 1873–75, p. 348.

611.—1873. Benevolent Endowment in the United States Treasury. <*Ann. Rec.*, 1873–75, p. 666.

647.—1874. Prepared Heads of Macas Indians. <*Ann. Rec.*, 1874–75, p. 297.

668.—1874. Exhibition of British Ethnology. <*Ann. Rec.*, 1874–75, p. 326.

670.—1874. Footprints in Solid Rock. <*Ann. Rec.*, 1874–75, p. 328.

671.—1874. Ancient Stone Fort in Indiana. <*Ann. Rec.*, 1874–75, p. 329.

672.—1874. Dall's Ethnological Explorations in Alaska. <*Ann. Rec.*, 1874–75, p. 329.

673.—1874. Trade among the Aborigines. <*Ann. Rec.*, 1874–75, p. 331.

693.—1874. Discovery of the Aleut Mummies. <*Ann. Rec.*, 1874–75, p. 345.

708.—1874. Consumption of Marine Products in Washington. <*Ann. Rec.*, 1874–75, p. 420.

720.—1874. Spinal Column of the Sturgeon as an Article of Food. <*Ann. Rec.*, 1874–75, p. 447.

787.—1875. Mr. George Latimer's Archæological Collection from Porto Rico. <*Ann. Rec.*, 1875–76, p. 325.

788.—1875. Stone Knives with Handles, from the Pai-Utes. <*Ann. Rec.*, 1875–76, p. 526.

789.—1875. Archæology of the Mammoth Cave. <*Ann. Rec.*, 1875–76, p. 327.

790.—1875. Discovery in Newfoundland of the Great Auk. <*Ann. Rec.*, 1875–76, p. 339.

INDUSTRY AND ART.

4.—1844. On the application of bi-chromate of potassa to photographic purposes. <*Literary Record and Journal of the Linnæan Association of Pennsylvania College*, i, No. 2, Dec., 1844, pp. 17–19.

105.—1860. On the principal plants used as food by man.—Sketch of the plants chiefly used as food by man, in different parts of the world and at various periods.—(Translated from the German) <*Report of the Commissioner of Patents for the year 1859.—Agriculture*, 1860, pp. 299–362.

189.—1871. Killing Fish with Torpedoes in Florida. <*Ann. Rec.*, 1871–72, p. 267.

197.—1871. Fish-Guano Flour from Loffoden. <*Ann. Rec.*, 1871–72, pp. 342, 343.

201.—1871. Artificial Ice in Packing Fish. <*Ann. Rec.*, 1871–72, p. 355.

202.—1871. Preservation of Dead Salmon for an indefinite time. <*Ann. Rec.*, 1871–72, p. 356.

203.—1871. Importance of Killing freshly-captured Fish. <*Ann. Rec.*, 1871–72, p. 387.

204.—1871. Ship-Canal across Cape Cod. <*Ann. Rec.*, 1871–72, pp. 422, 423.

205.—1871. Ship-Canal across New Jersey. <*Ann. Rec.*, 1871–72, pp. 423, 424.

206.—1871. Oil from Birds. <*Ann. Rec.*, 1871–72, pp. 466, 467.

209.—1871. "Archives of Science." <*Ann. Rec.*, 1871–72, p. 604.

211.—1871. Fishing Steamer. <*Ann. Rec.*, 1871–72, p. 606.

222.—1871. General summary of scientific and industrial progress for the year 1871. <*Ann. Rec.*, 1871–72, pp. xvii–xxxii.

258.—1872. Dwarfed Human Head. <*Ann. Rec.*, 1872–73, p. 235.

300.—1872. "The Lens," a new Scientific Journal. <*Ann. Rec.*, 1872–73, p. 310.

384.—1872. Oil Works on Unalaschka. <*Ann. Rec.*, 1872–73, p. 436.

395.—1872. Utilization of Refuse Fish. <*Ann. Rec.*, 1872–73, p. 444.

403.—1872. Department Report on the Preparation of Timber. <*Ann. Rec.*, 1872–73, p. 485.

461.—1875. General Summary of Scientific and Industrial Progress during the year 1874. <*Ann. Rec.*, 1874–75, p. xix.

533.—1873. Fish Guano. <*Ann. Rec.*, 1873–75, p. 387.

534.—1873. Value of Sea-Weed Manure. <*Ann. Rec.*, 1873–75, p. 395.

585.—1873. Oil from Birds. <*Ann. Rec.*, 1873–75, p. 566.

587.—1873. Utilization of Old Fish Pickle. <*Ann. Rec.*, 1873–75, p. 502.

589.—1873. Action of Cod-Liver Oil. <*Ann. Rec.*, 1873–75, p. 623.

826.—1875. Effect of Polluted Water on Fishes. <*Ann. Rec.*, 1875–76, p. 416.

844.—1875. Manufacture of Cod-Liver Oil. <*Ann. Rec.*, 1875–76, p. 428.

855.—1875. New Fish Product. <*Ann. Rec.*, 1875–76, p. 435.

872.—1875. Index of Patents from 1790 to 1873. <*Ann. Rec.*, 1875–76, p. 579.

931.—1876. Proposed Utilization of Fish Bones. <*Ann. Rec.*, 1876–77, p. 372.

941.—1876. New use for the Scrap of the Moss-Bunker. <*Ann. Rec.*, 1876–77, p. 390.

942.—1876. Utilizing the Offal of Codfish on the Gulf of St. Lawrence. <*Ann. Rec.* 1879–77, p. 391.

EXPLORATION AND GEOGRAPHY.

24.—1852. An account of natural history explorations in the United States during 1851· <*Sixth Annual Report Smithsonian Institution for the year 1851*, 1852, pp. 52–56. Appendix A to Report of Assistant Secretary.

42.—1853. Account of scientific explorations and reports on explorations, made in America during the year 1852. <*Seventh Annual Report Smithsonian Institution for the year 1852*, 1853, pp. 58–65. Appendix A to Assistant Secretary's Report.

12.—1855. Report on American Explorations in the years 1853 and 1854. <*Ninth Annual Report Smithsonian Institution (1854)*, 1855, pp. 79–97.

68.—1856. [Report of the Assistant Secretary of the Smithsonian Institution for the·year 1855.] <*Tenth Annual Rep. Smithsonian Institution (1855)*, 1856, pp. 36–61.

71.—1857. [Report of the Assistant Secretary of the Smithsonian Institution for the year 1856.] <*Eleventh Annual Rep. Smithsonian Institution for the year 1856*, 1857, pp. 47–68.

81.—1858. [Report for 1857 of the Assistant Secretary of the Smithsonian Institution.] <*Ann. Rep. Smithsonian Institution for the year 1857*, 1858, pp· 38–54.

89.—1859. Report for 1858 of the Assistant Secretary of the Smithsonian Institution. <*Ann. Rep. Smiths. Inst. for the year 1858*, 1859, pp. 44–62.

98.—1859. S. F. BAIRD's Résumé of Ornithological Field Operations in progress in America, etc. <*Ibis*, i, 1859, pp. 334, 335.

102.—1860. [Report for 1859 of the Assistant Secretary of the Smithsonian Institution.] <*Ann. Rep. Smiths. Inst. for the year 1859*, 1860, pp. 54–78.

107.—1861. [Report for 1860 of the Assistant Secretary of the Smithsonian Institution.] <*Ann. Rep. Smithsonian Institution for the year 1860*, 1861, pp. 55–86.

American explorations of the year...pp. 66–72
Continuation of enumeration of collections in the Museum, Nos. 73–94 .. 75

109.—1862. [Report for 1861 of the Assistant Secretary of the Smithsonian Institution.] <*Ann. Rep. Smithsonian Institution for the year 1861*, 1860, pp. 48–67.

Explorations of the year...pp. 58–61

110.—1863. [Report for 1862 of the Assistant Secretary of the Smithsonian Institution.] <*Ann. Rep. Smithsonian Institution for the year 1862*, 1863, pp. 46–59.

111.—1863. [Notice of R. Kennicott's and J. Xantus's Movements in North America.] <*Ibis*, v, 1863, pp. 238, 239.

112.—1863. [Letter on J. Xantus's collections at Colima, Mexico.] <*Ibis*, v, 1863, p. 476.

113.—1864. [Report for 1863 of the Assistant Secretary of the Smithsonian Institution.] <*Ann. Rep. Smithsonian Institution for the year* 1863, 1864, pp. 44–63.

Explorations of the year...pp. 52–56

116.—1865. Report of the Assistant Secretary (of the Smithsonian Institution for the year 1864). <*Ann. Rep. Smithsonian Institution for the year* 1864, 1865, pp. 74–100.

Explorations of the year...pp. 81–84

117.—1866. Report of the Assistant Secretary, Spencer F. Baird, relative to exchanges, collections of Natural History, &c. <*Ann. Rep. Smithsonian Institution for the year* 1865, pp. 75–88.

126.—1868. Report of Prof. S. F. Baird (Assistant Secretary of the Smithsonian Institution, for the year 1867). <*Ann. Rep. Smithsonian Institution for the year* 1867, 1868, pp. 64–78.

The report on the additions to the Museum and on the explorations, hitherto given in the report of the Assistant Secretary, are this year and afterwards incorporated in the report of the Secretary.

128.—1869. Explorations and Collections in Natural History (etc.). <*Ann. Rep. Smithsonian Institution for the year* 1868, 1869 (*in Rep. of the Secretary,* pp. 22–41 +) and pp. 54–67.

134.—1871. Rocky Mountain Explorations. <*Annual Record of Science an dIndustry,* 1871–72, p. 131.

135.—1871. Explorations of Professor Powell. <*Ann. Rec.,* 1871–72, pp. 132, 133.

136.—1871. Explorations of Professor Cope. <*Ann. Rec.,* 1871–72, pp. 133, 134.

137.—1871. Headwaters of the Yellowstone. <*Ann. Rec.,* 1871–72, p. 137.

138.—1871. Raymond's Report on the Yukon. <*Ann. Rec.,* 1871–72, pp. 138, 139.

139.—1871. Explorations in Vineyard Sound [by Professor Verrill]. <*Ann. Rec.,* 1871–72, p. 140.

140.—1871. Explorations in the West Indies. <*Ann. Rec.,* 1871–72, pp. 141, 142.

141.—1871. Explorations of Dr. Habel (in South America). <*Ann. Rec.,* 1871, pp. 142, 143.

142.—1871. Explorations of Professor Hartt (in Brazil). <*Ann. Rec.,* 1871–72, pp. 146, 147.

192.—1871. Zoological Stations in the Gulf of Naples. <*Ann. Rec.,* 1871–72, pp. 274, 275.

193.—1871. Verrill's Exploration in New Jersey. <*Ann. Rec.,* 1871–72, p. 276.

194.—1871. Dr. Stimpson's Exploration in Florida. <*Ann. Rec.,* 1871–72, p. 377.

213.—1871. Return of Mr. Gwyn Jeffreys to England. <*Ann. Rec.,* 1871–72, p. 608.

219.—1872. Explorations and Collections (of the Smithsonian Institution in the year 1871). <*Ann. Rep. Smithsonian Institution for the year* 1871, 1872, *in Report of Secretary,* pp. 26–34 + 42–62.

220.—1872. [Instructions to Capt. C. F. Hall for collecting objects of Natural History on the Expedition toward the North Pole.] <*Ann. Rep. Smithsonian Institution for the year* 1871, 1872, pp. 379–381.

223.—1872. Yellowstone Park. <*Ann. Rec.,* 1872–73, p. 125.

(Notice of passage of law establishing National Park.)

224.—1872. Second Report of the Geological Survey of Indiana for 1870. <*Ann. Rec.,* 1872–73, p. 125.

225.—1872. Report of the Geological Survey of Ohio for 1870. <*Ann. Rec.,* 1872–73, p. 125.

226.—1872. Report of Professor Hayden's Explorations. <*Ann. Rec.,* 1872–73, p. 126.

227.—1872. Report of Mr. Clarence King—Vol. 5. <*Ann. Rec.,* 1872–73, p 127.

228.—1872. Progress of the Geological Survey of California. <*Ann. Rec.,* 1872–73, p. 128.

230.—1872. Explorations of Lieutenant Wheeler in 1871. <*Ann. Rec.*, 1872-73, p. 152.

231.—1872. Explorations of Major Powell in 1871. <*Ann. Rec.*, 1872-73, p. 152.

232.—1872. Professor Marsh's Explorations in 1871. <*Ann. Rec.*, 1872-73, p. 153.

233.—1872. Explorations of Dr. Stimpson. <*Ann. Rec.*, 1872-73, p. 154.

234.—1872. Explorations of Prof. Hartt in Brazil. <*Ann. Rec.*, 1872-73, p. 157.

235.—1872. Recent Explorations in the United States. <*Ann. Rec.*, 1872-73, p. 173.

236.—1872. Explorations of William H. Dall. <*Ann. Rec.*, 1872-73, p. 175.

237.—1872. Explorations of the Navy Department in the North Pacific. <*Ann. Rec.*, 1872-73, p. 180.

238.—1872. Explorations of the Challenger. <*Ann. Rec.*, 1872-73, p. 181.

239.—1872. Explorations of Professor Powell in 1872. <*Ann. Rec.*, 1872-73, p. 192.

240.—1872. Report of the Circumnavigating Committee of the Royal Society. <*Ann. Rec.* 1872-73, p. 193.

241.—1872. Explorations of the Portsmouth. <*Ann. Rec.*, 1872-73, p. 196.

242.—1872. Explorations by Professor Hayden in 1872. <*Ann. Rec.*, 1872-73, p. 197.

244.—1872. The Voyage of the Hassler. <*Ann. Rec.*, 1872-73, p. 204.

245.—1872. Exploration in Central America. <*Ann. Rec.*, 1872-73, p. 211.

246.—1872. Powell's Report. <*Ann. Rec.*, 1872-73, p. 212.

247.—1872. Grandidier on the Zoology of Madagascar. <*Ann. Rec.*, 1872-73, p. 213.

248.—1872. Visit of Abbé David to Thibet. <*Ann. Rec.*, 1872-73, p. 213.

249.—1872. Fulfillment of the Predictions of Professor Agassiz. <*Ann Rec.*, 1872-73, p. 214.

440.—1873. Third and Fourth Annual Report of the Geological Survey of Indiana for 1871, 1872. <*Ann. Rec.*, 1873-75, p. 202.

441.—1873. Report for 1872 on the Geology of New Jersey. <*Ann. Rec.*, 1873-75, p. 203.

442.—1873. Geological Survey of Canada for 1871-72. <*Ann. Rec.*, 1873-75, p. 202.

444.—1873. Geological Survey of Ohio. <*Ann. Rec.*, 1873-75, p. 204.

445.—1873. Final Report of the Geological Survey of Ohio. <*Ann. Rec.*, 1873-75, p. 205.

447.—1873. Explorations in the Gulf of St. Lawrence in 1872. <*Ann. Rec.*, 1873-75, p. 216.

448.—1873.—Fauna of the St. George's Bank and adjacent waters. <*Ann. Rec.* 1873-75, p. 218.

449.—1873.—Report of the German North Polar Expedition. <*Ann. Rec.*, 1873-75, p. 220.

450.—1873.—Fictitious Account of Pary's Explorations. <*Ann. Rec.*, 1873-75, p. 221.

451.—1873.—Report on the Yellowstone Park. <*Ann. Rec.*, 1873-75, p. 222.

452.—1873.—Explorations of Lieut. Wheeler in 1871. <*Ann. Rec.*, 1873-75, p. 223.

453.—1873.—Dr. Hayden's Surveys. <*Ann. Rec.*, 1873-75, p. 226.

454.—1873.—Sixth Annual Report of Dr. Hayden's Explorations. <*Ann. Rec.*, 1873-75, p. 232.

455.—1873.—Report of Major J. W. Barlow. <*Ann. Rec.*, 1873-75, p. 232.

456.—1873.—Final Report of Dr. Hayden's Explorations. <*Ann. Rec.*, 1873-75, p. 236.

457.—1873.—The History of the Polaris. <*Ann. Rec.*, 1873-75, p. 237.

458.—1873.—The Cruise of the "Challenger" in 1873. <*Ann. Rec.*, 1873-75, p. 243.

459.—1873.—Explorations of W. H. Dall in the Aleutian Islands. <*Ann. Rec.*, 1873-75, p. 246.

460.—1873. Dr. Hayden's Geological Explorations in 1873. <*Ann. Rec.*, 1873-75, p. 249.

461.—1873. Lieutenant Wheeler's Exploration in 1873. <*Ann. Rec.*, 1873-75, p. 251.

462.—1873. Explorations of Captain William A. Jones. <*Ann. Rec.*, 1873–75, p. 254.

463.—1873. Interoceanic Canal Explorations by the United States Navy. <*Ann. Rec.*, 1873–75, p. 255.

464.—1873. Natural-History Explorations of the Northern Boundary Survey. <*Ann. Rec.*, 1873–75, p. 257.

465.—1873. Explorations of Professor Powell. <*Ann. Rec.*, 1873–75, p. 258.

466.—1873. Yellowstone Expedition. <*Ann. Rec.*, 1873–75, p. 261.

467.—1873. Recent Explorations in Spitzenberg. <*Ann. Rec.*, 1873–75, p. 262.

506.—1873. Recent Explorations of Professor Cope. <*Ann. Rec.*, 1873–75, p. 319.

632.—1874. Explorations of Pinart in Alaska. <*Ann. Rec.*, 1874–75, p. 246.

633.—1874. Explorations of W. M. Gabb in Costa Rica. <*Ann. Rec.*, 1874–75, p. 246.

634.—1874. Professor Orton's Explorations. <*Ann. Rec.*, 1874–75, p. 248.

635.—1874. Horetzky on the Hudson Bay Territory. <*Ann. Rec.*, 1874–75, p. 256.

636.—1874. Professor Stoddard's Expedition to Colorado. <*Ann. Rec.*, 1874–75, p. 257.

637.—1874. Explorations of Professor Powell in 1874. <*Ann. Rec.*, 1874–75, p. 262.

638.—1874. Explorations in 1874 of Lieutenant G. M. Wheeler, United States Engineers. <*Ann. Rec.*, 1874–75, p. 267.

639.—1874. Explorations of Dr. Hayden in 1874. <*Ann. Rec.*, 1874–75, p. 275.

672.—1874. Dall's Ethnological Explorations in Alaska. <*Ann. Rec.*, 1874–75, p. 329.

771.—1875. "(Report on) additions to the Museum and the various operations connected with it during the past year." <*Ann. Rep. Smithsonian Institution for the year* 1874, 1875, pp. 27–44, 49–76.

777.—1875. The Saranac Exploring Expedition. <*Ann. Rec.*, 1875–76, p. 260.

778.—1875. Explorations under Dr. Hayden in 1875. <*Ann. Rec.*, 1875–76, p. 263.

779.—1875. Explorations under Major Powell in 1875. <*Ann. Rec.*, 1875–76, p. 286.

780.—1875. Explorations and Surveys under Lieutenant George M. Wheeler, U. S. Army, in 1875. <*Ann. Rec.*, 1875–76, p. 293.

781.—1875. Major Powell's Final Report. <*Ann. Rec.*, 1875–76, p. 298.

782.—1875. Reports of the Northern Boundary Surveys. <*Ann. Rec.*, 1875–76, p. 300.

868.—1875. Annual Report of the United States Geological and Geographical Survey of the Territories for 1873. <*Ann. Rec.*, 1875–76, p. 573.

870.—1875. Report of the Icelandic Commission to Alaska. <*Ann. Rec.*, 1875–76, p. 576.

884.—1876. United States Fish Commission at Wood's Hole, Mass., 1875. <*Ann. Rec.*, 1875–76, p. cxxiv.

892.—1876. The U. S. Fish Commission. <*Forest and Stream*, vi, 1876, p. 147.

896.—1876. Work accomplished by the Challenger. <*Ann. Rec.*, 1876–77, p. 240.*

897.—1875. Explorations made under the direction of F. V. Hayden in 1876. <*Ann. Rec.*, 1876–77, p. 242.

898.—1876. Exploration of the Rocky Mountain Region by J. W. Powell. <*Ann. Rec.*, 1876–77, p. 255.

1013.—1878. Natural History of the Howgate Expedition. <*Forest and Stream*, ix., 1878, p. 413.
Memorandum given to Mr. L. Kumlein.

SPECIAL PAPERS IN THE REPORT OF THE UNITED STATES COMMISSIONER OF FISHERIES.

[Report for 1871.] See Nos. 421 and 422, pp. vii–xl. [Analysis, see above, pp. 134–135.]

423.—1873. Memoranda of inquiry relative to the Food-Fishes of the United States. <*Rep. U. S. Comm. Fish and Fisheries*, part i, 1873, pp. 1–3.

424.—1873. Questions relative to the Food-Fishes of the United States. <*Rep. U. S. Comm. Fish and Fisheries*, part i, 1873, pp. 3-6.

425.—1873. Testimony in regard to the present condition of the Fisheries, taken in 1871. <*Rep. U. S. Comm. Fish and Fisheries*, part i, 1873, pp. 7-72.

426.—1873. Report of conference held [by the U. S. Commissioner of Fish and Fisheries] at Boston, October 5, 1871, with the Fishery Commissioners of Massachusetts and Rhode Island. <*Rep. U. S. Comm. Fish and Fisheries*, 1873, part i, pp. 125-131.

427.—1873. Draught of law proposed for the consideration of and enactment by the Legislatures of Massachusetts, Rhode Island, and Connecticut. <*Rep. U. S. Comm. Fish and Fisheries*, part i, 1873, pp. 132-134.

428.—1873. Notices in regard to the Abundance of Fish on the New England Coast in former times. <*Rep. U. S. Comm. Fish and Fisheries*, part i, 1873, pp. 148-172.

429.—1873. Statistics of Fish and Fisheries on the South Shore of New England. <*Rep. U. S. Comm. Fish and Fisheries*, part i, 1873, pp. 172-181.

430.—1873. Supplementary Testimony and Information relative to the Condition of the Fisheries of the South Side of New England, taken in 1872. <*Rep. U. S. Comm. Fish and Fisheries*, part i, 1873, pp. 182-195.

431.—1873. Natural History of some of the more important Food-fishes of the South Shore of New England.—I. The Scup, *Stenotomus argyrops*, (Linn.,) Gill. <*Rep. U. S. Comm. Fish and Fisheries*, part i, 1873, pp. 228-235.

432.—1873. [Natural History of some of the more important Food-fishes of the South Shore of New England.]—II. The Blue-Fish, *Pomatomus saltatrix*, (Linn.,) Gill. <*Rep. U. S. Comm. Fish and Fisheries*, part i, 1873, pp. 235-252.

433.—1873. Description of Apparatus used in capturing Fish on the Sea-coast and Lakes of the United States. <*Rep. U. S. Comm. Fish and Fisheries*, part i, 1873, pp. 253-274.

434.—1873. List of Patents granted by the United States to the end of 1872, for inventions connected with the Capture, Utilization or Cultivation of Fishes and Marine Invertebrates. <*Rep. U. S. Comm. Fish and Fisheries*, part i, 1873, pp. 275-280.

435.—1873. List of Fishes collected at Wood's Hole. <*Rep. U. S. Comm. Fish and Fisheries*, part i, 1873, pp. 823-827.

627 and **628.**—Report of Commission for 1872-3.—Letters 627 and 628, pp. I-XCII. [Analysis above. pp. 154-157.]

629.—1874. Temperatures in the Gulf of Mexico. <*Rep. U. S. Comm. Fish and Fisheries*, part ii, 1874, pp. 745-748.

630.—1874. Reports of Special Conferences (of the U. S. Commissioner of Fisheries) with the American Fish-Culturists' Association and State Commissioners of Fisheries. <*Rep. U. S. Comm. Fish and Fisheries*, part ii, 1874, pp. 757-773.

889-890.—Report of Commission for 1873-4 and 1874-5. See Nos. 889-90. pp. vii-xlv. Analysis, pp. 184-186.

1002-1003.—Report of Commission for 1875-76. See Nos. 1002-1003. pp. 1-32. Analysis, pp. 199-207.

1030-1031.—Report of Commission for 1877. See Nos. 1030-1031. pp. . Analysis, pp. 208-220.

1020.—Report of Commission for 1878.

CONTRIBUTIONS TO REPORTS OF VARIOUS GOVERNMENT SURVEYS.

28, 29, 30, 31, 32, 33.—In Stansbury's Exploration and Survey of the Valley of the Great Salt Lake of Utah, &c. Washington, 1852.

47.—In Sitgreaves' Report on Expedition down the Zuni and Colorado Rivers. Washington, 1853.

49 and 50.—In Marcy and McClellan's Exploration of the Red River of Louisiana. Washington, 1853.

64-5.—In Gillis'—The U. S. Naval Astronomical Expedition to the Southern Hemisphere during the years 1849, '50, '51, and '52. Washington, 1855.

66, 76, 78, 79, 92, 93, 94, 95, 96, 97.—In Explorations and Surveys for a Railroad Route from the Mississippi River to the Pacific Ocean. 1855–9.

86, 87, 88.—In Report of United States and Mexican Boundary Survey. 1859.

108.—In Report upon the Colorado River of the West. 1861.

882.—In Report of Simpson's Explorations across the Great Basin of Utah in 1859.

COLLECTING AND TAXIDERMY.

4.—1844. On the application of bi-chromate of potassa to photographic purposes. *< Literary Record and Journal of the Linnæan Association of Pennsylvania College*, i, No. 2, Dec., 1844, pp. 17–19.

8.—1846. Hints for preserving Objects of Natural History, prepared by Prof. S. F. Baird, for Dickinson College, Carlisle, Pa. Carlisle: Printed by Gilt & Hinckley, 1846. 8vo. pp. 12.

16.—1850. General Directions for Collecting and Preserving Objects of Natural History.

25.—1852. Directions for Collecting, Preserving, and Transporting Specimens of Natural History, prepared for the use of the Smithsonian Institution. (Seal.) Smithsonian Institution: Washington, 1 January, 1852. 8vo. pp. 23.

48.—1853. [Directions for making collections in Natural History, prepared for the use of the parties engaged in the Exploration of a route for the Pacific Railroad along the 49th parallel.] 4to, about 10 pp. Printed on thin blue paper.

73.—1857. Directions for collecting, preserving, and transporting specimens of natural history. Prepared for the use of the Smithsonian Institution. *<Eleventh Ann. Rep. Smithsonian Institution* (1856), 1857, pp. 235–253.

84.—1858. Registers of Periodical Phenomena. *<Directions for Meteorological Observations & the Registry of Periodical Phenomena, Smith's Misc. Coll.* (148), pp. 63–68.

90.—1859. Smithsonian Miscellaneous Collections. Directions for collecting, preserving, and transporting specimens of natural history. Prepared for the use of the Smithsonian Institution. [Seal of Smithsonian Institution.] [Third edition.] Washington: Smithsonian Institution. March, 1859. 8vo. pp. 40, 6 cuts.

No. 34, S. I. in Smithsonian Miscellaneous Collections, Vol. II, Art. VII.

106.—1861. Smithsonian Miscellaneous Collections.—Instructions in reference to collecting Nests and Eggs of North American Birds. 8vo. pp. 82. (No title-page.)

220.—1872. (Instructions to Capt. C. F. Hall for collecting objects of Natural History on the Expedition toward the North Pole.) *<Ann. Rep. Smithsonian Institution, for the year* 1871, 1872, pp. 379–381.

MUSEUMS, SOCIETIES, ZOOLOGICAL STATIONS, &C.

192.—1871. Zoological Stations in the Gulf of Naples. *<Ann. Rec.*, 1871–72, pp. 274, 275.

214.—1871. Destruction of the Chicago Academy of Science by Fire. *<Ann. Rec.*, 1871–72, p. 609.

299.—1872. Report of the Museum of Comparative Zoology. <*Ann. Rec.*, 1872–73, p. 309.

404.—1872. Report of the Zoological Society for 1871. <*Ann. Rec.*, 1872–73, p. 601.

405.—1872. Washington Meeting of the Natural Academy of Sciences. <*Ann Rec.*, 1872–73, p. 604.

406.—1872. Twenty-first Meeting of the American Association for the Advancement of Science. <*Ann Rec.*, 1872–73, p. 607.

407.—1872. Fifth Report of the Peabody Museum, Cambridge. <*Ann. Rec.*, 1872–73, p. 608.

408.—1872. Report of the Peabody Academy of Science for 1871. <*Ann. Rec.*, 1872–73, p. 608.

409.—1872. Bloomington Scientific Association. <*Ann. Rec.*, 1872–73, p. 609.

410—1872. Meeting of the American Philological Society. <*Ann. Rec.*, 1872–73, p. 609.

411.—1872. Circular of the Chicago Academy of Sciences. <*Ann. Rec.*, 1872–73, p. 610.

412.—1872. Regulations of the New York Museum of Natural History. <*Ann. Rec.*, 1872–3, p. 611.

415.—1872. Renewed Activity of the St. Louis Academy of Sciences. <*Ann. Rec.*, 1872–73, p. 612.

423.—1873. "Account of the additions to the National Museum and the various operations connected with it during 1872." <*Ann.,Rep., Smithsonian Institution for the year* 1872, 1873, pp. 43–52 and 55–62.

469.—1873. The Oldest Zoological Museum in America. <*Ann. Rec.*, 1873–75, p. 264.

470.—1873. Cincinnati Acclimation Society. <*Ann. Rec.*, 1873–75, p. 265.

472.—1873. Opening of the "Anderson School of Natural History." <*Ann. Rec.*, 1873–75, p. 266.

473.—1873. The Brighton Aquarium. <*Ann. Rec.*, 1873–75, p. 267.

474.—1873. The Zoological Gardens of London. <*Ann. Rec.*, 1873–75, p. 267.

475.—1873. The Godeffroy Museum, at Hamburg. <*Ann. Rec.*, 1873–75, p. 268.

476.—1873. Preservation of British Prehistoric Monuments. <*Ann. Rec.*, 1873–75, p. 276.

Suggestions for United States.

489.—1873. International Exhibition of Horns. <*Ann. Rec.*, 1873–75, p. 303.

498.—1873. The Cesnola Collection. <*Ann. Rec.*, 1873–75, p. 312.

502.—1873. Additions to Yale College Museum. <*Ann. Rec.*, 1873–75, p. 315.

515.—1873. Bird Collections in London. <*Ann. Rec.*, 1873–75, p. 336.

516.—1873. Brighton Aquarium. <*Ann. Rec.*, 1873–75, p. 336.

518.—1873. An Aquarium for Central Park. <*Ann. Rec.*, 1873–75, p. 337.

519.—1873. Report of the Central Park Menagerie. <*Ann. Rec.*, 1873–75, p. 338.

520.—1873. The Gardens of the Acclimation Society of Paris. <*Ann. Rec.*, 1873–75, p. 338.

536.—1873. Tenth Annual Report of the Massachusetts Agricultural College. <*Ann. Rec.*, 1873–75, p. 418.

538.—1873. British Exhibition of Fishing Products at Vienna. <*Ann. Rec.*, 1873–75, p. 427.

539.—1873. Exhibition of Fishery Products at Vienna. <*Ann. Rec.*, 1873–75, p. 429.

540.—1873. Fishery Models at the late Scandinavian Exhibition. <*Ann. Rec.*, 1873–75, p. 429.

590.—1873. Buffalo Society of Natural History. <*Ann. Rec.*, 1873–75, p. 653.

591.—1873. The Torrey Botanical Club. <*Ann. Rec.*, 1873–75, p. 654.

592.—1873. Minnesota Academy of Natural Science. <*Ann. Rec.*, 1873–75, p. 655.

593.—1873.—Agassiz Natural-History Club at Penikese. <*Ann. Rec.*, 1873–75, p. 655.

594.—1873.—The Sixth Annual Report of the Peabody Museum, Cambridge. <*Ann. Rec.*, 1873–75, p. 656.

595.—1873.—Condition of the Boston Natural-History Society, 1871–2. <*Ann. Rec.*, 1873–75, p. 656.

596.—1873.—Building of the New York Musenm of Natural-History. <*Ann. Rec.* 1873–75, p. 657.

597.—1873.—Appropriations for the New York State Cabinet of Natural-History. <*Ann. Rec.*, 1873–75, p. 657.

598.—1873.—Report of the National Academy of Sciences for 1872. <*Ann. Rec.*, 1873–75, p. 657.

599.—1873.—Nourse's History of the U. S. Naval Observatory. <*Ann. Rec.*, 1873–75, p. 657.

600.—1873.—Twenty-second Meeting of the American Association for the Advance-ment of Science. <*Ann. Rec.*, 1873–75, p. 659.

601.—1873.—The Centennial Exhibition. <*Ann. Rec.*, 1873–75, p. 664.

602.—1873.—American Department of the Vienna Exposition. <*Ann. Rec.*, 1873–75, p. 662.

603.—1873.—The Bache Fund. <*Ann. Rec.*, 1873–75, p. 662.

604.—1873.—Gift of Land to the California Academy of Sciences. <*Ann. Rec.*, 1873–75, p. 663.

605.—1873.—The James Lick Donation to the California Academy of Sciences. <*Ann. Rec.*, 1873–75, p. 664.

606.—1873. Woodward's Garden at San Francisco. <*Ann. Rec.*, 1873–75, p. 664.

608.—1873. Catalogue of the Army Medical Museum. <*Ann. Rec.*, 1873–75, p. 665.

609.—1873. National Photographic Institute. <*Ann. Rec.*, 1873–75, p. 666.

610.—1873. National Invitation to the National Statistical Congress. <*Ann. Rec.*, 1873–75, p. 666.

611.—1873. Benevolent Endowment in the United States Treasury. <*Ann. Rec.*, 1873–75, p. 666.

640.—1874. The Godeffroy Museum at Hamburg. <*Ann. Rec.*, 1874–75, p. 282.

666.—1874. Success of the Naples Zoological Station. <*Ann. Rec.*, 1874–75, p. 317.

667.—1874. Zoological Garden at Hamburg. <*Ann. Rec.*, 1874–75, p. 318.

721.—1874. New Survey of the State of Massachusetts. <*Ann. Rec.*, 1874–75, p. 573.

723.—1874. Twenty-third Annual Meeting of the American Association for the Advancement of Science. <*Ann. Rec.*, 1874–75, p. 574.

724.—1874. Seventh Annual Report of the Peabody Museum, Cambridge, Massa-chusetts. <*Ann. Rec.*, 1874–75, p. 575.

725.—1874. First Report of the Anderson School of Natural History at Penikese. <*Ann. Rec.*, 1874–75, p. 576.

726.—1874. The Penikese School. <*Ann. Rec.*, 1874–75, p. 578.

727.—1874. Opening of the Anderson School at Penikese. <*Ann. Rec.*, 1874–75, p. 579.

728.—1874. Report for 1873 of the Peabody Academy of Science, Salem. <*Ann. Rec.*, 1874–75, p. 579.

729.—1874. Report of the Museum of Comparative Zoology for 1873. <*Ann. Rec.*, 1874–75, p. 580.

730.—1874. Bulletin of the Museum of Comparative Zoology. <*Ann. Rec.*, 1874–75, p. 581.

731.—1874. Catalogues of the Museum of Comparative Zoology. <*Ann. Rec.*, 1874–75, p. 581.

732.—1874. Annual Meeting of the Trustees of the Museum of Comparative Zool-ogy. <*Ann. Rec.*, 1874–75, p. 582.

733.—1874. Report of the Bussey Institution. <*Ann. Rec.*, 1874–75, p. 582.

734.—1874. The "Torrey Memorial Cabinet." <*Ann. Rec.*, 1874–75, p. 584.

736.—1874. The Bulletin of the Science Department of Cornell University. $<Ann.$
Rec., 1874–75, p. 585.

737.—1874. Issue of "Proceedings" by the New York Lyceum of Natural History.
$<Ann. Rec.$, 1874–75, p. 585.

738.—1874. Publishing Fund of the Historical Society of Pennsylvania. $<Ann.$
Rec., 1874–75, p. 586.

739.—1874. Report of the Philadelphia Academy of Natural Sciences. $<Ann. Rec.$,
1874–75, p. 586.

740.—1874. Philadelphia National Museum. $<Ann. Rec.$, 1874–75, p. 587.

741.—1874. Report of the Zoological Society of Philadelphia. $<Ann. Rec.$, 1874–75,
p. 587.

742.—1874. The Zoological Society of Philadelphia. $<Ann. Rec.$, 1874–75, p. 588.

743.—1874. Reorganization of the Maryland Academy of Sciences. $<Ann. Rec.$,
1874–75, p. 589.

744.—1874. Botanical Conservatory of the Maryland Academy of Sciences. $<Ann.$
Rec., 1874–75, p. 590.

745.—1874. Recent Publications of the Smithsonian Institution. $<Ann. Rec.$, 1874–
75, p. 591.

746.—1874. United States Departmental Centennial Board. $<Ann. Rec.$, 1874–75, p.
593.

748.—1874. European Savans in American Institutions. $<Ann. Rec.$, 1874–75, p. 594.

749.—1874. Sale of Dr. Troost's Cabinet of Minerals and Antiquities. $<Ann. Rec.$,
1874–75, p. 594.

750.—1874. Report of the Zoological Society of London. $<Ann. Rec.$, 1874–75, p.
595.

751.—1874. Annual Return of the British Museum. $<Ann. Rec.$, 1874–75, p. 596.

752.—1874. Temporary Museum at the late Meeting of the British Association. $<Ann.$
Rec., 1874–75, p. 599.

753.—1874. Meeting of the French Association for the Advancement of Science. $<$
Ann. Rec., 1874–75, p. 601.

754.—1874. The "American Society" of Paris. $<Ann. Rec.$, 1874–75, p. 602.

787.—1875. Mr. George Latimer's Archæological Collection from Porto Rico. $<Ann.$
Rec., 1875–76, p. 325.

636.—1875. The New Westminster Aquarium. $<Ann. Rec.$, 1875–76, p. 422.

839.—1875. Change of Water in Aquaria. $<Ann. Rec.$, 1875–76, p. 426.

862.—1875. Yarmouth Aquarium. $<Ann. Rec.$, 1875–76, p. 440.

863.—1875. Report of the American Museum of Natural History, New York, for 1874.
$<Ann. Rec.$, 1875–76, p. 571.

864.—1875. Kirtland School of Natural Sciences. $<Ann. Rec.$, 1875–76, p. 571.

865.—1875. Normal School of Natural Sciences. $<Ann. Rec.$, 1875–76, p. 572.

966.—1875. First Annual Report of the Zoological Society of Cincinnati. $<Ann.$
Rec., 1875–76, p. 572.

867.—1875. First Annual Report of the Geological and Agricultural Survey of Texas.
$<Ann. Rec.$, 1875–76, p. 574.

869.—1875. Arrangements for a Botanical Garden in Chicago. $<Ann. Rec.$, 1875–76,
p. 574.

871.—1875. Bequest to the Cincinnati Society of Natural History. $<Ann. Rec.$,
1875–76, p. 577.

873.—1875. Sums voted by the British Parliament for Scientific Instruction. $<Ann.$
Rec., 1875–76, p. 580.

874.—1875. Annual Report of the Council of the Zoological Society of London.
$<Ann. Rec.$, 1875–76, p. 581.

878.—1875. Royal Society's Catalogue of Learned Societies and Scientific Papers.
$<Ann. Rec.$, 1875–76, p. 284.

876.—1875. Meeting of the American Fish-Culturists' Association. <*Ann. Rec.,*
1875–76, p. 585.

880.—1875. Loan Exhibition of Scientific Apparatus. <*Ann. Rec.*, 1875–76, p. 588.

881.—1875. Annual Report of the Peabody Museum of Archæology and Ethnology.
<*Ann. Rec.*, 1876, p. 589.

PUBLICATIONS AS AN OFFICER OF THE SMITHSONIAN INSTITUTION.

16.—1851. Report of the Assistant Secretary in charge of the natural history department [of the Smithsonian Institution] for the year 1850. <*Fifth
Annual Report of the Secretary of the Smithsonian Institution for the
years* 1850, 1851, pp. 41–50.

17.—1851. [Note prefatory to catalogues of specimens of Natural History collected
in the Mauvaises Terres and on the Upper Missouri, by T. A. Culbertson.] <*Fifth Annual Report Smithsonian Institution for the year* 1850,
p. 133.

23.—1852. Report of Assistant Secretary in charge of the Museum, &c. <*Fourth
Annual Report of the Secretary of the Smithsonian Institution for the
year* 1851, pp. 40–52 (Appendices, pp. 52–65).

24.—1852. An account of natural history explorations in the United States during
1851. <*Sixth Annual Report Smithsonian Institution for the year* 1851,
1852, pp. 52–56· Appendix A to Report of Assistant Secretary.

41.—1853. Report of the Assistant Secretary in charge of the Museum, &c. <*Seventh
Annual Report Smithsonian Institution for the year* 1852, 1853, pp. 45–58.
Appendices, pp. 58–73.

42.—1853. Account of scientific explorations and reports on explorations, made in
America during the year 1852. <*Seventh Annual Report Smithsonian
Institution for the year* 1852, 1853, pp. 58–65. Appendix A of Assist-
ant Secretary's Report.

54.—1874. Report of the Assistant Secretary (of the Smithsonian Institution) in
charge of publications, exchanges, and natural history. <*Eighth Annual Report Smithsonian Institution* (1853), 1854, pp. 34–37.

61.—1855. Report of the Assistant Secretary (of the Smithsonian Institution) for the
year 1854. <*Ninth Annual Report Smithsonian Institution* (1854), 1855,
pp. 31–46.

62.—1855. Report on American Explorations in the years 1853 and 1854. <*Ninth Annual Report Smithsonian Institution* (1854), 1855, pp. 79–97.

63.—1855. Report on the fishes observed on the coasts of New Jersey and Long Island
during the summer of 1854, by Spencer F. Baird, Assistant Secretary
of the Smithsonian Institution. <*Ninth Annual Report of the Smithsonian Institution* [*for* 1854], 1855, pp. 317–325+*337.

68.—1856. [Report of the Assistant Secretary of the Smithsonian Institution for the
year 1855.] <*Tenth Ann. Rep. Smithsonian Institution* (1855), 1856, pp.
36–61.

71.—1857. [Report of the Assistant Secretary of the Smithsonian Institution for the
year 1856.] <*Eleventh Ann. Rep. Smithsonian Institution for the year*
(1857), 1857, pp. 47–68.

72.—1857. [Name Tamias pallasii proposed for Sciurus striatus, Pallas, nec Linn.
<*Eleventh Ann. Rep. Smithsonian Institution*, p. 55.

73.—1857. Directions for collecting, preserving, and transporting specimens of
ntaural history. Prepared for the use of the Smithsonian Institution.
<*Eleventh Ann. Rep. Smithsonian Institution* (1856), 1857, pp. 235–253.

81.—1858. [Report for 1857 of the Assistant Secretary of the Smithsonian Institution.] <*Ann. Rep. Smithsonian Institution for the year* 1857, 1858, pp.
38–54.

89.—1859. Report for 1858 of the Assistant Secretary of the Smithsonian Institution. *<Ann. Rep. Smiths. Inst. for the year* 1858, 1859, pp. 44–62.

102.—1860. [Report for 1859 of the Assistant Secretary of the Smithsonian Institution.] *<Ann. Rep. Smiths. Inst. for the year* 1859, 1860, pp. 54–78.

107.—1861. [Report for 1860 of the Assistant Secretary of the Smithsonian Institution.] *<Ann. Rep. Smithsonian Institution for the year* 1860, 1861, pp. 55–86.

109.—1862. [Report for 1861 of the Assistant Secretary of the Smithsonian Institution.] *< Ann. Rep. Smithsonian Institution for the year* 1861, 1860, pp. 48–67.

110.—1863. [Report for 1862 of the Assistant Secretary of the Smithsonian Institution.] *< Ann. Rep. Smithsonian Institution for the year* 1862, 1863, pp. 46–59.

113.—1864. [Report for 1863 of the Assistant Secretary of the Smithsonian Institution.] *< Ann. Rep. Smithsonian Institution for the year* 1863, 1864, pp. 44–63.

116.—1865. Report of the Assistant Secretary (of the Smithsonian Institution for the year 1864). *< Ann. Rep. Smithsonian Institution for the year* 1864, 1865, pp. 74–100.

117.—1866. Report of the Assistant Secretary, Spencer F. Baird, relative to exchanges, collections of Natural History, &c. *<Ann. Rep. Smithsonian Institution for the year* 1865, pp. 75–88.

126.—1868. Report of Prof. S. F. Baird (Assistant Secretary of the Smithsonian Institution, for the year 1867). *<Ann. Rep. Smithsonian Institution for the year* 1867, 1868, pp. 64–78.

128.—1869. Explorations and Collections in Natural History (etc.). *<Ann. Rep. Smithsonian Institution for the year* 1868, 1869 (*in Rep. of the Secretary*, pp. 22–41 and pp. 54–67).

219.—1872. Explorations and Collections (of the Smithsonian Institution in the year 1871). *<Ann. Rep. Smithsonian Institution for the year* 1871, 1872, *in Report of Secretary*, pp. 26–34+and 42–62.

220.—1872. [Instructions to Capt. C. F. Hall for collecting objects of Natural History on the Expedition toward the North Pole.] *<Ann. Rep. Smithsonian Institution for the year* 1871, 1872, pp. 379–381.

420.—1873. "Account of the additions to the National Museum, and the various operations connected with it during 1872." *<Ann. Rep. Smithsonian Institution for the year* 1872, 1873, pp. 43–52 and 55–62.

626.—1874. "Report * * * of the addition to the Museum and the onerous operations connected with it during the year 1873." *<Ann. Rep. Smithsonian Institution for the year* 1873, 1874, pp. 36–63 and 58–69.

771.—1875. "Report (on) additions to the Museum and the various operations connected with it during the past year." *<Ann. Rep. Smithsonian Institution for the year* 1874, 1875, pp. 27–44, 49–76.

891.—1876. "Report from Prof. Spencer F. Baird, Assistant Secretary, * * * (on) additions to the Museum and the various operations connected with it during the year 1875." *<Ann. Rep. Smithsonian Institution for the year* 1875, 1876, pp. 46–57, and 72–98.

893.—1876. "An account of the proposed plan of exhibition by the Smithsonian Institution at the International Centennial Exhibition, and the extent to which the work has been carried on." *<Ann. Rep. Smithsonian Institution for the year* 1875, 1876, pp. 58–71.

992.—1877. Report of Prof. Spencer F. Baird on the additions, &c., to the Museum in 1876. *<Ann. Rep. Smithsonian Institution for* 1876, 1877, pp. 38–63, and 84–115.

993.—1877. Report of Professor Baird on the Centennial Exhibition of 1876. <*Ann. Rep. Smithsonian Institution for* 1876, 1877, pp. 64–83.

1022.—1879. Prefatory Note [to Henry's Researches in Sound, with special reference to fog-signaling]. <*Report Smithsonian Institution,* 1878, pp. 455, 456.

1028.—1879. Annual Report of the Board of Regents of the Smithsonian Institution showing the operations, expenditures, and condition of the Institution for the year 1878. Washington: Government Printing Office, 1879. 8vo. p. 575.

PUBLICATIONS AS COMMISSIONER OF FISHERIES.

CIRCULARS.

133.—1871. [Letter addressed to fishermen and others living on the shores of Lake Michigan, and announcing that fishes with metallic tags had been liberated at twenty points, and asking for information about the subsequent capture.] 8vo., 1 page. Serial mark E—♀. Dated October 30, 1871. [U. S. F. C., 1.]

216.—1872. Memoranda of Inquiry relative to the food-fishes of the United States. [Washington: Government Printing Office. 1872.] 8vo. 5 pp. No title-page. [U. S. F. C., 2].

217.—1872. Questions relative to the food-fishes of the United States. [Washington: Government Printing Office. 1872.] 8vo. 7 pp. No title-page. [U. S. F. C., 3.]

218.—1872. Letter to accompany No. 217, inviting information concerning food-fishes. 1 p., letter size. [U. S. F. C., 4.]

419.—1873. Statistics of the Menhaden Fisheries, etc. (E♀). [Questions addressed to fishermen, etc.]—Washington, December 20, 1873. 4to., letter form, 2 l. [U. S. F. C., 5.]

631.—1874. Statistics of the Menhaden Fisheries, etc. (12). [Questions addressed to fishermen, etc.]—Washington, December 20, 1873. 4to., letter form, 2 l. Reprint of U. S. F. C., 5 [U. S. F. C., 10].

770.—1875. (E) ♀. U. S. Commission of Fish and Fisheries. Statistics of the Fishery Marine. Circular. [U. S. F. C., 12.] [Foolscap size, 2 pp. Washington: Government Printing Office, 1875.] [U. S. F. C. 12.]
  The blank tables to accompany this circular were printed in uniform style, and are registered (U. S, F. C., 11). Prepared by G. Brown Goode.

984.—1877. (E) ♀. U. S. Commission of Fish and Fisheries. Questions relative to the Food-fishes of the United States. Foolscap, one sheet. 4 pp. Government Printing Office, 1877. [U. S. F. C. 15.]
  A new edition of the circular bearing the same title, previously issued. [U. S. F. C., 3= No. 217.]

985.—1877. (E) ♀. U. S. Commission of Fish and Fisheries. Statistics of the Mackerel Fisheries, etc. Circular. [1 page, foolscap. Washington: Government Printing Office, 1877.] [U. S. F. C., 16.] To accompany circular No. 984. [U. S. F. C., 15.]

986.—1877. (E) ♀. U. S. Commission of Fish and Fisheries. Statistics of the Cod Fisheries, etc. Circular. [1 page, foolscap. Washington: Government Printing Office, 1877.] [U. S. F. C., 17.] To accompany circular No. 984. [U. S. F. C., 15.]

987.—1877. (E) ♀. U. S. Commission of Fish and Fisheries. Statistics of the Mullet Fisheries, etc. Circular. [1 page, foolscap. Washington: Government Printing Office, 1877.] [U. S. F. C., 18.] To accompany circular No. 984. [U. S. F. C., 15.]

988.—1877. (E) $\frac{p}{2}$. U. S. Commission of Fish and Fisheries. Statistics of Coast and River Fisheries. [4 pp., foolscap. Government Printing Office, 1877, Oct.] [U. S. F. C., 19.]

1005.—1878. (E) $\frac{p}{2}$. U. S. Commission of Fish and Fisheries. Questions relative to the Cod and the Cod Fisheries. [Foolscap size. 4 pp. Washington: Government Printing Office, 1878.] [U. S. F. C., 28.]

1006.—1878. $\frac{E}{101} \cdot \frac{p}{10}$. U. S. Commission of Fish and Fisheries. Questions relative to the Alewife and the Alewife Fisheries. [Foolscap size. 4 pp. Washington: Government Printing Office, 1878.] [U. S. F. C., 29.]

1007.—1878. $\frac{E}{101} \cdot \frac{p}{10}$. U. S. Commission of Fish and Fisheries. Questions relative to the Smelt and Smelt Fisheries. [Foolscap size. 4 pp. Washington: Government Printing Office, 1878.] [U. S. F. C., 30.]

1007.—1878. (E)$\frac{p}{11}$. U. S. Commission of Fish and Fisheries. Questions relative to the Mackerel and Mackerel Fisheries. [Foolscap size. 4 pp. Washington: Government Printing Office, 1879.] [U. S. F. C., 32.]

REPORTS.

I.

421.—1873. 42d Congress, 2d session. Senate Mis. Doc. No. 61. United States Commission on Fish and Fisheries. Part 1. Report on the condition of the sea fisheries of the south coast of New England in 1871 and 1872, by Spencer F. Baird, Commissioner. Washington: Government Printing Office. 1873. 8vo. pp. (5) 6–4 (1). [U. S. F. C.—6.]

The report of the Commissioner, without supplementary papers, pp. xlvii, was issued separately in advance, paged in Arabic numbers.

422.—1873. 42d Congress, 2d session. Senate Mis. Doc. No. 61. United States Commission of Fish and Fisheries. Part 1. Report on the condition of the sea fisheries of the south coast of New England in 1871 and 1872, by Spencer F. Baird, Commissioner. With supplementary papers. Washington: Government Printing Office. 1873. 8vo. pp. xlvii, 852. plates xxxviii, with 38 leaves explanatory to plates, 2 maps. [U. S. F. C.—7.]

Report in full—for contents see above, pp. 134–138, above.
This report includes papers numbered 423, 424, 425, 426, 427, 428, 429, 430, 431, 432, 433, 434, 435, for titles of which see *Chronological Catalogue*, pp. 138–140 above.

627.—1874. United States Commission of Fish and Fisheries. Part II. Report of the Commissioner for 1872 and 1873. A.—Inquiry into the decrease of the food-fishes. B.—The propagation of food-fishes in the waters of the United States. Washington: Government Printing Office. 1874. 8vo. pp. (v) vi–viii, (1) 2–92. [U. S. F. C., No. 8.]

Report without supplementary papers.

628.—1874. United States Commission of Fish and Fisheries. Part II. Report of the Commissioner for 1872 and 1873. A.—Inquiry into the decrease of the food-fishes. B.—The propagation of food-fishes in the waters of the United States. With supplementary papers. Washington: Government Printing Office. 1874. 8vo. pp. cii, 808, pls. xxxvii, 4 maps. [U. S. F. C., No. 9.]

Full Report—for contents see above pages 154–165. above.
Contains title 630.

889.—1876. United States Commission of Fish and Fisheries. Part III. Report of the Commissioner for 1873–4 and 1874–5. A.—Inquiry into the decrease of the food-fishes. B.—The propagation of food-fishes in the waters of the United States. Washington: Government Printing Office. 1876. 8vo. pp. (v) vi–xlvi. [U. S. F. C., 13.]

Report without supplementary papers.

890.—1876. United States Commission of Fish and Fisheries. Part III. Report of the Commissioner for 1873-4 and 1874-5. A.—Inquiry into the decrease of the food-fishes. B.—The propagation of food-fishes in the waters of the United States. Washington: Government Printing Office. 1876. 8vo. pp. lii, 777. [U. S. F. C., 14.]

Full Report—for contents see pages 184-190, above.

1002.—3878. United States Commission of Fish and Fisheries. Part IV. Report of the Commissioner for 1875-1876. A.—Inquiry into the decrease of the food-fishes. B.—The propagation of food-fishes in the waters of the United States. Washington: Government Printing Office, 1878. 8vo. pp. ix, 50*, 1029, plates vi (Hist. of Whale Fishery). [U. S. F. C., 26.]

Report without supplementary papers.

1003.—1878. United States Commission of Fish and Fisheries. Part IV. Report of the Commissioner for 1875-1876. A.—Inquiry into the decrease of the food-fishes. B.—The propagation of food-fishes in the waters of the United States. Washington: Government Printing Office. 1878. 8vo. pp. ix, 50*, 1029, plates vi (Hist. of Whale Fishery). [U. S. F. C., 27.]

Full Report, for contents see pp. 199-204, above.

1030.—1879. United States Commission of Fish and Fisheries. Part V. Report of the Commissioner for 1877. A.—Inquiry into the decrease of food-fishes. B.—The propagation of food-fishes in the waters of the United States. Washington: Government Printing Office. 1879. 8vo. pp. 48. [U. S. F. C., 37.]

Report without supplement.

1031.—1879. United States Commission of Fish and Fisheries. Part V. Report of the Commissioner for 1877. A.—Inquiry into the decrease of food-fishes. B.—The propagation of food-fishes in the waters of the United States. Washington: Government Printing Office. 1879. 8vo. pp. 48, 972. [U. S. F. C., 38.]

Full Report, for contents see pp.— 208-215, above.

1051.—1880. United States Commission of Fish and Fisheries. Part VI. Report of the Commissioner for 1878. A.—Inquiry into the decrease of food-fishes, B.—The propagation of food-fishes in the waters of the United States. Washington: Government Printing Office 1880. 8vo. pp. lxiv.

(Report of the Commissioner without supplementary papers.)

1952.—1880. United States Commission of Fish and Fisheries. Part VI. Report of the Commissioner for 1878. A.—Inquiry into the decrease of food-fishes. B.—The Propagation of food-fishes in the waters of the United States. Washington: Government Printing Office. 1880. 8vo. pp. lxiv.

Full Report—for table of contents. See above, pp. 217-226, above.

ANNUAL RECORD OF SCIENCE AND INDUSTRY.

(I.)

221.—1872. Annual Record of Science and Industry for 1871. Edited by Spencer F. Baird, with the assistance of eminent men of science. [Cut.] New York: Harper & Brothers, publishers, Franklin Square, 1872 [8vo. pp. xxxii, 634.] Titles 134 to 215, 222.

(II.)

417.—1873. Annual Record of Science and Industry for 1872. Edited by Spencer F. Baird, with the assistance of eminent men of science. [Cut.] New York: Harper & Brothers, Publishers, Franklin Square. 1873. (8vo. pp. lxviii, 650.) Titles 223-416.

(III.)

756.—1875. Annual Record of Science and Industry for 1873. Edited by Spencer F. Baird, with the assistance of Eminent Men of Science. [Cut.] New York: Harper & Brother, Publishers, Franklin Square. 1875. (8vo. pp. cxxxii, 714. Titles 439–613, 757–60.

(IV.)

762.—1875. Annual Record of Science and Industry for 1874. Edited by Spencer F. Baird, with the assistance of Eminent Men of Science. [Cut.] New York: Harper & Brother, Publishers, Franklin Square. 1875. (8vo. pp. cciv, 665.) Titles 632–761, 763–66.

(V.)

883.—1876. Annual Record of Science and Industry for 1875. Edited by Spencer F. Baird, with the assistance of eminent men of science. [Cut.] New York: Harper & Brothers, Publishers, Franklin Square. 1876. (8vo. pp. ccxc. 656.) Titles 777–881, 884, 7.

(VI.)

970.—1877. Annual Record of Science and Industry for 1876. Edited by Spencer F. Baird, with the assistance of Eminent Men of Science. [Cut.] New York: Harper & Brothers, publishers, Franklin Square. 1877. [8vo. pp. ccxxxvi, 609.] Titles 896–969, 971–74.

(VII.)

1004.—1878. Annual Record of Science and Industry for 1877. Edited by Spencer F. Baird, with the assistance of eminent men of science. [Cut.] New York: Harper & Brothers, Publishers, Franklin Square. 1878. 8vo. pp. xiv, 480.—Preface dated March 1, 1878.

(VIII.)

1020.—1879. Annual Record of Science and Industry for 1878. Edited by Spencer F. Baird; with the assistance of eminent men of science. [Cut.] New York: Harper & Brothers, Publishers, Franklin Square. 1879. 8vo. pp. xvii (i), 715. Preface dated March 1, 1879.

III. LIST OF SPECIES DISCUSSED AND ILLUSTRATED.

NOTE.

Under each of the classes mentioned in the analysis the genera and species are arranged alphabetically according to their generic and specific designations. These lists will be of importance chiefly to persons familiar with the synonymy of the several groups. The importance of this collation of references to descriptions and illustrations prepared by Professor Baird will be evident, since his work carried him over the entire field of North American vertebrate zoology, and every mammal, bird, and reptile known at the time of his researches was exhaustively treated, as well as a large number of fishes. These lists have also a definite value as constituting a key to the major p rtion of the descriptive work accomplished in the National Museum during the first twenty years of its history. The names of all the species illustrated by engravings or lithographs are printed in SMALL CAPS; the list thus serves as an index to the very numerous illustrations prepared under the supervision of Professor Baird.

It has not been thought necessary to specially index each genus of which a diagnosis has been given. It may be understood that each genus of which the species are discussed in "(76) Pacific Railroad Survey, vol. viii," and "(101) Mammals of North America," in "(78) Pacific Railroad Survey, vol. ix," or (632½) Birds of North America, 1874," or in "(39) Catalogue of North American Reptiles" is discussed and briefly diagnosed in those works. Discussions of genera in works other than these are especially indexed.

ANALYSIS.

MAMMALS.

ALCE AMERICANUS.
　(76) P. R. R. Surv., vol. viii, p. 632, figg. 1-2; (101) Mammals of N. A., 1859. Horns, adult, woodcut, fig. 1 young, fig. 2, p. 632.

ALCES AMERICANA.
　(22) Rep. Comm. Pat. for 1851, p. 112, pl. viii.

ANTILOCAPRA AMERICANA.
　(22) Rep. of Comm. of Patents for 1851, p. 121, plate 1; (76) P. R. R. Surv., vol. viii, p. 666, pll. xvi, xxx, figg. 23-24, p. 668; (86) Mex. Bound. Surv., vol. ii, p. 51; (101) Mammals of N. A., 1859. Animal, pl. xvi. Various horns, pl. xxx. Muzzle and hoof, woodcut, figg. 23-24, p. 668.

Aplocerus montanus.
　(76) P. R. R. Surv., vol. viii, p. 671.

APLODONTIA LEPORINA.
　(76) P. R. R. Surv., vol. viii, p. 353, pl. xx, fig. 4; (101) Mammals of N. A., 1859. Details, pl. xx, fig. 4. Skull, pl. xlix, fig. 2.

ARCTOMYS FLAVIVENTER.
　(76) P. R. R. Surv., vol. viii, p. 343, pl. xlvii, fig. 1; (101) Mammals of N. A., 1859. Skull, pl. xlvii, fig. 1.

Arctomys Lewisi.
　(76) P. R. R. Surv., vol. viii, p. 347.

ARCTOMYS MONAX.
　(76) P. R. R. Surv., vol. viii, p. 339, pl. xlix, fig. 1; (101) Mammals of N. A., 1859. Skull, pl. xlix, fig. 1.

Arctomys pruinosus.
　(76) P. R. R. Surv., vol. viii, p. 345.

Arvicola albo-rufescens.
　(76) P. R. R. Surv., vol. viii, p. 549.

ARVICOLA AUSTERA.
　(76) P. R. R. Surv., vol. viii, p. 539, pl. liv, No. 1587; (101) Mammals of N. A., 1859. Teeth, pl. liv, No. 1587.

Arvicola borealis.
　(76) P. R. R. Surv., vol. viii, p. 549.

Arvicola Breweri.
　(76) P. R. R. Surv., vol. viii, p. 525.

Arvicola californica.
　(76) P. R. R. Surv., vol. viii, p. 532.

ARVICOLA CINNAMONEA.
　(76) P. R. R. Surv., vol. viii, p. 541, pl. liv, No. 1714; (101) Mammals of N. A., 1859. Teeth, pl. liv, No. 1714.

Arvicola Dekayi.
(76) P. R. R. Surv., vol. viii, p. 549.
Arvicola Drummondii.
(76) P. R. R. Surv., vol. viii, p. 550.
Arvicola edax.
(76) P. R. R. Surv., vol. viii, p. 531.
Arvicola Gapperi.
(76) P. R. R. Surv., vol. viii, p. 518.
Arvicola Haydenii.
(76) P. R. R. Surv., vol. viii, p. 543.
Arvicola hirsutus.
(76) P. R. R. Surv., vol. viii, p. 550.
Arvicola longirostris.
(76) P. R. R. Surv., vol. viii, p. 530.
Arvicola modesta. •
(76) P. R. R. Surv., vol. viii, p. 535; (93) P. R. R. Surv., vol. x, p. 9.
ARVICOLA MONTANA.
(76) P. R. R. Surv., vol. viii, p. 528, pl. xxi, fig. 2; (101) Mammals of N. A., 1859. Details, pl. xxi, fig. 2.
Arvicola nasuta.
(76) P. R. R. Surv., vol. viii, p. 550.
Arvicola occidentalis.
(76) P. R. R. Surv., vol. viii, p. 534.
Arvicola oneida.
(76) P. R. R. Surv., vol. viii, p. 551.
Arvicola oregoni.
(76) P. R. R. Surv., vol. viii, p. 537.
ARVICOLA PINETORUM.
(76) P. R. R. Surv., vol. viii, p. 544, pl. liv, No. 1719; (101) Mammals of N. A., 1859. Teeth, pl. liv, No. 1719.
Arvicola Richardsonii.
(76) P. R. R. Surv, vol. viii, p. 551.
Arvicola riparia.
(76) P. R. R. Surv, vol. viii, p. 522.
Arvicola rubricatus.
(76) P. R. R. Surv., vol. viii, p. 551.
Arvicola rufidorsum.
(76) P. R. R. Surv., vol. viii, p. 526.
Arvicola texiana.
(76) P. R. R. Surv., vol. viii, p. 552.
ARVICOLA TOWNSENDII.
(76) P. R. R. Surv., vol. viii, p. 527; (101) Mammals of N. A., 1859. Teeth, pl. liv, No. 1595.
Arvicola xanthognathus.
(76) P. R. R. Surv., vol. viii, p. 552.
Auchenia llama.
(64) Gillis, Naval Astr. Exp., ii, p. 159; (65) Ibid., ii, p. 170.
Balæna antarctica.
(65) Gillis, Naval Astr. Exp., ii, p. 171.
BASSARIS ASTUTA.
(76) P. R. R. Surv., vol. viii, p. 147; (86) Mex. Bound. Surv., vol. ii, p. 18; (101) Mammals of N. A., 1859. Skull, pl. lxxiv, fig. 2.
Bison americanus.
(22) Rep. of Comm. of Patents for 1851, p. 124, pl. 7, figure 2nd.
BLARINA ANGUSTICEPS.
(76) P. R. R. Surv., vol. viii, p. 47, pl. xxx; (101) Mammals of N. Amer., 1859. Details, pl. xxx, fig. 7.

BLARINA BERLANDIERI.
(76) P. R. R. Surv., vol. viii, p. 53, pl. xxviii; (86) Mex. Bound. Surv., vol. ii, p. 5; (101) Mammals of N. Amer., 1859. Animal and skull, pl. xxviii, No. 2159.
BLARINA BREVICAUDA.
(76) P. R. R. Surv., vol. viii, p. 42, pl. xxx; (101) Mammals of N. Amer., 1859. Details, pl. xxx, fig. 5.
BLARINA CAROLINENSIS.
(76) P. R. R. Surv., vol. viii, pl. xxx, p. 45; (101) Mammals of N. Amer., 1859. Details, pl. xxx, fig. 8.
BLARINA CINEREA.
(76) P. R. R. Surv., vol. viii, p. 45, pl. xxx; (101) Mammals of N. Amer., 1859. Details, pl. xxx, figg. 9, 10.
BLARINA EXILIPES.
(76) P. R. R. Surv., vol. viii, p. 51, pl. xxviii; (101) Mammals of N. Amer., 1859. Animal and skull, pl. xxvii, No. 2157.
BLARINA TALPOIDES.
(76) P. R. R. Surv., vol. viii, p. 36, pll. xviii, xxx; (101) Mammals of N. Amer., 1859. Details of external form, pl. xviii, fig. 4, pl. xxx, fig. 6.
BOS AMERICANUS.
(76) P. R. R. Surv., vol. viii, p. 682, figg. 34, 35, p. 683; (86) U. S. and Mex. Bound. Surv., vol. ii, p. 52; (101) Mammals of N. A., 1859. Muzzle and hoof, woodcut, figg. 34, 35, p. 683.
Canis Azaræ.
(64) Gillis, Naval Astr. Exp., ii, p. 154; (65) Ibid., ii, p. 164.
Canis fulvipes.
(65) Gillis, Naval Astr. Exp. ii, p. 164.
CANIS LATRANS.
(76) P. R. R. Surv., vol. viii, p. 113; (86) Mex. Bound. Surv., vol. ii, p. 16, pl. xvi; (101) Mammals of N. Amer., 1859. Skull, pl. lxxvi.
Canis magellanicus.
(64) Gillis, Naval Astr. Exp., ii, p. 154; (65) Ibid., ii, p. 164.
Canis occidentalis, ater.
(76) P. R. R. Surv., vol. viii, p. 113.
CANIS OCCIDENTALIS GRISEO-ALBUS.
(76) P. R. R. Surv., vol. viii, p. 104, pl. xxxi. (101) Mammals of N. Amer., 1859. Skull, pl. xxxi.
Canis occidentalis mexicanus.
(76) P. R. R. Surv., vol. viii, p. 113; (86) Mex. Bound. Surv., vol. ii, p. 14.
Canis occidentalis nubilus.
(76) P. R. R. Surv., vol. viii, p, 111.
Canis occidentalis rufus.
(76) P. R. R. Surv., vol. viii, p. 113; (86) Mex. Bound Surv., vol. ii, p. 15.
CAPELLA RUPICAPRA.
(101) Mammals of N. A., 1859. Horns, pl. xxv, figg. 1854 and 882.
CAPRA AMERICANA.
(22) Rep. of Comm. of Patents for 1851, p. 120, plate 4.

Dydelphys elegans.

(64) Gillis, Naval Astr. Exp., ii, p. 155; (65) Gillis, Naval Astr. Exp., ii, p. 166.

Dysopes nasutus.

(65) Gillis, Naval Astr. Exp., ii, p. 163.

ELAPHUS CANADENSIS.

(22) Rep. of Comm. of Patents for 1851, p. 116, pl. 6.

Enhydra marina.

(76) P. R. R. Surv., vol. viii, p. 189.

ERETHIZON DORSATUS.

(76) P. R. R. Surv., vol. viii, p. 568, pl. lv, fig. 3; (101) Mammals of N. A., 1859. Upper view of skull, pl. lv, fig. 3.

ERETHIZON EPIXANTHUS.

(76) P. R. R. Surv., vol. viii, p. 569, pl. lv, figg. 1, 2.

(101) Mammals of N. A., 1859. Skull, pl. lv, figg. 1, 2.

FELIS CONCOLOR.

(65) Gillis, Naval Astr. Exp., ii, p. 164; (64) Gillis, Naval Astr. Exp., ii, p. 153; (65) Gillis, Naval Astr. Exp., ii, p. 164; (76) P. R. R. Surv., vol. viii, p. 83, pl. ii, fig. 2; (86) Mex. Bound. Surv., vol. ii, p. 5, pl. xi, figg. 1, 2; (101) Mammals of N. A., 1859. Animal very young, pl. ii, fig. 2; skull, adult, pl. lxxi, fig. 1; young, pl. lxxi, fig. 2.

FELIS EYRA.

(76) P. R. R. Surv., vol. viii, p. 88; (86) Mex. Bound. Surv., vol. ii, p. 10, pl. 2, fig. 7; pl. xiii, fig. 2; (101) Mammals of N. A., 1859. Animal, pl. lxii, fig. 1; Skull, pl. lxxiii, fig. 2.

Felis guigna.

(65) Gillis, Naval Astr. Exp., ii, p. 164.

Felis onca.

(76) P. R. R. Surv., vol. viii, p. 86.

Felis onza.

(86) Mex. Bound. Surv., vol. ii, p. 6.

Felis pajeros.

(65) Gillis, Naval Astr. Exp., ii, p. 164.

FELIS PARDALIS.

(76) P. R. R. Surv., vol. viii, p. 87; (86) Mex. Bound. Surv., vol. ii, p. 8, pll. xii, xiii, fig. 1; (101) Mammals of N. A., 1859. Skull, adult, pl. lxxii; young, pl. lxxiii, fig. 1.

FELIS YAGUARUNDI.

(76) P. R. R. Surv., vol. viii, p. 88; (86) Mex. Bound. Surv., vol. ii, p. 12, pl. xiv, fig. 1; (101) Mammals of N. A., 1859. Skull, pl. lxxiv, fig. 1.

FIBER ZIBETHECUS.

(29) Stansbury's Surv. Salt Lake [App. C], p. 312; (76) P. R. R. Surv., vol. viii, p. 561, pl. liv, No. 626; (86) U. S. and Mex. Bound. Surv., vol. ii, p. 45, pl. xxiv, fig. 3; (101) Mammals of N. A., 1859. Teeth, pl. liv, No. 626.

Galictis vittata.

(64) Gillis, Naval Astr. Exp., ii, p. 155; (65) Gillis, Naval Astr. Exp., ii, p. 165.

GEOMYS BREVICEPS.

(60) Pr. Ac. Nat. Sci. Phila., 1854, p. 335; (76) P. R. R. Surv., vol. viii, p. 378, pl. lii, fig. 2; (101) Mammals of N. A., 1859. Skull, pl. lii, fig. 2.

GEOMYS BURSARIUS.

(76) P. R. R. Surv., vol. viii, p. 372, pl. xxii, fig. 1, pl. 1, fig. 2; (101) Mammals of N. A., 1859. Details, pl. xxii, fig. 1; skull, pl. l, fig. 2.

GEOMYS CASTANOPS.

(76) P. R. R. Surv., vol. viii, p. 384, pl. x, fig. 2, pl. 1, fig. 1; (93) P. R. R. Surv., vol. x, p. 8, pl. x, fig. 2; (101) Mammals of N. A., 1859. Animal, pl. x, fig. 2; skull, pl. l, fig. 2.

GEOMYS CLARKII.

(59) Pr. Ac. Nat. Sci. Phila., 1854, p. 334; (76) P. R. R. Surv., vol. viii, p. 383; (86) Mex. Bound. Surv., vol. ii, p. 41, pl. ix, fig. 1, pl. xxiii, fig. 1; (101) Mammals of N. A., 1859. Animals and details, pl. lxix, fig. 1; skull, pl. lxxxiii, fig. 2.

GEOMYS HISPIDUS.

(76) P. R. R. Surv., vol. viii, p. 386, pl. xxii, fig. 4; (101) Mammals of N. A., 1859. Details, pl. xxii, fig. 4.

Geomys mexicanus.

(76) P. R. R. Surv., vol. viii, p. 387.

GEOMYS PINETIS.

(76) P. R. R. Surv., vol. viii, p. 380, pl. xxii, fig. 3; (101) Mammals of N. A., 1859. Details, pl. xxii, fig. 3.

Gulo luscus.

(29) Stansbury's Surv. Salt Lake [App. C], p. 311; (76) P. R. R. Surv., vol. viii, p. 181.

Habrocoma Bennetti.

(65) Gillis, Naval Astr. Exp., ii, p. 168.

Habrocomia Cuvieri.

(65) Gillis, Naval Astr. Exp., ii, p. 168.

Hesperomys austerus.

(60) Pr. Ac. Nat. Sci. Phila., 1854, p. 336; (76) P. R. R. Surv., vol. viii, p. 466.

HESPEROMYS BOYLII.

(60) Pr. Ac. Nat. Sci. Phila., 1854, p. 335; (76) P. R. R. Surv., vol. viii, p. 471, pl. viii, fig. 3, pl. lii, fig. 3.

Hesperomys brachyotis.

(65) Gillis, Naval Astr. Exp., ii, p. 169.

Hesperomys californicus.

(76) P. R. R. Surv., vol. viii, p. 478.

Hesperomys campestris.

(76) P. R. R. Surv., vol. viii, p. 485.

Hesperomys cognatus.

(76) P. R. R. Surv., vol. viii, p. 469.

Hesperomys Darwinii.

(65) Gillis, Naval Astr. Exp., ii, p. 169.

Hesperomys eremicus.

(76) P. R. R. Surv., vol. viii, p. 479; (86) U. S. and Mex. Bound. Surv., vol. ii, p. 44.

Hesperomys Gambelii.

(76) P. R. R. Surv., vol. viii, p. 464; (97) P. R. R. Surv., vol. x, p. 82.

Hesperomys gossypinus.

(76) P. R. R. Surv., vol. viii, p. 469.

Hesperomys leucogaster.

(76) P. R. R. Surv., vol. viii, p. 480.

Hesperomys leucopus.

(76) P. R. R. Surv., vol. viii, p. 459.

Hesperomys longicaudatus.

(65) Gillis, Naval Astr. Exp., ii, p. 170.

Hesperomys longipilis.
(65) Gillis, Naval Astr. Exp., ii, p. 169.
Hesperomys lutescens.
(65) Gillis, Naval Astr. Exp., ii, p. 170.
Hesperomys michiganensis.
(76) P. R. R. Surv., vol. viii, p. 476.
Hesperomys myoides.
(76) P. R. R. Surv., vol. viii, p. 472.
Hesperomys Nuttalli.
(76) P. R. R. Surv., vol. viii, p. 467.
HESPEROMYS PALUSTRIS.
(76) P. R. R. Surv., vol. viii, p. 482, pl. lii, fig. 4;
(101) Mammals of N. A., 1859. Skull, pl.
lii, fig. 4.
Hesperomys renggeri.
(65) Gillis, Naval Astr. Exp., ii, p. 169.
Hesperomys ? rupestris.
(65) Gillis, Naval Astr. Exp., ii, p. 169.
Hesperomys sonoriensis.
(76) P. R. R. Surv., vol. viii, p. 474; (86) Mex.
Bound. Surv., vol. ii, p. 43.
HESPEROMYS TEXANUS.
(76) P. R. R. Surv., vol. viii, p. 464, pl. viii,
fig. 1, pl. lii, fig. 5; (86) Mex. Bound. Surv.,
vol. ii, p. 43; (101) Mammals of N. A.,
1859. Animal, pl. viii, fig. 1; skull, pl. liii,
fig. 5.
Hesperomys xanthorhinus.
(65) Gillis, Naval Astr. Exp., ii, p. 169.
JACULUS HUDSONIUS.
(76) P. R. R. Surv., vol. viii, p. 430, pl. xxi,
fig. 5; (93) P. R. R. Surv., vol. x, p. 8; (101)
Mammals of N. A., 1859. Details, pl. xxi,
fig. 5.
Lagidium criniger.
(65) Gillis, Naval Astr. Exp., ii, p. 167.
Lagidium Cuvieri.
(64) Gillis, Naval Astr. Exp., ii, p. 156; (65)
Gillis, Naval Astr. Exp., ii, p. 167.
Lagidium pallipes.
(65) Gillis, Naval Astr. Exp., ii, p. 167.
Lagomys princeps.
(76) P. R. R. Surv., vol. viii, p. 619.
Lepus americanus.
(76) P. R. R. Surv., vol. viii, p. 579.
LEPUS AQUATICUS.
(76) P. R. R. Surv., vol. viii, p. 612, pl. lix, fig. 1;
(101) Mammals of N. A., 1859. Skull, pl.
lix, fig. 1.
LEPUS ARTEMESIA.
(76) P. R. R. Surv., vol. viii, p. 602; (86) Mex.
Bound. Surv., vol. ii, p. 48, pl. xxv, fig. 2;
(101) Mammals of N. A., 1859. Skull, pl.
lxxxv, fig. 2.
LEPUS AUDUBONII.
(76) P. R. R. Surv., vol. viii, p. 606, pll. xiii, lviii,
fig. 2; (101) Mammals of N. A., 1859. Ani-
mals, pl. xiii; skull, pl. lviii, fig. 2.
Lepus Bachmani.
(76) P. R. R. Surv., vol. viii, p. 606; (86) Mex.
Bound. Surv., vol. ii, p. 48.
LEPUS CALIFORNICUS.
(76) P. R. R. Surv., vol. viii, p. 594, pl. lvii, fig. 2;
(86) Mex. Bound. Surv., vol. ii, p. 47; (101)
Mammals of N. A., 1859. Skull, pl. lvii,
fig. 2.

LEPUS CALLOTIS.
(76) P. R. R. Surv., vol. viii, p. 585, pl. lvii, fig. 1;
(86) Mex. Bound. Surv., vol. ii, p. 45, pl. xxv,
fig. 1; (101) Mammals of N. A., 1859. Skull,
pl. lvii, fig. 1.
LEPUS CALLOTIS FLAVIGULARIS.
(101) Mammals of N. A., 1859. Skull, pl.
lxxxv, fig. 1.
LEPUS CAMPESTRIS.
(76) P. R. R. Surv., vol. viii, p. 585, pl. lvi, fig. 2;
(101) Mammals of N. A., 1859. Skull, pl.
lvi, fig. 2.
LEPUS GLACIALIS.
(76) P. R. R. Surv., vol. viii, p. 577, pl. lvi, fig. 1;
(101) Mammals of N. A., 1859. Skull, pl.
lvi, fig. 1.
Lepus Nuttalli.
(76) P. R. R. Surv., vol. viii, p. 617.
LEPUS PALUSTRIS.
(76) P. R. R. Surv., vol. viii, p. 615, pl. lix, fig.
2; (101) Mammals of N. A., 1859. Skull,
pl. lix, fig. 2.
LEPUS SYLVATICUS.
(76) P. R. R. Surv., vol. viii, p. 597, pl. lviii, fig.
1; (86) Mex. Bound. Surv., vol. ii, p. 47;
(101) Mammals of N. A., 1859. Skull, pl.
lviii, fig. 1.
Lepus texianus.
(76) P. R. R. Surv., vol. viii, p. 617.
Lepus Trowbridgii.
(60) Pr. Ac. Nat. Sci. Phila., 1854, p. 333; (76)
P. R. R. Surv., vol. viii, p. 610, pl. xiv; (101)
Mammals of N. A., 1859. Animal, pl. xiv.
LEPUS WASHINGTONII.
(60) Pr. Ac. Nat. Sci. Phila., 1854, p. 333; (76)
P. R. R. Surv., vol. viii, p. 583, pl. xv; (101)
Mammals of N. A., 1859. Animal, pl. xv.
Litra huidobria.
(65) Gillis, Naval Astr. Exp., ii, p. 165.
LUTRA CALIFORNICA.
(76) P. R. R. Surv., vol. viii, p. 187, pl. xix, fig.
8; (101) Mammals of N. A., 1859. De-
tails, pl. xix, fig. 8.
LUTRA CANADENSIS.
(76) P. R. R. Surv., vol. viii, p. 184, pl. xix, fig.
7; pl. xxxviii; (101) Mammals of N. A.,
1859. Details, pl. xix, fig. 7. Skull, pl.
xxxviii.
Lutra felina.
(65) Gillis, Naval Astr. Exp., ii, p. 165.
Lynx canadensis.
(76) P. R. R. Surv., vol. viii, p. 99.
LYNX FASCIATUS.
(76) P. R. R. Surv., vol. viii, p. 96, pl. ii, fig. 1;
(101) Mammals of N. A., 1859. Animal,
pl. ii, fig. 1.
LYNX MACULATUS.
(101) Mammals of N. A., 1859. Skull, adult,
pl. lxxv, fig. 1; young, pl. lxxv, fig. 2.
Lynx rufus.
(76) P. R. R. Surv., vol. viii, p. 90; (86) Mex.
Bound. Surv., vol. ii, p. 13.
LYNX RUFUS MACULATUS.
(76) P. R. R. Surv., vol. viii, p. 93; (86) Mex.
Bound. Surv., vol. ii, p. 13, pl. xv, figg. 1, 2;
(97) P. R. R. Surv., vol. x, p. 81.

Macrorhinus leoninus.
(65) Gillis, Naval Astr. Exp., ii, p. 166. •
MACROTUS CALIFORNICA.
(86) Mex. Bound. Surv., vol. ii, p. 4, pl. 1, fig. 2.
MACROTUS CALIFORNICUS.
(82) Pr. Ac. Nat. Sci. Phila., 1858, p. 116; (101) Mammals of N. A., 1859. Animal, pl. lxi, fig. 2.
Meles labradoria.
(29) Stansbury's Surv. Salt Lake [App. C], p. 311.
MEPHITIS BICOLOR.
(76) P. R. R. Surv., vol. viii, p. 197, pll. xxix, lx, fig. 3; (86) Mex. Bound. Surv., vol. ii, p. 321, pl. xvii; (101) Mammals of N. A., 1859. Animal, pl. xxix; skull, pl. lx, fig. 3, pl. lxxvii, fig. 3.
Mephitis chilensis.
(65) Gillis, Naval Astr. Exp., ii, p. 165.
MEPHITIS MEPHITICA.
(76) P. R. R. Surv., vol. viii, p. 195, pl. lx, fig. 2; (101) Mammals of N. A., 1859, pl. lx, fig. 1.
MEPHITIS MESOLEUCA.
(76) P. R. R. Surv., vol. viii, p. 192, pl. xix, fig. 1, pl. xxxix; (86) Mex. Bound. Surv., vol. ii, p. 19; (101) Mammals of N. A., 1859. Details, pl. xix, fig. 1; skull, pl. xxxix, fig. 3.
Mephitis mesomelas.
(76) P. R. R. Surv., vol. viii, p. 199.
Mephitis Molinæ.
(65) Gillis, Naval Astr. Exp., ii, p. 165.
Mephitis occidentalis.
(76) P. R. R. Surv., vol. viii, p. 194.
Mephitis patagonica.
(65) Gillis, Naval Astr. Exp., ii, p. 165.
MEPHITIS VARIANS.
(76) P. R. R. Surv., vol. viii, p. 193, pl. lx, fig. 2; (86) Mex. Bound. Surv., vol. ii, p. 19; (101) Mammals of N. A., 1859. Skull, pl. lx, fig. 2.
Mus decumanus.
(76) P. R. R. Surv., vol. viii, p. 438.
Mus musculus.
(76) P. R. R. Surv., vol. viii, p. 443.
Mus rattus.
(76) P. R. R. Surv., vol. viii, p. 439.
MUS TECTORUM.
(76) P. R. R. Surv., vol. viii, p. 441, pl. lii, fig. 6; (86) Mex. Bound. Surv., vol. ii, p. 42; (101) Mammals of N. A., 1859. Skull, pl. liii, fig. 6.
MUSTELA AMERICANA.
(76) P. R. R. Surv., vol. viii, p. 152, pl. xxxvi, fig. 2, pl. xxxvii, fig. 1; (101) Mammals of N. A., 1859. Skull, pl. xxxvi, fig. 2, and pl. xxxvii, fig. 1.
MUSTELA PENNANTII.
(76) P. R. R. Surv., vol. viii, p. 149, pl. xxxvi, fig. 1; (101) Mammals of N. A., 1859. Skull and gum folds, pl. xxxvi, fig. 1.
Myodes Cooperii.
(76) P. R. R. Surv., vol. viii, p. 558.
Myodes obensis.
(76) P. R. R. Surv., vol. viii, p. 559.

Myodes torquatus.
(76) P. R. R. Surv., vol. viii, p. 558.
Myopotamus coypus.
(64) Gillis, Naval Astr. Exp., ii, p. 157; (65) Ibid , p. 169.
NEOSOREX NAVIGATOR.
(76) P. R. R. Surv., vol. viii, p. 11, pl. xxvi; (101) Mammals of N. A., 1859. Animal and skull, pl. xxvi, No. 629.
NEOTOMA CINEREA.
(76) P. R. R. Surv., vol. viii, p. 499, pl. liii, fig. 5, pl. liv, No. 1694; (101) Mammals of N. A., 1859. Skull, pl. liii, fig. 5; teeth, pl. liv, No. 1694.
NEOTOMA FLORIDANA.
(76) P. R. R. Surv., vol. viii, p. 487, pl. lii, fig. 2; (101) Mammals of N. A., 1859. Skull, pl. lii, fig. 2.
NEOTOMA FUSCIPES.
(76) P. R. R. Surv., vol. viii, p. 495, pl. liii, fig. 1, pl. liv, No. 936; (101) Mammals of N. A., 1859. Skull, pl. liii, fig. 1; teeth, pl. liv, No. 936.
NEOTOMA MAGISTER.
(76) P. R. R. Surv., vol. viii, p. 499, pl. liii, fig. 4; (101) Mammals of N. A., 1859. Lower jaw, pl. liii, fig. 4.
NEOTOMA MEXICANA.
(59) Pr. Ac. Nat. Sci. Phila., 1854, p. 333; (76) P. R. R. Surv., vol. viii, p. 490, pl. liv; (86) Mex. Bound. Surv., vol. ii, p. 44, pl. x, fig. 3, pl. xxiv, fig. 1; (101) Mammals of N. A., 1859. Teeth, pl. liv; details, pl. lxx, fig. 3; skull, pl. lxxxiv.
NEOTOMA MICROPUS.
(59) Pr. Ac. Nat. Sci. Phila., 1854, p. 333; (76) P. R. R. Surv., vol. viii, p. 492; (86) U. S. and Mex. Bound. Surv., vol. ii, p. 44, pl. xxiv, fig. 2; (101) Mammals of N. A., 1859. Skull, pl. lxxxiv, fig. 2.
NEOTOMA OCCIDENTALIS.
(60) Pr. Ac. Nat. Sci. Phila., 1854, p. 335; (76) P. R. R. Surv., vol. viii, p. 496, pl. ix, fig. 2, pl. xxi, fig. 4, pl. liii, fig. 3; (101) Mammals of N. A., 1859. Animal, pl. ix, fig. 2; details, pl. xxi, fig. 4; skull, pl. liii, fig. 3.
Nycticejus macrotis.
(65) Gillis, Naval Astr. Exp., ii, p. 163.
Nycticejus varius.
(65) Gillis, Naval Astr. Exp., ii, p. 163.
Octodon Birdgesii.
(65) Gillis, Naval Astr. Exp., ii, p. 168.
Octodon degus.
(65) Gillis, Naval Astr. Exp., ii, p. 168.
Otaria flavescens.
(65) Gillis, Naval Astr. Exp., ii, p. 166.
Otaria jubata.
(65) Gillis, Naval Astr. Exp., ii, p. 166.
Otaria porcina.
Otaria ursina.|
(65) Gillis, Naval Astr. Exp., ii, p. 166.
OVIBOS MOSCHATUS.
(22) Rep. of Comm. of Patents for 1851, p. 121, pl. vii, fig. 1; (76) P. R. R. Surv, vol. viii, p. 680, fig. 33, p. 681; (101) Mammals of N. A., 1859. Muzzle, wood-cut, fig. 33, p. 681.

OVIS MONTANA.

(22) Rep. of Comm. of Patents for 1851, p. 123, pl. 3; (29) Stansbury's Surv. Salt Lake [App. C], p. 312; (76) P. R. R. Surv., vol. viii, p. 673, figg. 24, 25, p. 674, figg. 26, 29, p. 675, figg. 30, 32, p. 677; (101) Mammals of N. A., 1859. Muzzle and hoof, wood-cut, figg. 24, 25, p. 674; horns, male and female, wood-cut, figg. 26–29, p. 675; horns of male, figg. 30–32, p. 675.

Oxymicterus megalonyx.

(65) Gillis, Naval Astr. Exp., ii, p. 169.

Oxymicterus scalops.

(65) Gillis, Naval Astr. Exp., ii, p. 169.

Perognathus fasciatus.

(76) P. R. R. Surv., vol. viii, p. 420.

PEROGNATHUS FLAVUS.

(59) Pr. Ac. Nat. Sci. Phila., 1854, p. 334.

(76) P. R. R. Surv., vol. viii, p. 423, pl. viii, fig. 2, pl. xxi, fig. 3; (86) Mex. Bound. Surv., vol. ii, p. 42, pl. x, figg. 4, 5; (93) P. R. R. Surv., vol. x, p. 8.

(101) Mammals of N. A., 1859, p. 423. Animal, pl. viii, fig. 2; details, pl. xxi, fig. 3, and pl. lxx, figg. 4 and 5.

PEROGNATHUS HISPIDUS.

(76) P. R. R. Surv., vol. viii, p. 421, pl. li, fig. 4; (86) U. S. and Mex. Bound. Surv., vol. ii, p. 42, pl. ix, fig. 2, pl. xxiii, fig. 6; (101) Mammals of N. A., 1859. Skull, pl. li, fig. 4, and pl. lxxxiii, fig. 6. Animal, pl. lxix, fig. 2.

PEROGNATHUS MONTICOLA.

(76) P. R. R. Surv., vol. viii, p. 422, pl. li, fig. 3; (101) Mammals of N. A., 1859. Skull, pl. li, fig. 3.

Perognathus parvus.

(76) P. R. R. Surv., vol. viii, p. 425; (97) Ibid. vol. x, p. 82.

PEROGNATHUS PENICILLATUS.

(76) P. R. R. Surv., vol. viii, p. 418, pl. xx, fig. 5; (86) Mex. Bound. Surv., vol. ii, p. 42; (101) Mammals of N. A., 1859. Details, pl. xx, fig. 5.

Physeter macrocephalus.

(65) Gillis, Naval Astr. Exp., ii, p. 171.

Pistorius erminea.

(29) Stansbury's Sur. Salt Lake [App. C], p. 311.

Pistorius vison.

(29) Stansbury's Sur. Salt Lake [App. C], p. 311.

PROCYON HERNANDEZII.

(76) P. R. R. Surv., vol. viii, p. 212, pl. xl; (86) Mex. Bound. Surv., vol. ii, p. 22, pl. xviii; (101) Mammals of N. Amer., 1859. Skull, rather young, pl. lx; adult, pl. lxxvii.

Procyon Hernandezii mexicana.

(76) P. R. R. Surv., vol. viii, p. 215; (86) Mex. Bound. Surv., vol. ii, p. 22.

Procyon lotor.

(76) P. R. R. Surv., vol. viii, p. 209.

Procyon psora.

(76) P. R. R. Surv., vol. viii, p. 215.

Pseudostoma castanops.

(30) Stansbury's Sur. Salt Lake [App. C], p. 313.

Pteromys alpinus.

(76) P. R. R. Surv., vol. viii, p. 289.

Pteromys Hudsonius.

(76) P. R. R. Surv., vol. viii, p. 288.

Pteromys oregonensis.

(76) P. R. R. Surv., vol. viii, p. 290.

Pteromys volucella.

(76) P. R. R. Surv., vol. viii, p. 286.

PUTORIUS CICOGNANII.

(76) P. R. R. Surv., vol. viii, p. 161, pl. xix, fig. 4; (101) Mammals of N. A., 1859. Details, pl. xix, fig. 4.

PUTORIUS FRENATUS.

(76) P. R. R. Surv., vol. viii, p. 173, pl. xix, fig. 5; (86) Mex. Bound. Surv., vol. ii, p. 19, pl. ii, fig. 2, pl. xvii, figg. 1, 2; (101) Mammals of N. A., 1859. Details, pl. xix, fig. 5; details of young, pl. lxii, fig. 2; skull of adult, pl. lxxvii, fig. 1; of young, fig. 2.

Putorius Kaneii.

(76) P. R. R. Surv., vol. viii, p. 172.

Putorius longicauda.

(76) P. R. R. Surv., vol. viii, p. 169.

Putorius nigrescens.

(76) P. R. R. Surv., vol. viii, p. 180.

Putorius nigripes.

(76) P. R. R. Surv., vol. viii, p. 180.

PUTORIUS NOVEBORACENSIS.

(76) P. R. R. Surv., vol. viii, p. 166, pl. xxxvi, fig. 3; (101) Mammals of N. A., 1859. Skull, pl. xxxvi, fig. 3.

Putorius pusillus.

(76) P. R. R. Surv., vol. viii, p. 159.

PUTORIUS RICHARDSONII.

(76) P. R. R. Surv., vol. viii, p. 164, pl. xix, figg. 2–6; (101) Mammals of N. A., 1859. Details, pl. xix, figg. 2–6.

PUTORIUS VISON.

(76) P. R. R. Surv., vol. viii, p. 177, pl. xxxvii, figg. 2, 3; (101) Mammals of N. A., 1859. Skull, pl. xxxvii, fig. 2, adult, fig. 3, young.

PUTORIUS XANTHOGENYS.

(76) P. R. R. Surv., vol. viii, p. 176, pl. iii, fig. 1; (101) Mammals of N. A., 1859. Animal, pl. iii, fig. 1. Details, pl. xix, fig. 3.

RANGIFER CARIBOU.

(76) P. R. R. Surv., vol. viii, p. 633, figg. 3, 4, 5, 6, p. 634; (101) Mammals of N. A., 1859. Horns, adult, wood-cut, fig. 3; young, figg. 4, 5, 6, p. 634.

RANGIFER GROENLANDICUS.

(76) P. R. R. Surv., vol. viii, p. 634, figg. 7, 8, p. 635; (101) Mammals of N. A., 1859. Horns, adult, male, wood-cut, fig. 7; adult, female ? fig. 8, p. 635.

Reithroden carolinensis.

(76) P. R. R. Surv., vol. viii, p. 452.

Reithrodon chinchilloides.

(65) Gillis, Naval Astr. Exp., ii, p. 170.

Reithrodon humilis.

(76) P. R. R. Surv., vol. viii, p. 448.

Reithrodon longicauda.

(76) P. R. R. Surv., vol. viii, p. 451.

REITHRODON.

(101) Mammals of N. A., 1859, p. 451. Details, pl. lxvii, fig. 4, and pl. lxx, fig. 6.

REITHRODON MEGALOTIS.
(76) P. R. R. Surv., vol. viii, p. 451; (86) U. S. and Mex. Bound. Surv., vol. ii, p. 43, pl. vii, fig. 4, pl. x, fig. 6, pl. xxiv, fig. 4; (101) Mammals of N. A., 1859. Skull, pl. lxxxiv, fig. 4.

REITHRODON MONTANUS.
(60) Pr. Ac. Nat. Sci. Phila., 1854, p. 335; (76) P. R. R. Surv., vol. viii, p. 449, pl. liv, No. 1306; (93) Ibid., v., vol. x, p. 9; (101) Mammals of N. A., 1859. Teeth, pl. liv, No. 1306.

SCALOPS AQUATICUS.
(76) P. R. R. Surv., vol. viii, p. 60, pl. xvii, fig. 1; (101) Mammals of N. A., 1859. Details of external form, pl. xvii, fig. 1.

Scalops argentatus.
(76) P. R. R. Surv., vol. viii, p. 63.

SCALOPS BREWERI.
(76) P. R. R. Surv., vol. viii, p. 68, pll. xvii, xxx; (101) Mammals of N. A., 1859. Details of external form, pl. xvii, figg. 3, 4, pl. xxx, fig. 2.

Scalops latimanus.
(76) P. R. R. Surv., vol. viii, p. 65.

SCALOPS TOWNSENDII.
(76) P. R. R. Surv., vol. viii, p. 65, pll. xvii, xxx; (101) Mammals of N. A., 1859. Details of external form, pl. xvii, fig. 5, pl. xxx, fig. 1.

SCALOPS TOWNSENDII CALIFORNICUS.
(76) P. R. R. Surv., vol. viii, p. 65, pll. xvii, xxx; (101) Mammals of N. A, 1859. Details of external form, pl. xvii, figg. 2, 6, pl. xxx, fig. 3.

Schizodon fuscus.
(65) Gillis, Naval Astr. Exp., ii, p. 168.

Sciurus Aberti.
(76) P. R. R. Surv., vol. viii, p. 267.

SCIURUS CAROLINENSIS.
(76) P. R. R. Surv., vol. viii, p. 256, pl. xlv, fig. 2; (101) Mammals of N. A., 1859. Skull, pl. xlv, fig. 2.

Sciurus carolinensis mexicanus.
(76) P. R. R. Surv., vol. viii, p. 263.

Sciurus castanotus.
(59) Pr. Ac. Nat. Sci. Phila., 1854, p. 334.

SCIURUS CASTANONOTUS.
(76) P. R. R. Surv., vol. viii, p. 266; (86) Mex. Bound. Surv., vol. ii, p. 35, pll. v, xxi, fig. 2; (101) Mammals of N. A., 1859. Animal, pl. lxv; skull.

SCIURUS CINEREUS.
(76) P. R. R. Surv., vol. viii, p. 248, pl. xlviii, fig. 2; (101) Mammals of N. A., 1859. Skull, pl. xlviii, fig. 2.

Sciurus Colliæi.
(76) P. R. R. Surv., vol. viii, p. 280.

SCIURUS DOUGLASSII.
(76) P. R. R. Surv., vol. viii, p. 275, pl. xx, fig. 1, pl. xlv, fig. 3; (101) Mammals of N. A., 1859. Details, pl. xx, fig. 1; skull, pl. xlv, fig. 3.

SCIURUS DOUGLASSII SUCKLEYI.
(76) P. R. R. Surv., vol. viii, p. 275, pl. vii; (101) Mammals of N. A., 1859. Animal, pl. vii.

Sciurus ferruginiventris.
(76) P. R. R. Surv., vol. viii, p. 281.

Sciurus fossor.
(76) P. R. R. Surv., vol. viii, p. 264; (97) Ibid, vol. x, p. 81.

SCIURUS FREMONTII.
(76) P. R. R. Surv., vol. viii, p. 272, pl. vi; (93) Ibid, vol. x, p. 7, pl. xi; (101) Mammals of N. A., 1859. Animal, pl. vi.

SCIURUS HUDSONIUS.
(76) P. R. R. Surv., vol. viii, p. 260, pl. xlvi, fig. 1; (101) Mammals of N. A., 1859. Skull, pl. xlvi, fig. 1.

Sciurus lanigerus.
(76) P. R. R. Surv., vol. viii, p. 280.

Sciurus leporinus.
(76) P. R. R. Surv., vol. viii, p. 280.

SCIURUS LIMITIS.
(59) Pr. Ac. Nat. Sci. Phila. 1854, p. 334; (76) P. R. R. Surv., vol. viii, p. 256; (86) Mex. Bound. Surv., vol. ii, p. 34, pll. iv, xxi, fig. 1; (101) Mammals of N. A., 1859. Animal, pl. lxiv; skull, pl. lxxxi, fig. 1.

Sciurus ludovicianus.
(76) P. R. R. Surv., vol. viii, p. 251; (86) Mex. Bound. Surv., vol. ii, p. 35.

Sciurus mustelinus.
(76) P. R. R. Surv., vol. viii, p. 280.

Sciurus nigrescens.
(76) P. R. R. Surv., vol. viii, p. 250.

Sciurus Richardsonii.
(76) P. R. R. Surv., vol. viii, p. 273.

Sciurus Suckleyi.
(60) Pr. Ac. Nat. Sci. Phila., 1854, p. 333.

Sciurus vulpinus.
(76) P. R. R. Surv., vol. viii, p. 246.

SIGMODON BERLANDIERI.
(59) Pr. Ac. Nat. Sci. Phila., 1854, p. 333.
(76) P. R. R. Surv., vol. viii, p. 504, pl. liii, figg. 6, 7; (86) Mex. Bound. Surv., vol. ii, p. 44, pl. vi, fig. 2, pl. x. fig. 2; (101) Mammals of N. A., 1859. Skull, pl. liii, figg. 6, 7, pl. lxxxi, fig. 3; animal, pl. lxvi, fig. 2; details, pl. lxx, fig. 2.

Sigmodon hispidus.
(76) P. R. R. Surv., vol. viii, p. 502.

SOREX COOPERI.
(76) P. R. R. Surv., vol. viii, p. 27, pl. xxvi; (101) Mammals of N. A., 1859. Animal and skull, pl. xxvi, No. 2047.

Sorex fimbripes.
(76) P. R. R. Surv., vol. viii, p. 55.

SOREX FORSTERI.
(76) P. R. R. Surv., vol. viii, p. 22; pl. xxx, fig. 4; (101) Mammals of N. A., 1859. Details, pl. xxx, fig. 4.

Sorex Harlani.
(76) P. R. R. Surv., vol. viii, p. 56.

SOREX HAYDENI.
(76) P. R. R. Surv., vol. viii, p. 29, pl. xxvii; (101) Mammals of N. A., 1859. Animal and skull, pl. xxvii, No. 1685.

SOREX HOYI.
(76) P. R. R. Surv., vol. viii, p. 32, pl. xxviii; (101) Mammals of N. A., 1859. Animal and skull, pl. xxviii, No. 1688.

SOREX PACHYURA.
- (101) Mammals of N. A., 1859. Animal and skull, pl. xxvii, No. 1674; (76) P. R. R. Surv., vol. viii, p. 20, pl. xxvii.

Sorex palustris.
(76) P. R. R. Surv., vol. viii, p. 55.

Sorex parvus.
(76) P. R. R. Surv., vol. viii, p. 56.

Sorex personatus.
(76) P. R. R. Surv., vol. viii, p. 30.

SOREX PLATYRHINUS.
(76) P. R. R. Surv., vol. viii, p. 25, pl. xxviii; (101) Mammals of N. A., 1859. Animal and skull, pl. xxviii, No. 1699.

Sorex Richardsonii.
(76) P. R. R. Surv., vol. viii, p. 24.

SOREX SUCKLEYI.
(76) P. R. R. Surv., vol. viii, p. 18, pl. xxvii; (101) Mammals of N. A., 1859. Animal and skull, pl. xxvii, No. 1677.

SOREX THOMPSONI.
(76) P. R. R. Surv., vol. viii, p. 34, pl. xxvii; (101) Mammals of N. A., 1859. Animal and skull, pl. xxvii, No. 1686.

SOREX TROWBRIDGII.
(76) P. R. R. Surv., vol. viii, p. 13, pl. xxvi: (101) Mammals of N. A., 1859. Animal and skull, pl. xxvi.

SOREX VAGRANS.
(76) P. R. R. Surv., vol. viii, p. 15, pl. xviii, figg. 5, 6, pl. xxvi; (101) Mammals of N. A., 1859. Details of external form, pl. xviii, figg. 5, 6; animal and skull, pl. xxvi, No. 1675.

Spalacopus Poeppigii.
(64) Gillis, Naval Astr. Exp., ii, p. 157; (65) Ibid., ii, p. 168.

SPERMOPHILUS BEECHEYI.
(60) Pr. Ac. Nat. Sci. Phila. 1854, p. 334; (76) P. R. R. Surv., vol. viii, p. 307, pl. iii, fig. 2, pl. xlvi, fig. 3; (97) Ibid., vol. x, p. 81; (101) Mammals of N. A., 1859. Animal, pl. iii, fig. 2; skull, pl. xlvi, fig. 3.

SPERMOPHILUS COUCHII.
(59) Pr. Ac. Nat. Sci. Phila., 1854, p. 334; (76) P. R. R. Surv., vol. viii, p. 311; (86) Mex. Bound. Surv., vol. ii, p. 38, pl. xxi, fig. 3; (101) Mammals of N. A., 1859. Skull, pl. lxxxi, fig. 3.

SPERMOPHILUS DOUGLASSII.
(76) P. R. R. Surv., vol. viii, p. 309, pl. xlv, fig. 1; (101) Mammals of N. A., 1859. Skull, pl. xlv, fig. 1.

SPERMOPHILUS FRANKLINI.
(76) P. R. R. Surv., vol. viii, p. 314, pl. lxvi, fig. 4; (101) Mammals of N. A., 1859. Skull, pl. lxvi, fig. 4.

SPERMOPHILUS GRAMMURA.
(101) Mammals of N. A. Animal, pl. iv, fig. 1; details, pl. lxvii, fig. 1: skull, pl. lxxxii, fig. 2.

SPERMOPHILUS GRAMMURUS.
(60) Pr. Ac. Nat. Sci. Phila., 1854, p. 334; (76) P. R. R. Surv., vol. viii, p. 310, pl. iv, fig. 1; (86) Mex. Bound. Surv., vol. ii, p. 38, pl. vii, fig. 1, pl. xxii, fig. 1.

Spermophilus Gunnisoni.
(60) Pr. Ac. Nat. Sci. Phila., 1854, p. 334.

SPERMOPHILUS HARRISII.
(76) P. R. R. Surv., vol. viii, p. 313, pl. xlviii, fig. 3; (97) Ibid., vol. x, p. 82; (101) Mammals of N. A., 1859. Skull, pl. xlviii, fig. 3.

SPERMOPHILUS LATERALIS.
(76) P. R. R. Surv., vol. viii, p. 312, pl. xx, fig. 3; pl. lxv. fig. 5; (101) Mammals of N. A., 1859. Details, pl. xx, fig. 3; skull, pl. xiv, fig. 5.

Spermophilus macrourus.
(76) P. R. R. Surv., vol. viii, p. 327.

SPERMOPHILUS MEXICANUS.
(76) P. R. R. Surv., vol. viii, p. 319; (86) Mex. Bound. Surv., vol. ii, p. 39, pl. xxii, fig. 2; (101) Mammals of N. A., 1859. Skull, pl. lxxxii, fig. 2.

Spermophilus Parryi.
(76) P. R. R. Surv., vol. viii, p. 323.

Spermophilus Richardsonii.
(76) P. R. R. Surv., vol. viii, p. 325.

SPERMOPHILUS SPILOSOMA.
(59) Pr. Ac. Nat. Sci., Phila. 1854, p. 334; (76) P. R. R. Surv., vol. viii, p. 321; (86) Mex. Bound. Surv., vol. ii, p. 39, pl. vii, fig. 3, pl. xxii, fig. 3; (101) Mammals of N. A., 1859. Details, pl. lxvii, fig. 3; skull, pl. lxxxix, fig. 3.

SPERMOPHILUS TERETICAUDA.
(76) P. R. R. Surv., vol. viii, p. 315; (86) Mex. Bound. Surv., vol. ii, p. 38, pl. vii, fig. 2, pl. xxi, fig. 4; (101) Mammals of N. A., 1859. Details, pl. lxvii, fig. 2; skull, pl. lxxxi, fig. 4.

Spermophilus Townsendi.
(76) P. R. R. Surv., vol. viii, p. 326.

Spermophilus tridecem-lineatus.
(29) Stansbury's Surv. Salt Lake [App. C], p. 312. (76) P. R. R. Surv., vol. viii, p. 316.

Stenoderma chilensis.
(65) Gillis, Naval Astr. Exp., ii, p. 163.

Stenorhynchus leptonyx.
(65) Gillis, Naval Astr. Exp., ii, p. 166.

TALPA EUROPŒA.
(76) P. R. R. Surv., vol. viii, p. 68, pl. xvii, fig. 7; (101) Mammals of N. Amer., 1859. Muzzle, pl. xvii, fig. 7.

Tamias Cooperi.
(60) Pr. Ac. Nat. Sci. Phila. 1854, p. 334.

TAMIAS TOWNSENDII COOPERI.
(76) P. R. R. Surv., vol. viii, p. 301, pl. 6, fig. 2.

TAMIAS DORSALIS.
(59) Pr. Ac. Nat. Sci. Phila., 1854, p. 334; (76) P. R. R. Surv., vol. viii, p. 300; (86) Mex. Bound. Surv., vol. ii, p. 37, pl. 6, fig. 1; (101) Mammals of N. A., 1859. Animal, pl. lxvi, fig. 1.

TAMIAS QUADRIVITTATUS.
(93) P. R. R. Surv., vol. x, p. 7; (76) Ibid., vol. viii, p. 297, pl. xx, fig. 2; (101) Mammals of N. A., 1859. Details, pl. xx, fig. 2.

TAMIAS STRIATUS.
(76) P. R. R. Surv., vol. viii, p. 292, pl. xlvi, fig. 2; (101) Mammals of N. A., 1859. Skull, pl. xlvi, fig. 2.

TAMIAS TOWNSENDII.
  (76) P. R. R. Surv., vol. viii, p. 301, pl. xlv, fig. 4; (101) Mammals of N. A., 1859. Skull, pl. xlv, fig. 4.

TAMIAS TOWNSENDII COOPERI.
  (101) Mammals of N. A., 1859. Animal, pl. v, fig. 2.

TARANDUS ARCTICUS.
  (22) Rep. of Comm. of Patents for 1851, p. 105, pl. 2.

Tarandus furcifer.
  (22) Rep. of Comm. of Patents for 1851, p. 109.

Tarandus hastalis.
  (22) Rep. of Comm. of Patents for 1851, p. 108.

TAXIDEA AMERICANA.
  (76) P. R. R. Surv., vol. viii, p. 202, pl. xxxix; (101) Mammals of N. A., 1859. Upper jaw, pl. xxxix, fig. 2.

TAXIDEA BERLANDIERI.
  (76) P. R. R. Surv., vol. viii, p. 205, pl. xxxix, fig. 7; (86) Mex. Bound. Surv., vol. ii, p. 321; (101) Mammals of N. A., 1859. Skull, pl. xxxix, fig. 7.

THOMOMYS BOREALIS.
  (76) P. R. R. Surv., vol. viii, p. 396, pl. xxii, fig. 2; (101) Mammals of N. A., 1859. Details, pl. xxii, fig. 2.

Thomomys bottœ.
  (60) Pr. Ac. Nat. Sci. Phila. 1854, p. 335.

THOMOMYS BULBIVORUS.
  (76) P. R. R. Surv., vol. viii, p. 389, pll. xli, l, fig. 3, pl. lii, fig. 1; (97) P. R. R. Surv., vol. x, p. 82.

THOMOMYS BULBIVORUS.
  (101) Mammals of N. A., 1859. Animal, with details of external form, pl. xi; skull, pl. l, fig. 3, pl. liii, fig. 1.

Thomomys Douglassii.
  (76) P. R. R. Surv., vol. viii, p. 394.

THOMOMYS FULVUS.
  (76) P. R. R. Surv., vol. viii, p. 402, pl. xii, fig. 2; (86) U. S. and Mex. Round. Surv., vol. ii, p. 41; (101) Mammals of N. A., 1859. Animal, with details, pl. xii, fig. 2.

THOMOMYS LATICEPS.
  (60) Pr. Ac. Nat. Sci. Phila. 1854, p. 335; (76) P. R. R. Surv., vol. viii, p. 392, pl. xii, fig. 1; (101) Mammals of N. A., 1859. Animal, with details, pl. xii, fig. 1.

THOMOMYS RUFESCENS.
  (60) Pr. Ac. Nat. Sci. Phila., 1854, p. 335; (76) P. R. R. Surv., vol. viii, p. 397, pl. x, fig. 1; (93) P. R. R. Surv., vol. x, p. 8, pl. x, fig. 1; (101) Mammals of N. A., 1859. Animal, pl. x, fig. 1.

Thomomys talpoides.
  (76) P. R. R. Surv., vol. viii, p. 403.

THOMOMYS UMBRINUS.
  (59) Pr. Ac. Nat. Sci. Phila., 1854, 334; (76) P. R. R. Surv., vol. viii, p. 399; (86) U. S. and Mex. Bound. Surv., vol. ii, p. 41, pll. viii, x, fig. 1, pl. xxxiii, fig. 5; (101) Mammals of N. A., 1859. Animal and details, pl. lxviii; details, pl. lxx, fig. 1; skull, pl. lxxxiii, fig. 5.

UROTRICHUS GIBBSII.
  (76) P. R. R. Surv., vol. viii, p. 76, pl. xviii, fig. 3, pl. xxviii; (101) Mammals of N. A., 1859. Details of external form, pl. xviii, fig. 3; animal and skull, pl. xxviii, No. 602.

URSUS AMERICANUS.
  (76) P. R. R. Surv., vol. viii, p. 225, pl. xliii, figg. 1–9; (101) Mammals of N. A., 1859. Skull, pl. xliii, figg. 10–13.

Ursus AMERICANUS cinnamoneus.
  (76) P. R. R. Surv., vol. viii, p. 228.

URSUS ARCTOS.
  (101) Mammals of N. A., 1859. Skull rather immature, pl. xliii, figg. 1–9.

URSUS CINNAMONEUS.
  (86) Mex. Bound. Surv., vol. ii, p. 29, pl. xix; (101) Mammals of N. A., 1859. Skull, pl. lxxix.

URSUS HORRIBILIS.
  (76) P. R. R. Surv., vol. viii, p. 219, pll. xli, xlii; (101) Mammals of N. A., 1859. Skull very old, pl. xli; skull rather young, pl. xlii.
  (86) Mex. Bound. Surv., vol. ii, p. 24, pl. xx; (101) Mammals of N. A., 1859. Skull, pl. lxxx.

URSUS MARITIMUS.
  (76) P. R. R. Surv., vol. viii, p. 229, pl. xliv; (101) Mammals of N. A., 1859. Skull, pl. xliv.

Vespertilio chilensis.
  (65) Gillis, Naval Astr. Exp., ii, p. 163.

VESPERTILIO PALLIDUS.
  (86) Mex. Bound. Surv., vol. ii, p. 4, pl. 1, fig. 1; (97) P. R. R. Surv., vol. x, p. 81; (101) Mammals of N. A., 1859. Animal, pl. lxi, fig. 1.

Vespertilio velatus.
  (65) Gillis, Naval Astr. Exp., ii, p. 163.

Vulpes fulvus argentatus.
  (76) P. R. R. Surv., vol. viii, p. 128.

Vulpes fulvus decussatus.
  (76) P. R. R. Surv., vol. viii, p. 127.

VULPES FULVUS FULVUS.
  (76) P. R. R. Surv., vol. viii, p. 124, pl. xxxii; (101) Mammals of N. A., 1859. Skull, pl. xxxii.

Vulpes lagopus.
  (76) P. R. R. Surv., vol. viii, p. 137.

VULPES LITTORALIS.
  (76) P. R. R. Surv., vol. viii, p. 143, pl. xxxv, fig. 2; (101) Mammals of N. A., 1859. Animal, pl. i; skull, pl. xxxv, fig. 2.

VULPES MACROURA.
(101) Mammals of N. A., 1859. Skull, pl. xxxiii.

Vulpes Macrourus.
(29) Stansbury's Surv. Salt Lake [App. C], p. 309; (76) P. R. R. Surv., vol. viii, p. 130, pl. xxxiii.

Vulpes Utah.
(34) Pr. Ac. Nat. Sci. Phila., 1852-3 (1854), p. 124.

VULPES VELOX.
(76) P. R. R. Surv., vol. viii, p. 133, pl. xxxiv; (101) Mammals of N. A, 1859. Skull, pl. xxxiv.

VULPES VIRGINIANUS.
(76) P. R. R. Surv., vol. viii, p. 138, pl. xxxv, fig. 1; (86)' Mex. Bound. Surv., vol. ii, p. 16; (101) Mammals of N. A., 1859. Skull, pl. xxxv, fig. 1.

BIRDS.

Acanthylis Vauxii.
(32) Stansbury's Surv. Salt Lake [App. C], p. 327.

Accipiter Cooperi.
(78) P. R. R. Surv., vol. ix, p. 16.

Accipiter Cooperi.
(87) U. S. and Mex. Bound. Surv., vol. ii, p. 3.

Accipiter fuscus.
(31) Stansbury's Surv. Salt Lake [App. C], p. 314; (78) P. R. R. Surv, vol. ix, p. 18; (87) U. S. and Mex. Bound. Surv., vol. ii, p. 3.

Accipiter mexicanus.
(78) P. R. R. Surv., vol. ix, p. 17.

Actiturus Bartramius.
(31a) Stansbury's Surv. Salt Lake [App. C], p. 326; (78) P. R. R. Surv., vol. ix, p. 737.

Actodromas Bonapartii.
(78) P. R. R. Surv., vol. ix, p. 722.

Actodromas maculata.
(78) P. R. R. Surv., vol. ix, p. 720.

Actodromas Wilsonii.
(78) P. R. R. Surv., vol. ix, p. 721.

Adamastor cinereus.
(78) P. R. R. Surv., vol. ix, p. 835.

Ædon.
(115) Review of N. A. Birds, Aug., 1864, Part I, p. 138.

Ædon aztecus.
(115) Review of N. A. Birds, Aug., 1864, Part I, p. 138.

Aegialeus melodus.
(78) P, R. R. Surv., vol. ix, p. 695.

Aegialeus semipalmatus.
(78) P. R. R. Surv., vol. ix. p. 694.

Aegialitis melodus.
(78) P. R. R. Surv., vol. ix, p. 695.

Aegialitis montanus.
(78) P. R. R. Surv., vol. ix, p. 693.

Aegialitis nivosa.
(78) P. R. R. Surv., vol. ix, p. 695.

AEGIALITIS NIVOSUS.
(104) Birds of N. A., 1860, p. 695, pl. xc, figg. 1, 2.

Aegialitis semipalmatus.
(78) P. R. R. Surv., vol. ix, p. 694.

Aegialitis vociferus.
(78) P. R. R. Surv., vol. ix, p. 692; (87) U. S. and Mex. Bound. Surv., vol. ii, p. 25; (99) Pr. Ac. Nat. Sci. Phila., 1859, p. 306.

Aegialitis Wilsonius.
(78) P. R. R. Surv., vol. ix, p. 693.

Ægiothus.
(632½) Birds of N. A., 1874, vol. i, p. 491.

Aegiothus canescens.
(78) P. R. R. Surv., vol. ix, p. 429; (632½) Birds of N. A., 1874, vol. i, p. 498, pl. xxii, fig. 2.

ÆGIOTHUS FLAVIROSTRIS BREWSTERI.
(632½) Birds of N. A., 1874, vol. i, p. 501, pl. xxii, fig. 6.

Aegiothus linaria.
(78) P. R. R. Surv., vol. ix, p. 428.

ÆGIOTHUS LINARIUS.
(632½) Birds of N. A., 1874, vol. i, p. 493, pl. xxii, figg. 3, 5 (cuts, p. 491).

ÆSALON LITHOFALCO COLUMBARIUS.
(632½) Birds of N. A., 1874, vol. iii, p. 144 (cut, p. 146).

Æsalon lithofalco Richardsoni.
(632½) Birds of N. A., 1874, vol. iii, p. 148.

Æsalon lithofalco Suckleyi.
(632½) Birds of N. A., 1874, vol. iii, p. 147.

Aestrelata meridionalis.
(78) P. R. R. Surv., vol. ix, p. 827.

Agelainæ.
(632½) Birds of N. A., 1874, vol. ii, p. 148.

Agelaius.
(632½) Birds of N. A., 1874, vol. ii, p. 158.

Agelaius ——.
(99) Pr. Ac. Nat. Sci. Phila., 1859, p. 305.

Agelaius gubernator.
(78) P. R. R. Surv., vol. ix, p 529; (87) U. S. and Mex. Bound. Surv., vol. ii, p. 18.

AGELAIUS PHŒNICEUS.
(78) P. R. R. Surv., vol. ix, p. 526; (87) U. S. and Mex. Bound. Surv., vol. ii, p. 18; (632½) Birds of N. A., 1874, vol. ii, p. 159, pl. xxxiii, figg. 1, 2, 3 (cuts, p. 158).

AGELAIUS PHŒNICEUS GUBERNATOR.
(632½) Birds of N. A., 1874, vol. ii, p. 163, pl. xxxiii. figg. 4, 8.

Agelaius tricolor.
(78) P. R. R. Surv., vol. ix, p. 530; (87) U. S. and Mex. Bound. Surv, vol. ii, p. 18; (632½) Birds of N. A., 1874, vol. ii, p. 165, pl. xxxiii, figg. 5, 6, 7.

Agelaius xanthocephalus.
(31a) Stansbury's Surv. Salt Lake [App. C], p. 326.

Agrodoma Spraguei.
(32) Stansbury's Surv. Salt Lake [App. C], p. 329.

Aix sponsa.
(78) P. R. R. Surv., vol. ix, p. 785.

Alauda.
(632½) Birds of N. A., 1874, vol. ii, p. 135.

Alauda Arvensis.
(632½) Birds of N. A., 1874, vol. ii, p. 136, pl. xxxii, fig. 3.

Alauda rufa.
(32) Stansbury's Surv. Salt Lake [App. C], p. 331.

Alaudidæ.
(632½) Birds of N. A., 1874, vol. ii, p. 135.

Alca impennis.
(78) P. R. R. Surv., vol. ix, p. 900.

Alca torda.
(78) P. R. R. Surv., vol. ix, p. 901.

Alcedinidæ.
(632½) Birds of N. A., 1874, vol. ii, p. 391.

Ammodromus.
(632½) Birds of N. A., 1874, vol. i, p. 556.

AMMODROMUS CAUDACUTUS.
(78) P. R. R. Surv., vol. ix, p. 453; (632½) Birds of N. A., 1874, vol. i, p. 557, pl. xxv, fig. 7 (cuts, pp. 556, 557).

AMMODROMUS MARITIMUS.
(78) P. R. R. Surv., vol. ix, p. 454; (632½) Birds of N. A., 1874, vol. i, p. 560, pl. xxv, fig. 8; (632½) Birds of N. A., 1874, App., p. 513.

AMMODROMUS SAMUELIS.
(78) P. R. R. Surv., vol. ix, p. 455; (83) Pr. Bost. Soc. Nat. Hist., 1858, p. 379; (104) Birds of N. A., 1860, p. 455, pl. lxxi, fig. 1.

Ampelidæ.
(115) Review of N. A. Birds, May, 1866, Part I, p. 322; Ibid, p. 400; (632½) Birds of N. A., 1874, vol. i, p. 395.

Ampelinæ.
(115) Review of N. A. Birds, May, 1866, Part I, p. 401; Ibid, p. 403; (632½) Birds of N. A., 1874, vol. i, p. 395.

Ampelis.
(115) Review of N. A. Birds, May, 1866, Part I, p. 403; (632½) Birds of N. A., 1874, vol. i, p. 395.

Ampelis cedrorum.
(78) P. R. R. Surv., vol. ix, p. 318; (87) U. S. and Mex. Bound. Surv., vol. ii, p. 11; (115) Review of N. A. Birds, May, 1866, Part I, p. 407; (632½) Birds of N. A., 1874, vol. i, p. 401.

Ampelis garrula.
(115) Review of N. A. Birds, May, 1866, Part I, p. 403.

AMPELIS GARRULUS.
(78) P. R. R. Surv., vol. ix, p. 317; (632½) Birds of N. A., 1874, vol. i, p. 396, pl. xviii, fig. 1 (cut, p. 397).

Anas boschas.
(31) Stansbury's Surv. Salt Lake [App. C], p. 322; (78) P. R. R. Surv., vol. ix, p. 774; (87) U. S. and Mex. Bound. Surv., vol. ii, p. 26; (94) P. R. R. Surv., vol. x, p. 15.

Anas europhasianus.
(32) Stansbury's Surv. Salt Lake [App. C], p. 334.

Anas obscura.
(78) P. R. R. Surv., vol. x, p. 775.

Anorthura.
(115) Review of N. A. Birds, Sept., 1864, Part I, p. 144.

Anorthura hyemalis.
(78) P. R. R. Surv., vol. ix, p. 369.

Anous stolidus.
(78) P. R. R. Surv., vol. ix, p. 865.

Anser canadensis.
(31) Stansbury's Surv. Salt Lake [App. C], p. 321.

Anser cœrulescens.
(78) P. R. R. Surv., vol. ix, p. 761.

Anser erythropus.
(31) Stansbury's Surv. Salt Lake [App. C], p. 321.

Anser frontalis.
(78) P. R. R. Surv., vol. ix, p. 762.

Anser (Anser) Gambelii.
(78) P. R. R. Surv., vol. ix, p. 761.

Anser Gambelii.
(87) Mex. Bound. Surv., vol. ii, p. 26.

Anser hyperboreus.
(78) P. R. R. Surv., vol. ix, p. 760.

Anser nigricans.
(32) Stansbury's Surv. Salt Lake [App. C], p. 334.

Anthinæ.
(632½) Birds of N. A., 1874, vol. 1, p. 169.

Anthus.
(115) Review of N. A. Birds, Aug., 1864, Part I, p. 151; Ibid., p. 152; Ibid., Oct., 1864, Part I, p. 153.

Anthus.
(115) Review of N. A. Birds, Aug., 1864, Part I, p. 158.

Anthus.
(632½) Birds of N. A., 1874, vol. 1, p. 170.

Anthus bogotensis.
(115) Review of N. A. Birds, Aug., 1864, Part I, p. 157.

ANTHUS LUDOVICIANUS.
(78) P. R. R. Surv., vol. ix, p. 232; (87) Mex. Bound. Surv., vol. ii, p. 10; (115) Review of N. A. Birds, Aug., 1864, Part I, p. 153; (632½) Birds of N. A., 1874, vol. i, p. 171, pl. x, fig. 3 (cuts, pp. 170, 171); Ibid., App., p. 504.

ANTHUS PRATENSIS.
(115) Review of N. A. Birds, Aug., 1864, Part I, p. 155; (632½) Birds of N. A., 1874, vol. i, p. 173, pl. x, fig. 4.

Anthus rufus.
(115) Review of N. A. Birds, Aug., 1864, Part I, p. 156.

Anthus Spraguei.
(115) Review of N. A. Birds, Aug., 1864, Part I, p. 155.

Antrostomus.
(632½) Birds of N. A., 1874, vol. ii, p. 408.

ANTROSTOMUS CAROLINENSIS.
(78) P. R. R. Surv., vol. ix, p. 147; (632½) Birds of N. A., 1874, vol. ii, p. 410, pl. xlvi, fig. 1 (cut, p. 410); (632½) Birds of N. A., 1874, App., p. 520.

ANTROSTOMUS NUTTALLI.
(32) Stansbury's Surv. Salt Lake [App. C], p. 327; (78) P. R. R. Surv., vol. ix, p. 149; (87) Mex. Bound. Surv., vol. ii, p. 6; (632½) Birds of N. A., 1874, vol. ii, p. 417, pl. xlvi, fig. 3 (cut, p. 408).

ANTROSTOMUS VOCIFERUS.
 (78) P. R. R. Surv., vol. ix, p. 148; (632½) Birds of N. A., 1874, vol. ii, p. 413, pl. xlvi, fig. 2 (cut, p. 416).
Aphriza virgata.
 (78) P. R. R. Surv., vol. ix, p. 698.
Apobapton brachypterus.
 (78) P. R. R. Surv., vol. ix, p. 917.
Apobapton Kittlitzii.
 (78) P. R. R. Surv., vol. ix, p. 917.
Apobapton marmoratus.
 (78) P. R. R. Surv., vol. ix, p. 915.
Apobapton Temminckii.
 (78) P. R. R. Surv., vol. ix, p. 916.
Apobapton Wrangelii.
 (78) P. R. R. Surv., vol. ix, p. 917.
Aquila.
 (632½) Birds of N. A., 1874, vol. iii, p. 312.
Aquila canadensis.
 (78) P. R. R. Surv., vol. ix, p. 41.
AQUILA CHRYSAËTUS CANADENSIS.
 (632½) Birds of N. A., 1874, vol. iii, p. 314 (cuts, pp. 312, 316, 317).
Aramus giganteus.
 (78) P. R. R. Surv., vol. ix, p. 657.
Archibuteo.
 (632½) Birds of N. A., 1874, vol. iii, p. 291.
ARCHIBUTEO FERRUGINEUS.
 (32) Stansbury's Surv. Salt Lake [App. C], p. 327; (78) P. R. R. Surv., vol. ix, p. 34; (87) Mex. Bound. Surv., vol. ii, p. 4; (632½) Birds of N. A., 1874, vol. iii, p. 300 (cuts, p. 298).
Archibuteo lagopus.
 (78) P. R. R. Surv., vol. ix, p. 32.
ARCHIBUTEO LAGOPUS SANCTI-JOHANNIS.
 (632½) Birds of N. A., 1874, vol. iii, p. 304 (cuts, pp. 298, 307, 308, 312).
Archibuteo Sancti-Johannis.
 (78) P. R. R. Surv., vol. ix, p. 33.
Ardea herodias.
 (78) P. R. R. Surv., vol. ix, p. 668; (87) U. S. & Mex. Bound. Surv., vol. ii, p. 24.
ARDEA WÜRDEMANNII.
 (78) P. R. R. Surv., vol. ix, p. 669; (104) Birds of N. A., 1860, p. 661, pl. lxxxvi.
Ardenna major.
 (78) P. R. R. Surv., vol. ix, p. 833.
Ardetta exilis.
 (78) P. R. R. Surv., vol. ix, p. 673.
Arquatella maritima.
 (78) P. R. R. Surv., vol. ix, p. 717.
Astur.
 (632½) Birds of N. A., 1874, vol. iii, p. 236.
Astur atricapillus.
 (78) P. R. R. Surv., vol. ix, p. 15.
ASTUR PALUMBARIUS ATRICAPILLUS.
 (632½) Birds of N. A., 1874, vol. iii, p. 237 (cuts, pp. 236, 241).
Asturina.
 (632½) Birds of N. A., 1874, vol. iii, p. 244.
ASTURINA NITIDA.
 (78) P. R. R. Surv., vol. ix, p. 35; (87) U. S. & Mex. Bound. Surv., vol. ii, p. 4; (104) Birds of N. A., 1860, p. 35, pl. lxiv. Adult and young.

ASTURINA NITIDA PLAGIATA.
 (632½) Birds of N. A., 1874, vol. iii, p. 246 (cuts, pp. 244, 247).
Athene cunicularia.
 (78) P. R. R. Surv., vol. ix, p. 60; (87) Mex. Bound. Surv., vol. ii, p. 5; (94) P. R. R. Surv., vol. x, p. 13.
Athene hypugæa.
 (31) Stansbury's Surv. Salt Lak [App.C],p. 314; (78). P. R. R. Surv., vol. ix, p. 59; (87) Mex. Bound. Surv., vol. ii, p. 5.
Atthis.
 (632½) Birds of N. A., 1874, vol. ii, p. 464.
Atthis anna.
 (78) P. R. R. Surv., vol. ix, p. 137.
ATTHIS COSTÆ.
 (78) P. R. R. Surv., vol. ix, p. 138; (104) Birds of N. A., 1860, p. 138, pl. xix. Male and female.
ATTHIS HELOSIA.
 (632½) Birds of N. A., 1874, vol. ii, p. 465, pl. xlvii, fig. 6 (cut, p. 464).
Atticora.
 (115) Review of N. A. Birds, May, 1865, Part I, p. 270; Ibid., p. 271; Ibid., p. 305.
Atticora cyanoleuca.
 (115) Review of N. A. Birds, May, 1865, Part I, p. 308.
Atticora cyanoleuca montana.
 (115) Review of N. A. Birds, May, 1865, Part I, p. 310.
Atticora fasciata.
 (115) Review of N. A. Birds, May, 1865, Part I, p. 306.
Atticora fucata.
 (115) Review of N. A. Birds, May, 1865, Part I, p. 308.
Atticora melanoleuca.
 (115) Review of N. A. Birds, May, 1865, Part I, p. 310.
Atticora murina.
 (115) Review of N. A. Birds, May, 1865, Part I, p. 311.
Atticora patagonica.
 (115) Review of N. A. Birds, May, 1865, Part I, p. 311.
Atticora pileata.
 (115) Review of N. A. Birds, May, 1865, Part I, p. 307.
Atticora tibialis.
 (115) Review of N. A. Birds, May, 1865, Part I, p, 307.
Audubonia occidentalis.
 (78) P. R. R. Surv., vol. ix, p. 670.
Auriparus.
 (115) Review of N. A. Birds, Aug., 1864, Part I, p. 85; (632½) Birds of N. A., 1874, vol. i, p. 111.
AURIPARUS FLAVICEPS.
 (115) Review of N. A. Birds, Aug., 1864, Part I, p. 85; (632½) Birds of N. A., 1874, vol. i, p. 112 (cuts, p. 112).
Aythya americana.
 (78) P. R. R. Surv., vol. ix, p. 793; (87) Mex. Bound. Surv., vol. ii, p. 27; (94) P. R. R. Surv., vol. x, p. 16.

Aythya vallisneria.
  (78) P. R. R. Surv., vol. ix, p. 704.
Basilenterus.
  (115) Review of N. A. Birds, Apr., 1865, Part
    I, p. 227; Ibid., p. 238; Ibid., May, 1865,
    Part I, p. 241.
Basilenterus Belli.
  (115) Review of N. A. Birds, May, 1865, Part
    I, p. 250.
Basilcuterus bivittatus.
  (115) Review of N. A. Birds, May, 1865, Part
    I, p. 241.
Basileuterus chrysogaster.
  (115) Review of N. A. Birds, May, 1865, Part I,
    p. 244.
Basilenterus cinereicollis.
  (115) Review of N. A. Birds, May, 1865, Part I,
    p. 244.
Basilenterus coronatus.
  (115) Review of N. A. Birds, May, 1865, Part I,
    p. 244.
Basilenterus culicivorus.
  (115) Review of N. A. Birds, May, 1865, Part I,
    p. 245.
Basileuterus Delattrii.
  (115) Review of N. A. Birds, May, 1865, Part I,
    p. 249.
Basilenterus hypoleucus.
  (115) Review of N. A. Birds, May, 1865, Part I,
    p. 241.
Basileuterus leucoblepharum.
  (115) Review of N. A. Birds, May, 1865, Part I,
    p. 244.
Basileuterus melanogenys.
  (115) Review of N. A. Birds, May, 1865, Part I,
    p. 248.
Basileuterus mesochrysus.
  (115) Review of N. A. Birds, May, 1865, Part I,
    p. 250.
Basileuterus nigricristatus.
  (115) Review of N. A. Birds, May, 1865, Part I,
    p. 251.
Basileuterus rufifrons.
  (115) Review of N. A. Birds, May, 1865, Part I,
    p. 248.
Basileuterus semicervinus.
  (115) Review of N. A. Birds, May, 1865, Part I,
    p. 244.
Basilenterus stragulatus.
  (115) Review of N. A. Birds, May, 1865, Part I,
    p. 244.
Basilenterus superciliosus.
  (115) Review of N. A. Birds, May, 1865, Part I,
    p. 244.
Basilenterus uropygialis.
  (115) Review of N. A. Birds, May, 1865, Part I,
    p. 246.
Basilenterus vermivorus.
  (115) Review of N. A. Birds, May, 1865, Part I,
    p. 241.
Basileuterus viridicatus.
  (115) Review of N. A. Birds, May, 1865, Part I,
    p. 244.
BATHMIDURUS MAJOR.
  (78) P. R. R. Surv., vol. ix, p. 166; (104) Birds of
    N. A., 1860, p. 166, pl. xlvii, fig. 2, female.

Bernicia (Bernicla) brenta.
  (78) P. R. R. Surv., vol. ix, p. 767.
Bernicla canadensis.
  (78) P. R. R. Surv., vol. ix, p. 764; (87) Mex.
    Bound. Surv., vol. ii, p. 26.
Bernicla Hutchinsii.
  (78) P. R. R. Surv., vol. ix, p. 765.
Bernicla leucopareia.
  (78) P. R. R. Surv., vol. ix, p. 765.
Bernicla leucopsis.
  (78) P. R. R. Surv., vol. ix, p. 768.
Bernicla (Bernicia) nigricans.
  (78) P. R. R. Surv., vol. ix, p. 767.
Bewickii.
  (115) Review of N. A. Birds, Aug., 1864, Part
    I, p. 126.
Blasipus Heermanni.
  (78) P. R. R. Surv., vol. ix, p. 848; (99) Pr. Acad.
    Nat. Sci. Phila., 1859, p. 306.
Bonasa.
  (632½) Birds of N. A., 1874, vol. iii, p. 446.
Bonasa Sabinii.
  (78) P. R. R. Surv., vol. ix, p. 631.
Bonasa umbelloides.
  (78) P. R. R. Surv., vol. ix, p. 630.
Bonasa umbellus.
  (78) P. R. R. Surv., vol. ix, p. 630.
Bonasa umbellus sabina.
  (632½) Birds of N. A., 1874, vol. iii, p. 454.
BONASA UMBELLUS UMBELLOIDES.
  (632½) Birds of N. A., 1874, vol. iii, p. 453, pl. lxi,
    fig. 10.
BONASA UMBELLUS UMBELLUS.
  (632½) Birds of N. A., 1874, vol. iii, p. 448, pl. lxi,
    figg. 3, 9 (cuts, pp. 448, 449).
Botaurus lentiginosus.
  (31) Stansbury's Surv. Salt Lake [App. C],
    p. 320; (78) P. R. R. Surv., vol. ix, p. 674;
    (87) Mex. Bound. Surv., vol. ii, p. 24.
Brachyotus Cassinii.
  (78) P. R. R. Surv., vol. ix, p. 54.
Brachyrhamphus antiquus.
  (78) P. R. R. Surv., vol. ix, p. 916.
Brachyrhamphus brachypterus.
  (32) Stansbury's Surv. Salt Lake [App. C],
    p. 335; (78) P. R. R. Surv., vol. ix, p. 917.
Brachyrhamphus hypoleucus.
  (99) Pr. Acad. Nat. Sci. Phila., 1859, p. 306.
Brachyrhamphus Kittlitzii.
  (78) P. R. R. Surv., vol. ix, p. 917.
Brachyrhamphus marmoratus.
  (78) P. R. R. Surv., vol. ix, p. 915.
Brachyrhamphus Temminckii.
  (78) P. R. R. Surv., vol. ix, p. 916.
Brachyrhamphus Wrangelli.
  (32) Stansbury's Surv. Salt Lake [App. C],
    p. 335; (78) P. R. R. Surv., vol. ix, p. 917.
Brunneicollis.
  (115) Review of N. A. Birds, Aug., 1864, Part I,
    p. 138.
Bubo.
  (632½) Birds of N. A., 1874, vol. iii, p. 60.
Bubo virginianus.
  (87) Mex. Bound. Surv., vol. ii, p. 4; (99) Pr.
    Acad. Nat. Sci. Phila., 1859 (1860), p. 302.

Bubo virginianus arcticus.
> (78) P. R. R. Surv., vol. ix, p. 49; (632½) Birds of N. A., 1874, vol. iii, p. 64.

Bubo virginianus atlanticus.
> (78) P. R. R. Surv., vol. ix, p. 49.

Bubo virginianus magellanicus.
> (78) P. R. R. Surv., vol. ix, p. 49.

Bubo virginianus pacificus.
> (78) P. R. R. Surv., vol. ix, p. 49; (632½) Birds of N. A., 1874, vol. iii, p. 65.

BUBO VIRGINIANUS VIRGINIANUS.
> (632½) Birds of N. A., 1874, vol. iii, p. 62 (cuts, pp. 62, 63, 66, 98, 99, 100-1).

Bucephala albeola.
> (78) P. R. R. Surv., vol. ix, p. 797; (87) Mex. Bound. Surv., vol. ii, p. 27.

Bucephala americana.
> (78) P. R. R. Surv., vol. ix, p. 796; (94) P. R. R. Surv., vol. x, p. 16.

Bucephala islandica.
> (78) P. R. R. Surv., vol. ix, p. 796.

Budytes.
> (632½) Birds of N. A., 1874, vol. i, p. 167.

BUDYTES FLAVA.
> (129) Chicago Acad. Sci., 1869, p. 312, pl. xxx, fig. 1; (632½) Birds of N. A., 1874, vol. i, p. 167, pl. x, fig. 2 (cuts, pp. 167, 168).

Buteo.
> (632½) Birds of N. A., 1874., vol. iii, p. 254.

Buteo Bairdii.
> (78) P. R. R. Surv., vol. ix, p. 21.

Buteo borealis.
> (31) Stansbury's Surv. Salt Lake [App. C], p. 314; (78) P. R. R. Surv., vol. ix, p. 25; (87) Mex. Bound. Surv., vol. ii, p. 3; (632½) Birds of N. A., 1874, vol. iii, p. 281.

BUTEO BOREALIS BOREALIS.
> (632½) Birds of N. A., 1874, vol. iii, p. 282 (cuts, pp. 255, 288).

Buteo borealis calurus.
> (632½) Birds of N. A., 1874, vol. iii, p. 286.

Buteo borealis Krideri.
> (632½) Birds of N. A., 1874, vol. iii, p. 284.

Buteo borealis lucasanus.
> (632½) Birds of N. A., 1874, vol. iii, p. 285.

BUTEO CALURUS.
> (78) P. R. R. Surv., vol. ix, p. 22: (87) Mex. Bound. Surv., vol. ii, p. 3; (94) P. R. R. Surv., vol. x, p. 11, pl. xiv; (104) Birds of N. A., 1860, p. 22, pl. xiv.

BUTEO COOPERI.
> (78) P. R. R. Surv., vol. ix, p. 31; (104) Birds of N. A., 1860, p. 31, pl. xvi; (632½) Birds of N. A., 1874, vol. iii, p. 295 (cuts, pp. 295, 296).

BUTEO ELEGANS.
> (78) P. R. R. Surv., vol. ix, p. 28; (87) Mex. Bound. Surv., vol. ii, p. 3; (104) Birds of N. A., 1860, p. 28, pl. ii, adult; pl. iii, young.

BUTEO FULIGINOSUS.
> (104) Birds of N. A., 1860, p. 30, pl. xv, fig. 1.

BUTEO HABLANI.
> (78) P. R. R. Surv., vol. ix, p. 24; (632½) Birds of N. A., 1874, vol. iii, p. 292 (cut, p. 293).

Buteo insignatus.
> (78) P. R. R. Surv., vol. ix, p. 23.

Buteo lineatus.
> (78) P. R. R. Surv., vol. ix, p. 28.

BUTEO LINEATUS ELEGANS.
> (632½) Birds of N. A., 1874, vol. iii, p. 277 (cut, p. 281).

Buteo lineatus lineatus.
> (632½) Birds of N. A., 1874, vol. iii, p. 275.

Buteo montanus.
> (78) P. R. R. Surv., vol. ix, p. 26; (87) Mex. Bound. Surv., vol. ii, p. 3; (94) P. R. R. Surv., vol. x, p. 12.

BUTEO OXYPTERUS.
> (78) P. R. R. Surv., vol. ix, p. 30; (94) P. R. R. Surv., vol. x, p. 11, pl. xv; (104) Birds of N. A., 1860, p. 30, pl. xv, fig. 2.

BUTEO PENNSYLVANICUS.
> (78) P. R. R. Surv., vol. ix, p. 29; (632½) Birds of N. A., 1874, vol. iii, p. 259 (cut, p. 261).

BUTEO SWAINSONI.
> (78) P. R. R. Surv., vol. ix, p. 19; (87) Mex. Bound. Surv., vol. ii, p. 3; (94) P. R. R. Surv., vol. x, p. 11, pll. xii, xiii; (104) Birds of N. A., 1860, p. 19, pll. xii, xiii.

BUTEO SWAINSONI OXYPTERUS.
> (632½) Birds of N. A., 1874, vol. iii, p. 266 (cut p. 267).

BUTEO SWAINSONI SWAINSONI.
> (632½) Birds of N. A., 1874, vol. iii, p. 263 (cuts, pp. 255, 264, 269, 270).

BUTEO ZONOCERCUS.
> (632½) Birds of N. A., 1874, vol. iii, p. 272 (cuts, pp. 255, 271-272, 274).

Butorides brunnescens.
> (78) P. R. R. Surv., vol. ix, p. 676.

Butorides virescens.
> (78) P. R. R. Surv., vol. ix, p. 676; (87) Mex. Bound. Surv., vol. ii, p. 24.

Calamospiza.
> (632½) Birds of N. A., 1874, vol. ii, p. 60.

CALAMOSPIZA BICOLOR.
> (78) P. R. R. Surv., vol. ix, p. 492; (87) Mex. Bound. Surv., vol. ii, p. 16; (99) Pr. Acad. Nat. Sci. Phila., 1859, p. 304; (632½) Birds of N. A., 1874, vol. ii, p. 61, pl. xxix, figg. 2, 3 (cuts, pp. 60, 61).

Calidris arenaria.
> (78) P. R. R. Surv., vol. ix, p. 723; (99) Pr. Acad. Nat. Sci. Phila., 1859, p. 306.

Callichelodon.
> (115) Review of N. A. Birds, May, 1865, Part I, p. 271; *Ibid.*, p. 293; *Ibid.*, p. 303.

Callipepla.
> (632½) Birds of N. A., 1874, vol. iii, p. 487.

Callipepla Douglassii.
> (32) Stansbury's Surv. Salt Lake [App. C], p. 334.

Callipepla elegans.
> (32) Stansbury's Surv. Salt Lake [App. C], p. 334.

Callipepla Gambeli.
> (31a) Stansbury's Surv. Salt Lake [App. C], p. 326; (32) *Ibid.*, p. 334.

Callipepla picta.
> (32) Stansbury's Surv. Salt Lake [App. C], p. 334.

CALLIPEPLA SQUAMATA.
 (32) Stansbury's Surv. Salt Lake [App. C], p.
 334; (78) P. R. R. Surv., vol. ix, p. 646; (87)
 Mex. Bound. Surv., vol. ii, p. 23; (632½)
 Birds of N. A., 1874, vol. iii, p. 487, pl. lxiii,
 fig. 6 (cuts p. 488).
Calypte.
 (632½) Birds of N. A., 1874, vol. ii, p. 453.
CALYPTE ANNA.
 (632½) Birds of N. A., 1874, vol. ii, p. 454, pl.
 xlvii, fig. 7 (cuts, p. 454).
CALYPTE COSTAE.
 (632½) Birds of N. A., 1874, vol. ii, p. 457, pl.
 xlvii, fig. 8 (cuts, p. 453).
Campephilus.
 (632½) Birds of N. A., 1874, vol. ii, p. 494.
Campephilus imperialis.
 (78) P. R. R. Surv., vol. ix, p. 82.
CAMPEPHILUS PRINCIPALIS.
 (78) P. R. R. Surv., vol. ix, p. 81; (632½) Birds
 of N. A., 1874, vol. ii, p. 496, pl. xlix, figg. 1,
 2 (cuts, pp. 495, 497).
Campylorhynchus.
 (115) Review of N. A. Birds, Aug., 1864, Part
 I, p. 94; Ibid., p. 96; (632½) Birds of N. A.,
 1874, vol. i, p, 131.
CAMPYLORHYNCHUS AFFINIS.
 (99) Pr. Acad. Nat. Sci. Phila., 1859, p. 304;
 (115) Review of N. A. Birds, Aug., 1864,
 Part I, p. 100; (632½) Birds of N. A., 1874,
 vol. i, p. 133, pl. viii, fig. 6.
Campylorhynchus albibrunneus.
 (115) Review of N. A. Birds, Aug., 1864, Part
 I, p. 96.
Campylorhynchus balteatus.
 (115) Review of N. A. Birds, Aug., 1864, Part
 I, p. 103.
CAMPYLORHYNCHUS BRUNNEICAPILLUS.
 ·(78) P. R. R. Surv., vol. ix, p. 355; (87) U. S.
 and Mex. Bound. Surv., vol. ii, p. 13; (115)
 Review of N. A. Birds, Aug., 1864, Part
 I, p. 99; (632½) Birds of N. A., 1874, vol. i,
 p. 132, pl. viii, fig. 5 (cuts, pp. 131, 132);
 Ibid., App., p. 503.
Campylorhynchus capistratus.
 (115) Review of N. A. Birds, Aug., 1864, Part
 I, p. 104.
Campylorhynchus gularis.
 (115) Review of N. A. Birds, Aug., 1864, Part
 I, p. 109.
Campylorhynchus guttatus.
 (115) Review of N. A. Birds, Aug., 1864, Part
 I, p. 108.
Campylorhynchus humilis.
 (115) Review of N. A. Birds, Aug., 1864, Part
 I, p. 107.
Campylorhynchus jocosus.
 (115) Review of N. A. Birds, Aug., 1864, Part
 I, p. 106.
Campylorhynchus megalopterus.
 (115) Review of N. A. Birds, Aug:, 1864, Part
 I, p. 101.
Campylorhynchus nigriceps.
 (115) Review of N. A. Birds, Aug., 1864, Part
 I, p. 109.

Campylorhynchus pallescens.
 (115) Review of N. A. Birds, Aug., 1864, Part
 I, p. 101.
Campylorhynchus rufinucha.
 (115) Review of N. A. Birds, Aug., 1864, Part
 I, p 105.
Campylorhynchus zonatus.
 (115) Review of N. A. Birds, Aug., 1864, Part
 I, p. 104.
Canace.
 (632½) Birds of N. A., 1874, vol. iii, p. 415.
CANACE CANADENSIS CANADENSIS.
 (632½) Birds of N. A., 1874, vol. iii, p. 416, pl.
 lix, figg. 5, 6, pl. lxi, fig. 5 (cut, p. 419).
CANACE CANADENSIS FRANKLINI.
 (632½) Birds of N. A., 1874, vol. iii, p. 419, pl.
 lix, fig. 3 (cut, p. 419).
Canace obscurus fuliginosus.
 (632½) Birds of N. A., 1874, vol. iii, p. 425.
CANACE OBSCURUS OBSCURUS.
 (632½) Birds of N. A., 1874, vol. iii, p. 422, pl. lix,
 figg. 1, 2 (cuts, pp. 421, 422).
CANACE OBSCURA RICHARDSONI.
 (632½) Birds of N. A., 1874, vol. iii, p. 427, pl.
 lix, fig. 4.
Caprimulgidæ.
 (632½) Birds of N. A., 1874, vol. ii, p. 398.
Caprimulginæ.
 (632½) Birds of N. A., 1874, vol. ii, p. 398.
Cardellina.
 (115) Review of N. A. Birds, Apr., 1865, Part
 I, p. 236; Ibid., p. 238; Ibid., May, 1865,
 Part I, p. 263.
Cardellina rubra.
 (78) P. R. R. Surv., vol. ix, p. 296; (115) Review of N. A. Birds, May, 1865, Part I,
 p. 264.
Cardellina rubrifrons.
 (115) Review of N. A. Birds, May, 1865, Part
 I, p. 264.
Cardellina versicolor.
 (115) Review of N. A. Birds, May, 1865, Part
 I, p. 265.
Cardinalis.
 (632½) Birds of N. A., 1874, vol. ii, p. 98.
Cardinalis igneus.
 (99) Pr. Acad. Nat. Sci., Phila., 1859, p. 305.
Cardinalis sinuatus.
 (32) Stansbury's Surv. Salt Lake [App. C],
 p. 331.
CARDINALIS VIRGINIANUS.
 (78) P. R. R. Surv., vol. ix, p. 509; (87) Mex
 Bound. Surv., vol. ii, p. 17; (632½) Birds
 of N. A., 1874, vol. ii, p. 100, pl. xxx, figg.
 6, 7 (cuts, pp. 98, 100).
CARDINALIS VIRGINIANUS IGNEUS.
 (632½) Birds of N. A., 1874, vol. ii, p. 103, pl. xxx,
 fig. 10; Ibid., App., p. 516.
Carpodacus.
 (632½) Birds of N. A., 1874, vol. i, p. 459.
CARPODACUS CALIFORNICUS.
 (78) P. R. R. Surv., vol. ix, p. 413; (104) Birds
 of N. A., 1860, p. 413, pl. lxxii, figg. 2, 3.
 Male and female.

CARPODACUS CASSINII.
(56) Pr. Acad. Nat. Sci. Phila., 1854, p. 119;
(78) P. R. R. Surv., vol. ix, p. 414; (104)
Birds of N. A., 1860, p. 414, pl. xxvii, fig. 1.
Male; (632½) Birds of N. A., 1874, vol. i, p.
460, pl. xxi, figg. 4, 5.

Carpodacus familiaris.
(32) Stansbury's Surv. Salt Lake [App. C],
p. 331.

Carpodacus frontalis.
(78) P. R. R. Surv., vol. ix, p. 415; (87) Mex.
Bound. Surv., vol. ii, p. 14; (99) Pr. Acad.
Nat. Sci. Phila., 1859, p. 304; (632½) Birds
of N. A., 1874, vol. i, p. 465.

CARPODACUS FRONTALIS FRONTALIS.
(632½) Birds of N. A., 1874, vol. i, p. 466, pl. xxi,
figg. 3, 6 (cuts, pp. 459, 461).

CARPODACUS FRONTALIS RHODOCOLPUS.
(632½) Birds of N. A., 1874, vol. i, p. 468, pl. xxi,
fig. 9.

Carpodacus haemorrhous.
(78) P. R. R. Surv., vol. ix, p. 417.

Carpodacus obscurus.
(32) Stansbury's Surv. Salt Lake [App. C], p.
331.

CARPODACUS PURPUREUS.
(78) P. R. R. Surv., vol. ix, p. 412; (632½) Birds
of N. A., 1874, vol. i, p. 462, pl. xxi, figg.
7, 8.

CARPODACUS PURPUREUS CALIFORNICUS.
(632½) Birds of N. A., 1874, vol. i, p. 465, pl. xxi,
figg. 10, 11.

Carpodagus purpureus.
(6) Lit. Rec. and Journ. Linnæan, Ass. Penn.
Col., Oct., 1845, p. 254.

Cataractes arra.
(78) P. R. R. Surv., vol. ix, p. 915.

Cataractes lomvia.
(78) P. R. R. Surv., vol. ix, p. 913.

Cataractes ringvia.
(78) P. R. R. Surv., vol. ix, p. 914.

Catharista.
(632½) Birds of N. A., 1874, vol. iii, p. 350.

CATHARISTA ATRATA.
(632½) Birds of N. A., 1874, vol. iii, p. 351 (cuts,
pp. 350, 352, 355, 356).

Cathartes atratus.
(78) P. R. R. Surv., vol. ix, p. 5.

Cathartes aura.
(78) P. R. R. Surv., vol. ix, p. 4; (87) Mex.
Bound. Surv., vol. ii, p. 3.

Cathartes Burrovianus.
(78) P. R. R. Surv., vol. ix, p. 6.

Cathartes californianus.
(78) P. R. R. Surv., vol. ix, p. 5,

Cathartidæ.
(632½) Birds of N. A., 1874, vol. viii, p. 335.

Catharus.
(115) Review of N. A. Birds, June, 1864, Part
I, p. 4; Ibid., p. 6; Ibid., p. 7.

Catharus dryas.
(115) Review of N. A. Birds, June, 1864, Part
I, p. 10.

Catharus Frantzii.
(115) Review of N. A. Birds, June, 1864, Part
I, p. 9.

Catharus melpomene.
(115) Review of N. A. Birds, June, 1864, Part
I, p. 7.

Catharus mexicanus.
(115) Review of N. A. Birds, June, 1864, Part
I, p. 11.

Catharus occidentalis.
(115) Review of N. A. Birds, June, 1864, Part
I, p. 8.

Catherpes.
(115) Review of N. A. Birds, Aug., 1864, Part
I, p. 94; Ibid., p. 110; (632½) Birds of N.
A., 1874, vol. i, p. 137.

Catherpes mexicanus.
(78) P. R. R. Surv., vol. ix, p. 356; (87) Mex.
Bound. Surv., vol. ii, p. 13; (115) Review
of N. A. Birds, Aug., 1864, Part I, p. 111.

CATHERPES MEXICANUS CONSPERCUS.
(632½) Birds of N. A., 1874, vol. i, p. 139, pl. viii,
fig. 4 (cuts, p. 138); Ibid., App., p. 503.

Centrocercus.
(632½) Birds of N. A., 1874, vol. iii, p. 428.

CENTROCERCUS UROPHASIANUS.
(78) P. R. R. Surv., vol. ix, p. 624; (94) P. R. R.
Surv., vol. x, p. 14; (632½) Birds of N. A.,
1874, vol. iii, p. 429, pl. lx, figg. 2, 4, pl. lxi,
fig. 6 (cuts, pp. 430, 431).

Centronyx.
(632½) Birds of N. A., 1874, vol. i, p. 530.

CENTRONYX BAIRDII.
(78) P. R. R. Surv., vol. ix, p. 441; (632½) Birds
of N. A., 1874, vol. i, p. 531, pl. xxv, fig. 3
(cuts, p. 531); Ibid., App., p. 510.

Centrophanes lapponicus.
(78) P. R. R. Surv., vol. ix, p. 433.

Centrophanes melanomus.
(78) P. R. R. Surv., vol. ix, p. 436.

Centrophanes ornatus.
(78) P. R. R. Surv., vol. ix, p. 435.

Centrophanes pictus.
(78) P. R. R. Surv., vol. ix, p. 434.

Centurus.
(632½) Birds of N. A., 1874, vol. ii, p. 553.

CENTURUS AURIFRONS.
(632½) Birds of N. A., 1874, vol. ii, p. 557, pl. lii,
figg. 3, 6.

CENTURUS CAROLINUS.
(78) P. R. R. Surv., vol. ix, p. 109; (632½) Birds
of N. A., 1874, vol. ii, p. 554, pl. lii, figg. 1, 4
(cuts, p. 555).

Centurus elegans.
(32) Stansbury's Surv. Salt Lake [App. C], p. 333.

CENTURUS FLAVIVENTRIS.
(32) Stansbury's Surv. Salt Lake [App. C], p.
333; (78) P. R. R. Surv., vol. ix, p. 110; (87)
Mex. Bound. Surv., vol. ii, p. 5, pl. iv; (104)
Birds of N. A., 1860, p. 110, pl. xliii.

Centurus Santacruzii.
(32) Stansbury's Surv. Salt Lake [App. C], p. 333.

CENTURUS UROPYGIALIS.
(56) Pr. Acad. Nat. Sci. Phila., 1854, p. 121; (78)
P. R. R. Surv., vol. ix, p. 111; (87) Mex.
Bound. Surv., vol. ii, p. 6; (99) Pr. Acad.
Sci. Phila., 1859 (1860), p. 302; (104) Birds
of N. A., 1860, p. 111, pl. xxxvi; (632½)
Birds of N. A., 1874, vol. ii, p. 558, pl. lii,
figg. 2, 5; Ibid., App., p. 521.

Cerorhina monocerata.
(78) P. R. R. Surv., vol. ix, p. 905.
Cerorhina Suckleyi.
(78) P. R. R. Surv., vol. ix, p. 906.
Certhia.
(115) Review of N. A. Birds, Aug., 1864, Part I,
p. 89; (632½) Birds of N. A., 1874, vol. i, p. 124.
Certhia americana.
(78) P. R. R. Surv., vol. ix, p. 372; (115) Review
of N. A. Birds, Aug., 1864, Part I, p. 89.
CERTHIA FAMILIARIS AMERICANA.
(632½) Birds of N. A., 1874, vol. i, p. 125, pl. viii,
fig. 11 (cuts, p. 124).
Certhia familiaris mexicana.
(632½) Birds of N. A., 1874, vol. i, p. 128.
CERTHIA MEXICANA.
(78) P. R. R. Surv., vol. ix, p. 378; (104) Birds of
N. A., 1860, p. 373, pl. lxxxiii, fig. 2; (115) Review of N. A. Birds, Aug., 1864, Part I, p. 90.
Certhiadæ.
(115) Review of N. A. Birds, Aug., 1864, Part
I, p. 89; (632½) Birds of N. A., 1874, vol. i,
p. 124.
Certhiola.
(632½) Birds of N. A., 1874, vol. i, p. 425.
CERTHIOLA BAHAMENSIS.
(115) Review of N. A. Birds, Aug., 1864, Part
I, p. 163, fig. 2; (632½) Birds of N. A., 1874,
vol. i, p. 428, pl. xix, fig. 5.
Certhiola barbadensis.
(632½) Birds of N. A., 1874, App., p. 508.
Certhiola caboti.
(632½) Birds of N. A., 1874, App., p. 508.
CERTHIOLA FLAVEOLA.
(78) P. R. R. Surv., vol. ix, p. 924; (104) Birds
of N. A., 1860, p. 924, pl. lxxxiii, fig. 3.
Certhiola frontalis.
(632½) Birds of N. A., 1874, App., p. 508.
Certhiola newtoni.
(632½) Birds of N. A., 1874, App., p. 508.
Ceryle.
(632½) Birds of N. A., 1874, vol. ii, p. 391.
CERYLE ALCYON.
(78) P. R. R. Surv., vol. ix, p. 158; (87) Mex.
Bound. Surv., vol. ii, p. 7; (632½) Birds of
N. A., 1874, vol. ii, p. 392, pl. xlv, fig. 6
(cuts, pp. 392, 393, 397).
CERYLE AMERICANA.
(32) Stansbury's Surv. Salt Lake [App. C],
p. 327; (78) P. R. R. Surv., vol. ix, p. 159;
(87) Mex. Bound. Surv., vol. ii, p. 7, pl. vii;
(104) Birds of N. A., 1860, p. 159, pl. xlv.
CERYLE AMERICANA CABANISI.
(632½) Birds of N. A., 1874, vol. ii, p. 396, pl.
xlv, fig. 9.
Chætura.
(632½) Birds of N. A., 1874, vol. ii, p. 431.
CHÆTURA PELAGICA.
(632½) Birds of N. A., 1874, vol. ii, p. 432, pl.
xlv, fig. 7 (cuts, pp. 421, 431, 432).
Chætura pelasgia.
(78) P. R. R. Surv,, vol. ix, p. 144.
CHÆTURA VAUXII.
(78) P. R. R. Surv., vol. ix, p. 145; (104) Birds
of N. A., 1860, p. 145, pl. xviii, fig. 2; (532½)
Birds of N. A., 1874, vol. ii, p. 435, pl. xlv,
fig. 8; Ibid., App., p. 521.

Chæturinæ.
(632½) Birds of N. A., 1874, vol. ii, p. 427.
Chamæa..
(115) Review of N. A. Birds, July, 1864, Part
I, p. 76; (632½) Birds of N. A., 1874, vol. i,
p. 83.
CHAMÆA FASCIATA.
(78) P. R. R. Surv., vol. ix, p. 370; (115) Review
of N. A. Birds, July, 1864, Part I, p. 76;
(632½) Birds of N. A., 1874, vol. i, p. 84, pl.
vi, fig. 8 (cuts, p. 83); Ibid., App., p. 502;
(32) Stansbury's Surv. Salt Lake [App. C],
p. 331.
Chamæadæ.
(115) Review of N. A. Birds, July, 1874, Part
I, p. 75; (632½) Birds of N. A., 1874, vol. i,
p. 83.
Chamæpelia.
(632½) Birds of N. A., 1874, vol. iii, p. 389.
CHAMÆPELIA PASSERINA.
(78) P. R. R. Surv., vol. ix, p. 606; (87) Mex.
Bound. Surv., vol. ii, p. 22; (99) Pr. Acad.
Nat. Sci. Phila., 1859' p. 305; (632½) Birds
of N. A., 1874, vol. iii, p. 389, pl. lviii, fig. 6
(cuts, pp. 389, 390); Ibid., App., p. 522.
Charadrius virginicus.
(78) P. R. R. Surv., vol. ix, p. 690; (87) Mex.
Bound. Surv., vol. ii, p. 25.
Charadrius vociferus.
(31) Stansbury's Surv. Salt Lake [App. C], p.
319.
Chaulelasmus streperus.
(78) P. R. R. Surv., vol. ix, p. 882; (87) Mex.
Bound. Surv., vol. ii, p. 27.
Chelidon.
(115) Review of N. A. Birds, May, 1865, Part
I, p. 271.
Chen caerulescens.
(78) P. R. R. Surv., vol. ix, p. 761.
Chen hyperboreus.
(78) P. R. R. Surv., vol. ix, p. 760.
Chenalopex impennis.
(78) P. R. R. Surv., vol. ix, p. 900.
Chloephaya canagica.
(78) P. R. R. Surv., vol. ix, p. 768.
Chloroceryle americana.
(78) P. R. R. Surv., vol. ix, p. 159.
CHLOROPHANES ATRICAPILLA.
(115) Review of N. A. Birds, Aug., 1864, Part
I, p. 163, fig. 3.
Chondestes.
(632½) Birds of N. A., 1874, vol. i, p. 562.
CHONDESTES GRAMMACA.
(78) P. R. R. Surv., vol. ix, p. 456; (87) Mex.
Bound. Surv., vol. ii, p. 15; (99) Pr. Acad.
Nat. Sci. Phila., 1859, p. 304; (632½) Birds
of N. A., 1874, vol. i, p. 562 (cuts, pp. 562,
563), vol. ii, pl. xxxi, fig. 1.
Chordeiles.
(632½) Birds of N. A., 1874, vol. ii, p. 400.
CHORDEILES ACUTIPENNIS TEXENSIS.
(632½) Birds of N. A., 1874, vol. ii, p. 406, pl.
xlvi, fig. 5.
Chordeiles brasilianus.
(32) Stansbury's Surv. Salt Lake [App. C].
p. 327.

CHORDEILES HENRYI.
(78) P. R. R. Surv., vol. ix, p. 153, 922; (87)
Mex. Bound. Surv., vol. ii, p. 7; (94)
P. R. R. Surv., vol. x, p. 13, pl. xvii;
(104) Birds of N. A., 1860, p. 153, pl.
xvii.

Chordeiles popetue.
(78) P. R. R. Surv., vol. ix, p. 151.

CHORDEILES POPETUE HENRYI.
(632½) Birds of N. A., 1874, vol. ii, p. 404, pl.
xlvi, fig. 4.

Chordeiles popetue minor.
(632¼) Birds of N. A., 1874, App., p. 520.

CHORDEILES POPETUE POPETUE.
(632½) Birds of N. A., 1874, vol. ii, p. 401 (cuts,
pp. 399, 401).

CHORDEILES TEXENSIS.
(78) P. R. R. Surv., vol. ix, p. 154; (87) U. S.
and Mex. Bound. Surv., vol. ii, p. 7, pl.
vi; (99) Pr. Acad. Nat. Sci. Phila., 1859
(1860), p. 303; (104) Birds of N. A., 1860,
p. 154, pl. xliv; (632½) Birds of N. A., 1874,
App., p. 520.

Chroicocephalus atricilla.
(78) P. R. R. Surv., vol. ix. p, 850; (87) Mex.
Bound. Surv., vol. ii, p. 27.

CHROICOCEPHALUS CUCULLATUS.
(78) P. R. R. Surv., vol. ix, p. 851; (104) Birds
of N. A., 1860, p. 851, pl. xciii, fig. 1.

CHROICOCEPHALUS FRANKLINII.
(78) P. R. R. Surv., vol. ix, p. 851; (104) Birds
of N. A., 1860, p. 851, pl. xciii, fig. 3.

Chroicocephalus minutus.
(78) P. R. R. Surv., vol. ix, p. 853.

Chroicocephalus philadelphia.
(78) P. R. R. Surv., vol. ix, p. 852.

Chrysomitris.
(632½) Birds of N. A., 1874, vol. i, p. 470.

CHRYSOMITRIS LAWRENCEII.
(32) Stansbury's Surv. Salt Lake [App. C], p.
330; (78) P. R. R. Surv., vol. ix, p. 424;
(632½) Birds of N. A., 1874, vol. i, p. 478, pl.
xxii, figg. 14, 15.

Chrysomitris magellanicus.
(78) P. R. R. Surv., vol. ix, p. 419.

CHRYSOMITRIS MEXICANA.
(78) P. R. R. Surv., vol. ix, p. 423; (87) Mex.
Bound. Surv., vol. ii, p. 14, pl. xvi, fig. 1;
(104) Birds of N. A., 1860, p. 424, pl. liv,
fig. 1.

CHRYSOMITRIS PINUS.
(78) P. R. R. Surv., vol. ix, p. 425; (632½) Birds
of N. A., 1874, vol. i, p. 480, pl. xxii, fig. 16,
(cut, p. 480).

Chrysomitris psaltria.
(78) P. R. R. Surv., vol. ix, p. 422; (632½) Birds
of N. A., 1874, App., p. 509.

CHRYSOMITRIS PSALTRIA ARIZONÆ.
(632½) Birds of N. A., 1874, vol. i, p. 476, pl. xxii,
fig. 11; Ibid., App., p. 509.

CHRYSOMITRIS PSALTRIA MEXICANA.
(632½) Birds of N. A., 1874, vol. i, p. 478, pl. xxii,
figg. 12, 13.

CHRYSOMITRIS PSALTRIA PSALTRIA.
(632½) Birds of N. A., 1874, vol. i, p. 474, pl. xxii,
figg. 9, 10.

Chrysomitris Stanleyi.
(78) P. R. R. Surv., vol. ix, p. 420.

Chrysomitris tristis.
(78) P. R. R. Surv., vol. ix, p. 421; (632½) Birds
of N. A., 1874, vol. i, p. 471, pl. xxii, figg.
7, 8 (cuts pp. 470, 472).

Chrysomitris Yarrelli.
(78) P. R. R. Surv., vol. ix, p. 421.

Chrysopoga typica.
(52) Stansbury's Surv. Salt Lake [App.C], p.330.

Ciceronia microceros.
(78) P. R. R. Surv., vol. ix, p. 908.

Ciceronia pusillus.
(78) P. R. R. Surv., vol. ix, p. 909.

Cichlerminia Bonapartii.
(115) Review of N. A. Birds, July, 1864, Part
I, p. 59.

Cichlopsis.
(115) Review of N. A. Birds, June, 1866, Part
I, p. 417; (115) Review of N. A. Birds,
June, 1866, Part I, p. 433.

Cichlopsis leucogonys.
(115) Review of N. A. Birds, June, 1866, Part
I, p. 434.

Cinclidæ.
Review of N. A. Birds, June, 1864, Part I, p. 3;
Ibid., p. 59; Ibid., July, 1864, Part I, p. 59;
(632½) Birds of N. A. 1874, vol. i, p. 55.

Cinclocerthia gutturalis.
(115) Review of N. A. Birds, July, 1864, Part
I, p. 59.

Cinclocerthia ruficauda.
(115) Review of N. A. Birds, July, 1864, Part
I, p. 59.

Cinclus.
(115) Review of N. A. Birds, July, 1864, Part I,
p. 59; (632½) Birds of N. A., 1874, vol. i, p. 55.

Cinclus mexicanus.
(115) Review of N. A. Birds, July, 1864, Part
I, p. 60; (632½) Birds of N. A., 1874, vol. i,
p. 56, pl. v, fig. 1 (cuts, p. 55).

Cinnicerthia.
(115) Review of N. A. Birds, Aug., 1864, Part
I, p. 94; Ibid., p. 111.

Cinnicerthia unibrunnea.
(115) Review of N. A. Birds, Aug., 1864, Part
I, p. 111.

Cinnicerthia unirufa.
(115) Review of N. A. Birds, Aug., 1864, Part I,
p. 111.

Circus.
(632½) Birds of N. A., 1874, vol. iii, p. 212.

CIRCUS CYANEUS HUDSONIUS.
(632½) Birds of N. A., 1874, vol. iii, p. 214 (cuts
pp. 212, 217).

Circus hudsonius.
(78) P. R. R. Surv., vol. ix, p. 38; (87) Mex.
Bound. Surv., vol. ii, p. 4; (94) P. R. R.
Surv., vol, p. 12.

Cistothorus.
(115) Review N. A. Birds, Aug., 1864, Part I,
p. 94; Ibid., p. 146; (632½) Birds of N. A.
1874, vol. i, p. 158.

Cistothorus elegans.
(115) Review of N. A. Birds, Aug., 1864, Part I, p. 146.
CISTOTHORUS PALUSTRIS.
(78) P. R. R. Surv., vol. ix, p. 364; (115) Review of N. A. Birds, Aug., 1864, Part I, p. 147; (632½) Birds of N. A., 1874, vol. i, p. 161, pl. ix, fig. 6 (cuts pp. 158, 160).
Cistothorus palustris paludicola.
(115) Review of N. A. Birds, Aug., 1864, Part I, p. 148.
Cistothorus (Cistothorus) stellaris.
(78) P. R. R. Surv., vol. ix, p. 365.
CISTOTHORUS STELLARIS.
(115) Review of N. A. Birds, Aug., 1864, Part I, p. 146; (632½) Birds of N. A., 1874, vol. i, p. 159, pl. ix, fig. 7; *Ibid.*, App., 9, p. 504 (cut).
Clangula albeola.
(31) Stansbury's Surv. Salt Lake [App. C], p. 324.
Coccothraustinæ.
(632½) Birds of N. A., 1874, vol. i, p. 446.
Coccothraustes ferreo-rostris.
(32) Stansbury's Surv. Salt Lake [App. C], p. 331.
Coccyginæ.
(632½) Birds of N. A., 1874, vol. ii, p. 470.
Coccygus.
(632½) Birds of N. A., 1874, vol. ii, p. 475.
COCCYGUS AMERICANUS.
(78) P. R. R. Surv., vol. ix, p. 76; (632½) Birds of N. A., 1874, vol. ii, p. 477, pl. xlviii, fig. 3 (cuts, pp. 476, 477).
COCCYGUS ERYTHROPHTHALMUS.
(78) P. R. R. Surv., vol. ix, p. 77; (632½) Birds N. A., 1874, vol. ii, p. 484, pl. xlviii, fig. 5.
COCCYGUS MINOR.
(78) P. R. R. Surv., vol. ix, p. 78; (632½) Birds of N. A., 1874, vol. ii, p. 482, pl. xlviii, fig. 4.
Cœrebidæ.
(632½) Birds of N. A., 1874, vol. i, p. 425.
Colaptes.
(632½) Birds of N. A., 1874, vol. ii, p. 573.
COLAPTES AURATUS.
(78) P. R. R. Surv., vol. ix, p. 118; (632½) Birds of N. A., 1874, vol. ii, p. 575, pl. lv, figg. 1, 2 (cut, p. 574).
Colaptes Ayresii.
(32) Stansbury's Surv. Salt Lake [App. C], p. 333.
COLAPTES CHRYSOIDES.
(78) P. R. R. Surv., vol. ix, p. 125; (87) Mex. Bound. Surv., vol. ii, p. 6; (99) Pr. Acad. Nat. Sci. Phila., 1859 (1860), p. 302; (632½) Birds of N. A., 1874, vol. ii, p. 583, pl. liv, figg. 1, 2.
Colaptes collaris.
(32) Stansbury's Surv. Salt Lake [App. C], p. 333.
COLAPTES HYBRIDUS.
(78) P. R. R. Surv., vol. ix, p. 122; (632½) Birds of N. A. 1873, vol. ii, p. 582, pl. liv, fig. 3.
Colaptes mexicanoides.
(32) Stansbury's Surv. Salt Lake [App. C], p. 333.

COLAPTES MEXICANUS.
(78) P. R. R. Surv., vol. ix, p. 120; (87) Mex. Bound. Surv., vol. ii, p. 6; (632½) Birds of N. A., 1874, vol. ii, p. 578, pl. lv, figg. 3, 4 (cut, p. 579).
Collurio.
(632½) Birds of N. A., 1874, vol. i, p. 412; (115) Review of N. A. Birds, June, 1866, Part I, p. 437.
COLLURIO BOREALIS.
(115) Review of N. A. Birds, June, 1866, Part I, p. 437; (632½) Birds of N. A., 1874, vol. i, p. 415, pl. xix, figg. 1, 2.
Collurio elegans.
(115) Review of N. A. Birds, June, 1866, Part I, p. 444.
Collurio excubitoroides.
(115) Review of N. A. Birds, June, 1866, Part I, p. 445.
COLLURIO LUDOVICIANUS.
(115) Review of N. A. Birds, June, 1866, Part I, p. 443; (632½) Birds of N. A, 1874, vol. i, p. 418, pl. xix, fig. 4.
COLLURIO LUDOVICIANUS EXCUBITOROIDES.
(632½) Birds of N. A., 1874, vol. i, p. 421, pl. xix, fig. 3 (cuts, pp. 412, 415, 421).
Collurio ludovicianus robustus.
(632½) Birds of N. A., 1874, vol. i, p. 420; *Ibid.*, App., p. 508.
Collyrio borealis.
(78) P. R. R. Surv., vol. ix, p. 324.
COLLYRIO ELEGANS.
(78) P. R. R. Surv., vol. ix, p. 328; (104) Birds of N. A., 1860, p. 328, pl. lxxv, fig. 1.
COLLYRIO EXCUBITOROIDES.
(78) P. R. R. Surv., vol. ix, p. 327; (87) Mex. Bound. Surv., vol. ii, p. 11; (104) Birds of N. A., 1860, p. 327, pl. lxxv, fig. 2.
Collyrio ludovicianus.
(78) P. R. R. Surv., vol. ix, p. 325.
Columba.
(632½) Birds of N. A., 1874, vol. iii, p. 358.
Columba (Columba) fasciata.
(78) P. R. R. Surv., vol. ix, p. 597.
COLUMBA FASCIATA.
(87) Mex. Bound. Surv., vol, ii, p. 21; (632½) Birds of N. A., 1874, vol. iii, p. 360, pl. lvii, fig. 2 (cuts, pp. 358, 361).
Columba (Columba) flavirostris.
(78) P. R. R. Surv., vol. ix, p. 598.
COLUMBA FLAVIROSTRIS.
(32) Stansbury's Surv. Salt Lake [App. C], p. 334; (87) Mex. Bound. Surv., vol. ii, p. 21, pl. xxiii; (104) Birds of N. A., 1860, p. 598, pl. lxi; (632½) Birds of N. A., 1874, vol. iii, p. 366, pl. lvii, fig. 5.
COLUMBA LEUCOCEPHALA.
(78) P. R. R. Surv., vol. ix, p. 599; (632½) Birds of N. A., 1874, vol. iii, p. 208, pl. lvii, fig. 3 (cut, p. 364).
Columba leucoptera.
(31-a) Stansbury's Surv. Salt Lake [App. C], p. 326.
Columba solitaria.
(32) Stansbury's Surv. Salt Lake [App. C], p. 333.

Columbidæ.
> (632½) Birds of N. A., 1874, vol. iii, p. 357.

Columbinæ.
> (632½) Birds of N. A., 1874, vol. iii, p. 357.

Colymbus arcticus.
> (78) P. R. R. Surv., vol. ix, p. 888.

Colymbus glacialis.
> (31) Stansbury's Surv. Salt Lake [App. C], p. 324.

Colymbus pacificus.
> (78) P. R. R. Surv., vol. ix, p. 889.

Colymbus septentrionalis.
> (78) P. R. R. Surv., vol. ix, p. 890.

Colymbus torquatus.
> (78) P. R. R. Surv., vol. ix, p. 888.

Comptolæmus labradorius.
> (78) P. R. R. Surv., vol. ix, p. 803.

Conirostrum ornatum.
> (32) Stansbury's Surv. Salt Lake [App. C], p. 327.

Contopus.
> (632¼) Birds of N. A., 1874, vol. ii, p. 350.

CONTOPUS BOREALIS.
> (78) P. R. R. Surv., vol. ix, p. 188; (632½) Birds of N. A., 1874, vol. ii, p. 353, pl. xliv, fig. 1 (cuts, pp. 350, 353).

CONTOPUS PERTINAX.
> (632½) Birds of N. A., 1874, vol. 2, p. 356, pl. xliv, fig. 2.

Contopus Richardsonii.
> (78) P. R. R. Surv., vol. ix, p. 189; (87) Mex. Bound. Surv., vol. ii, p. 9.

CONTOPUS VIRENS.
> (78) P. R. R. Surv., vol. ix, p. 190; (632½) Birds of N. A., 1874, vol. ii, p. 357, pl. xliv, fig. 3.

CONTOPUS VIRENS RICHARDSONI.
> (632½) Birds of N. A., 1874, vol. ii, p. 360, pl. xliv, fig. 4.

Conurus.
> (632½) Birds of N. A., 1874, vol. ii, p. 586.

CONURUS CAROLINENSIS.
> (78) P. R. R. Surv., vol. ix, p. 67; (632½) Birds of N. A., 1874, vol. ii, p. 587, pl. lvi, figg. 1, 2 (cuts, pp. 586, 587).

Corvidæ.
> (632½) Birds of N. A., 1874, vol. ii, p. 231.

Corvinæ.
> (632½) Birds of N. A., 1874, vol. ii, p. 231.

Corvus.
> (632½) Birds of N. A., 1874, vol. ii, p. 232.

CORVUS AMERICANUS.
> (104) Birds of N. A., 1860, p. 566, pl. xxiii; (632½) Birds of N. A., 1874, vol. ii, p. 243, pl. xxxvii, fig. 5.

CORVUS AMERICANUS FLORIDANUS.
> (78) P. R. R. Surv., vol. ix, p. 568; (632½) Birds of N. A., 1874, vol. ii, p. 247, pl. xxxvii, fig. 9.

CORVUS CACALOTL.
> (78) P. R. R. Surv., vol. ix, p. 563; (104) Birds of N. A., 1860, p. 563, pl. xx.

CORVUS CARNIVORUS.
> (78) P. R. R. Surv., vol. ix, p. 560; (87) Mex. Bound. Surv., vol. ii, p. 20; (94) P. R. R. Surv., vol. x, p. 14; (104) Birds of N. A., 1860, p. 560, pl. xxi.

CORVUS CAURINUS.
> (78) P. R. R. Surv., vol. ix, p. 569; (104) Birds of N. A., 1860, p. 569, pl. xxiv; (632½) Birds of N. A., 1874, vol. ii, p. 248, pl. xxxvii, fig. 3.

CORVUS CORAX CARNIVORUS.
> (632½) Birds of N. A., 1874, vol. ii, p. 234, pl. xxxvii, fig. 6 (cuts, pp. 232, 234).

CORVUS CRYPTOLEUCUS.
> (78) P. R. R. Surv., vol. ix, p. 565; (87) Mex. Bound. Surv., vol. ii, p. 20; (104) Birds of N. A., 1860, p. 565, pl. xxii; (632½) Birds of N. A., 1874, vol. ii, p. 242, pl. xxxvii, fig. 8; Ibid., App., p. 518.

CORVUS FLORIDANUS.
> (104) Birds of N. A., 1860, p. 568, pl. lxvii, fig. 1.

CORVUS OSSIFRAGUS.
> (78) P. R. R. Surv., vol. ix, p. 571; (104) Birds of N. A., 1860, p. 571, pl. lxvii, fig. 2; (632½) Birds of N. A., 1874, vol. ii, p. 251, pl. xxxvii, fig. 7.

Coturnicops noveboracensis.
> (78) P. R. R. Surv., vol. ix, p. 750.

Coturniculus.
> (632½) Birds of N. A., 1874, vol. i, p. 548.

COTURNICULUS HENSLOWI.
> (78) P. R. R. Surv., vol. ix, p. 451; (632½) Birds of N. A., 1874, vol. i, p. 550, pl. xxv, fig. 5.

COTURNICULUS LECONTEI.
> (632½) Birds of N. A., 1874, vol. i, p. 552, pl. xxv, fig. 6; Ibid App., p. 513.

Coturniculus Lecontii.
> (78) P. R. R. Surv., vol. ix, p. 452.

COTURNICULUS PASSERINUS.
> (78) P. R. R. Surv., vol. ix, p. 450; (87) Mex. Bound. Surv., vol. ii, p. 15; (632½) Birds of N. A., 1874, vol. i, p. 553, pl. xxv, fig. 4 (cuts, pp. 548, 550).

Coturniculus passerinus perpallidus.
> (632½) Birds of N. A., 1874, vol. i, p. 556; Ibid., App., p. 513.

Cotyle.
> (115) Review of N. A. Birds, May, 1865, Part I, p. 271; Ibid., p. 318; Ibid., p. 319; (632½) Birds of N. A., 1874, vol. i, p. 353.

COTYLE RIPARIA.
> (78) P. R. R. Surv., vol. ix, p. 313; (115) Review of N. A. Birds, May, 1865, Part I, p. 319; (632½) Birds of N. A. 1874, vol. i, p. 353, pl. xvi, fig. 14 (cuts, pp. 353, 354).

Cotyle serripennis.
> (78) P. R. R. Surv., vol. ix, p. 313; (87) Mex. Bound. Surv., vol. ii, p. 11.

Cracidæ.
> (632½) Birds of N. A., 1874, vol. iii, p. 397.

Craxirex unicinctus.
> (78) P. R. R. Surv., vol. ix, p. 46; (87) Mex. Bound. Surv., vol. ii, p. 4.

Creagrus furcatus.
> (78) P. R. R. Surv., vol. ix, p. 857.

Creciscus jamaicensis.
> (78) P. R. R. Surv., vol. ix, p. 749.

CREX PRATENSIS.
> (78) P. R. R. Surv., vol. ix, p. 751; (104) Birds of N. A., 1860, p. 751, pl. lxxxix, fig. 2.

Crotophaga.
> (632½) Birds of N. A., 1874, vol. ii, p. 486.

CROTOPHAGA ANI.
(78) P. R. R. Surv., vol. ix, p. 72 : (104) Birds
of N. A., 1860, p. 72, pl. lxxxiv, fig. 2;
(632½) Birds of N. A.,1874,vol. ii, p. 488, pl.
xlviii,fig. 2 (cuts, pp. 487, 487).
CROTOPHAGA RUGIROSTRIS.
(78) P. R. R. Surv., vol. ix, p. 71; (104) Birds of
of N. A., 1860, p. 71, pl. lxxxiv, fig. 1.
Cuculidæ.'
(632½) Birds of N. A., 1874, vol. ii, p. 470.
Culicivora atricapilla.
(32) Stansbury's Surv. Salt Lake [App. C], p.
328.
Culicivora plumbea.
(56) Proc. Acad. Nat. Sci. Phila., 1854, p. 118.
Cupidonia.
(632½) Birds of N. A., 1874, vol. iii, p. 439.
Cupidonia cupido.
(78) P. R. R. Surv., vol. ix, p. 628.
CUPIDONIA CUPIDO CUPIDO.
(632½) Birds of N. A., 1874, vol. iii, p. 440, pl. lxi,
figg. 1, 7 (cuts, p. 441).
Cupidonia cupido pallidicinctus.
(632½) Birds of N. A., 1874, vol. iii, p. 446.
Curvirostra americana.
(78) P. R. R. Surv., vol. ix, p. 426.
Curvirostra leucoptera.
(78) P. R. R. Surv., vol. ix, p. 427.
Cyanocitta.
(632½) Birds of N. A., 1874, vol. ii, p. 282.
CYANOCITTA CALIFORNICA.
(78) P. R. R. Surv., vol.ix, p. 584; (87) Mex.
Bound. Surv., vol. ii, p. 20; (99) Pr.
Acad. Nat. Sci. Phila., 1859, p. 305; (632½)
Birds of N. A., 1874, vol. ii, p. 288, pl. xl, fig.
1 (cuts, pp. 283, 288); Ibid., App., p. 518.
CYANOCITTA CALIFORNICA WOODHOUSEI.
(632½) Birds of N. A., 1874, vol. ii, p. 291, pl. xl,
fig. 3.
CYANOCITTA FLORIDANA.
(78) P. R. R. Surv., vol. ix, p. 586; (632½) Birds
of N. A., 1874, vol ii, p. 285, pl. xl, fig. 4.
Cyanocitta macrolopha.
(56) Pr. Acad. Acad. Sci. Phila., 1854, p. 118.
CYANOCITTA WOODHOUSII.
(78) P. R. R. Surv., vol. ix, p. 585; (87) Mex.
Bound. Surv., vol. ii, p. 20; pl. xxi; (104)
Birds of N. A., 1860, p. 585, pl. lix.
CYANOCITTA SORDIDA.
(78) P. R. R. Surv., vol. ix, p. 587; (87) Mex.
Bound. Surv., vol. ii, p. 21, pl. xxii, fig. 1;
(104) Birds of N. A., 1860, p. 587, pl. lx,
fig. 1.
CYANOCITTA ULTRAMARINA.
(78) P. R. R. Surv., vol. ix, p. 588; (87) Mex.
Bound. Surv., vol. ii, p. 21, pl. xxii, fig.
2; (104) Birds of N. A., 1860, p. 588, pl. lx,
fig. 2.
CYANOCITTA ULTRAMARINA ARIZONÆ.
(632½) Birds of N. A., 1874, vol. ii, p. 292, pl. xli,
fig. 2.
Cyanocitta ultramarina Couchi.
(632½) Birds of N. A., 1874, vol. ii, p. 293.
Cyanocorax Cassinii.
(32) Stansbury's Surv. Salt Lake [App. C], p
331.

Cyanocorax coronatus.
(32) Stansbury's Surv. Salt Lake [App. C], p.
331.
Cyanocorax luxuosus.
(32) Stansbury's Surv. Salt Lake [App. C]. p,
331.
Cyanopterus Rafflesii.
(32) Stansbury's Surv. Salt Lake [App. C], p.
334.
Cyanospiza.
(632½) Birds of N. A., 1874, vol. ii, p. 81.
CYANOSPIZA AMŒNA.
(78) P. R. R. Surv., vol ix, p. 504; (632½) Birds
of N. A., 1874, vol. ii, p. 84, pl. xxix, figg.
11, 12 (cuts, pp. 81, 84).
Cyanospiza ciris.
(78) P. R. R. Surv., vol. ix, p. 503: (87) Mex.
Bound. Surv., vol. ii, p. 17; (632½) Birds of
N. A., 1874, vol. ii, p. 87, pl. xxix, fig. 7, 8.
CYANOSPIZA CYANEA.
(78) P. R. R. Surv , vol. ix, p. 505; (632½) Birds
of N. A., 1874, vol. ii, p. 82, pl. xxix, figg.
13, 14.
CYANOSPIZA PARELLINA.
(78) P. R. R. Surv., vol. ix, p. 502; (87) Mex.
Bound. Surv., vol. ii, p. 17, pl. xviii, fig. 1;
(104) Birds of N. A., 1860, p. 502, pl. lvi,
fig. 1.
CYANOSPIZA VERSICOLOR.
(78) P. R. R. Surv., vol. ix, p. 503; (87) Mex.
Bound. Surv., vol. ii, p. 17, pl. xvii, fig. 2;
(99) Pr. Acad. Nat. Sci. Phila., 1859, p.
304; (104) Birds of N. A., 1860, p. 503, pl.
lvi, fig. 2; (632½) Birds of N. A., 1874, vol.
ii, p, 86, pl. xxix, figg. 9, 10.
Cyanura.
(632½) Birds of N. A., 1874, vol. ii, p. 271; Ibid.,
App., p. 518,
CYANURA CRISTATA.
(78) P. R. R. Surv., vol. ix, p. 580; (632½) Birds
of N. A., 1874, vol. ii, p. 273, pl. xlii, fig. 3
(cuts, pp. 271, 274).
Cyanura macrolophus.
(78) P. R. R. Surv., vol. ix, p. 582.
CYANURA STELLERI.
(78) P. R. R. Surv., vol. ix, p. 581; (632½) Birds
of N. A., 1874, vol. ii. p. 277. pl. xxxix, fig. 1.
CYANURA STELLERI FRONTALIS.
(632½) Birds of N. A., 1874, vol. ii, p. 279, pl.
xxxix, fig. 2.
CYANURA STELLERI MACROLOPHA.
(632½) Birds of N. A., 1874, vol. ii, p. 281, pl.
xxxix, fig. 3.
Cyclorhis.
(115) Review of N. A. Birds, May, 1866, Part I,
p. 324; (115) Review of N. A. Birds, May,
1866, Part I, p. 384.
Cyclorhis flavipectus.
(115) Review of N. A. Birds, May, 1866, Part
I, p. 389.
Cyclorhis flaviventris.
(115) Review of N. A. Birds, May, 1866, Part
I, p. 384.
Cyclorhis guianensis.
(115) Review of N. A. Birds, May, 1866, Part
I, p. 389.

Cyclorhis nigrirostris.
(115) Review of N. A. Birds, May, 1866, Part I, p. 394.

Cyclorhis ochrocephala.
(115) Review of N. A. Birds, May, 1866, Part I, p. 391.

Cyclorhis subflavescens.
(115) Review of N. A. Birds, May, 1866, Part I, p. 388.

Cyclorhis virenticeps.
(115) Review of N. A. Birds, May, 1866, Part I, p. 393.

Cyclorhis viridis.
(115) Review of N. A. Birds, May, 1866, Part I, p. 392.

Cygnus americanus.
(31) Stansbury's Surv. Salt Lake [App. C], p. 321; (78) P. R. R. Surv., vol. ix, p. 758; (94, P. R. R. Surv., vol. x., p. 15.

Cygnus buccinator.
(78) P. R. R. Surv., vol. ix, p. 758.

Cyphorinus.
(115) Review of N. A. Birds, Aug., 1864, Part I, p. 94; (115) Review of N. A. Birds, Aug., 1864, Part I, p. 112.

Cyphorinus Lawrencii.
(115) Review of N. A. Birds, Aug., 1864, Part I, p. 113.

Cypselidæ.
(632¼) Birds of N. A., 1874, vol. ii, p. 421.

Cypselinæ.
(632¼) Birds of N. A., 1874, vol. ii, p. 423.

Cypselus melanoleucus.
(56) Proc. Acad. Nat. Sci. Phila., 1884, p. 118.

Cyrtonyx.
(632¼) Birds of N. A., 1874, vol. iii, p. 491.

CYRTONYX MÁSSENA.
(32) Stansbury's Surv. Salt Lake [App. C], p. 334; (78) P. P. R. Surv., vol. ix, p. 647; (87) Mex. Bound. Surv., vol. ii, p. 23; (632¼) Birds of N. A., 1874, vol. iii, p. 492, pl. lxiv, figg. 3, 6 (cuts, pp. 491, 492).

Cyrtopelecanus erythrorhynchus.
(78) P. R. R. Surv., vol. ix, p. 868.

DACIUS CÁYANA.
(115) Review of N. A. Birds, Aug., 1864, Part I, p. 163, fig. 4.

Dafila acuta.
(31) Stansbury's Surv. Salt Lake [App. C], p. 323; (78) P. R. R. Surv., vol. ix, p. 776; (87) Mex. Bound. Surv., vol. ii, p. 26.

Daption capensis.
(78) P. R. R. Surv., vol. ix, p. 828.

Demiegretta ludoviciana.
(78) P. R. R. Surv., vol. ix, p. 663.

Demiegretta Pealii.
(78) P. R. R. Surv., vol. ix, p. 661.

Demiegretta rufa.
(78) P. R. R. Surv., vol. ix, p. 662; (87) Mex. Bound. Surv., vol. ii, p. 24.

Dendrocygna arborea?
(32) Stansbury's Surv. Salt Lake [App. C], p. 334.

DENDROCYGNA AUTUMNALIS.
(32) Stansbury's Surv. Salt Lake [App. C], p. 334; (78) P. R. R. Surv., vol. ix, p. 770; (87) Mex. Bound. Surv., vol. ii, p. 26; (104) Birds of N. A., 1860, p. 770, pl. lxiii, fig. 2.

DENDROCYGNA FULVA.
(78) P. R. R. Surv., vol. ix, p. 770; (104) Birds of N. A., 1860, p. 770, pl. lxiii, fig. 1.

Dendroica.
(115) Review of N. A. Birds, Apr., 1865, Part I, p. 182; Ibid., p. 201; (632½) Birds of N. A., 1874, vol. i, p. 215.

Dendroica Adelaidæ.
(115) Review of N. A. Birds, Apr., 1865, Part I. p. 212.

DENDROICA ÆSTIVA.
(78) P. R. R. Surv., vol. ix, p. 282; (87) Mex. Bound. Surv., vol. ii, p. 10; (115) Review of N. A. Birds., Apr., 1865, Part I, p. 195; (632½) Birds of N. A., 1874, vol. i, p. 222, pl. xiv, fig. 1.

DENDROICA AUDUBONII.
(78) P. R. R. Surv., vol. ix, p. 273; (87) Mex. Bound. Surv., vol. ii, p. 10; (115) Review of N. A. Birds, Apr., 1865, Part I, p. 188; (632½) Birds of N. A., 1874, vol. i, p. 229, pl. xiii, fig. 1 (cut, p. 215); (632½) Birds of N. A., 1874, App., p. 505.

Dendroica aureola.
(115) Review of N. A. Birds, Apr., 1865, Part I, p. 192.

DENDROICA BLACKBURNLÆ.
(78) P. R. R. Surv., vol. ix, p. 274; (115) Review of N. A. Birds, Apr., 1865, Part I, p. 189; (632½) Birds of N. A., 1874, vol. i, p. 237, pl. xiii, fig. 2; Ibid., App., p. 505.

Dendroica canadensis.
(78) P. R. R. Surv., vol. ix, p. 271.

Dendroica carbonata.
(78) P. R. R. Surv., vol. ix, p. 287; (115) Review of N. A. Birds, Apr., 1865, Part I, p. 207.

DENDROICA CASTANEA.
(78) P. R. R. Surv., vol. ix, p. 276; (115) Review of N. A. Birds, Apr., 1865, Part I, p. 189; (632½) Birds of N. A., 1874, vol. i, p. 251, pl. xiii, figg. 4, 5; Ibid., App., p. 505.

DENDROICA CHRYSOPAREIA.
(115) Review of N. A. Birds, Apr., 1865, Part I, pp. 183, 267; (632½) Birds of N. A., 1874, vol. i, p. 260, pl. xii, fig. 6.

DENDROICA CŒRULEA.
(78) P. R. R. Surv., vol. ix, p. 280; (115) Review of N. A. Birds, Apr., 1865, Part I, p. 191; (632½) Birds of N. A., 1874, vol. i, p. 235, pl. xiii, figg. 10, 11; Ibid., 11; App., p. 505.

Dendroica cœrulescens.
(115) Review of N. A. Birds, Apr., 1865, Part I, p. 186; (632½) Birds of N. A., 1874, vol. i, p. 254, pl. xii, figg. 10, 11.

DENDROICA CORONATA.
(78) P. R. R. Surv., vol. ix, p. 272; (115) Review of N. A. Birds, Apr., 1865, Part I, p. 187; (632½) Birds of N. A., 1874, vol. i, p. 227, pl. xii, fig. 9 (cut, p. 215).

DENDROICA DISCOLOR.
> (78) P. R. R. Surv., vol. ix, p. 290; (115) Review of N. A. Birds, Apr., 1865, Part I, p. 213; (632½) Birds of N. A., 1874, vol. i, p. 276, pl. xiv, fig. 6.

DENDROICA DOMINICA.
> (115) Review of N. A. Birds, Apr., 1865, Part I, p. 209; (632½) Birds of N. A., 1874, vol. i, p. 240, pl. xiv, figg. 6, 7; *Ibid.*, App., p. 505.

Dendroica dominica albilora.
> (632½) Birds of N. A., 1874, App., p. 505.

Dendroica eoa.
> (115)Review of N. A. Birds, Apr., 1865, Part I, p. 192.

DENDROICA GRACIÆ.
> (115) Review of N. A. Birds, Apr., 1865, Part I, p. 210; (632½) Birds of N. A., 1874, vol. i, p. 243, pl. xiv, fig. 10.

Dendroica Graciæ decora.
> (632½) Birds of N. A., 1874, App., p. 505.

Dendroica Gundlachi.
> (115) Review of N. A. Birds, Apr., 1865, Part I, p. 197.

DENDROICA KIRTLANDII.
> (78) P. R. R. Surv., vol. ix, p. 286; (115) Review N. A. Birds, Apr., 1865, Part I, p. 206; (632½) Birds of N. A., 1874, vol. i, p. 272, pl. xiv, fig. 5.

DENDROICA MACULOSA.
> (78) P. R. R. Surv., vol. ix, p. 284; (115) Review of N. A. Birds, Apr., 1865, Part I, p. 206; (632½) Birds of N. A., 1874, vol. i, p. 232, pl. xiv, fig. 2.

DENDROICA MONTANA.
> (78) P. R. R. Surv., vol. ix, p. 278; (115) Review of N. A. Birds, Apr., 1865, Part I, p. 190; (632½) Birds of N. A., 1874, vol. i, p. 271, pl. xiv, fig. 3.

DENDROICA NIGRESCENS.
> (78) P. R. R. Surv., vol. ix, p. 270; (115) Review of N. A. Birds, Apr., 1865, Part I, p. 186; (632½) Birds of N. A., 1874, vol. i, p. 258, pl. xii, fig. 8; *Ibid.*, App., p. 506.

DENDROICA OCCIDENTALIS.
> (78) P. R. R. Surv., vol. ix, p. 268; (115) Review N. A. Birds, Apr., 1865, Part I, p. 183; (632½) Birds of N. A., 1874, vol. i, p. 266, pl. xii, fig. 5; App., p. 506.

DENDROICA OLIVACEA.
> (115) Review of N. A. Birds, Apr., 1865, Part I, p. 205; (632½) Birds of N. A., 1874, vol. i, p. 258, pl. xiv, fig. 4.

DENDROICA PALMARUM.
> (78) P. R. R. Surv., vol. ix, p. 288; (115) Review of N. A. Birds, Apr., 1865, Part I, p. 207; (632½) Birds of N. A., 1874, vol. i, p. 273, pl. xiv, fig. 8.

DENDROICA PENNSYLVANICA.
> (78) P. R. R. Surv., vol. ix, p. 279; (115) Review of N. A. Birds, Apr., 1865, Part I, p. 191; (632½) Birds of N. A., 1874, vol. i, p. 245, pl. xiii, figg. 7, 8.

Dendroica petechia.
> (115) Review of N. A. Birds, Apr., 1865, Part I, p. 199.

Dendroica pharetra.
> (115) Review of N. A. Birds, Apr., 1865, Part I, p. 192.

DENDROICA PINUS.
> (78) P. R. R. Surv., vol. ix, p. 277; (115) Review of N. A. Birds, Apr., 1865, Part I, p. 190; (632½) Birds of N. A., 1874, vol. i, p. 268, pl. xiii, fig. 6.

Dendroica pityophila.
> (115) Review of N. A. Birds, Apr., 1865, Part I, p. 208.

Dendroica rufigula.
> (115) Review of N. A. Birds, Apr., 1865, Part I, p. 204.

DENDROICA STRIATA.
> (78) P. R. R. Surv., vol. ix, p. 280; (115) Review of N. A. Birds, Aug., 1864, Part I, p. 163, fig. 6; *Ibid.*, Apr., 1865, Part I, p. 192; (632½) Birds of N. A., 1874, vol. i, p. 248, pl. xiii, fig. 9.

Dendroica superciliosa.
> (78) P. R. R. Surv., vol. ix, p. 289; (87) Mex. Bound. Surv., vol. ii, p. 10.

Dendroica tigrina.
> (78) P. R. R. Surv., vol. ix, p. 286.

DENDROICA TOWNSENDII.
> (78) P. R. R. Surv., vol. ix, p. 269; (115) Review of N. A. Birds, Apr., 1865, Part I, p. 185; (632½) Birds of N. A., 1874, vol. i, p. 265, pl. xii, fig. 7; *Ibid.*, App., p. 506.

Dendroica Vieilloti.
> (115) Review of N. A. Birds, Apr., 1865, Part I, p. 203.

Dendroica Vieilloti Bryanti.
> (632½) Birds of N. A., 1874, App., p. 504.

DENDROICA VIRENS.
> (78) P. R. R. Surv., vol. ix, p. 267; (87) Mex. Bound. Surv., vol. ii, p. 10; (115) Review of N. A. Birds, Apr., 1865, Part I, p. 182; (632½) Birds of N. A., 1874, vol. i, p. 261, pl. xii, fig. 4.

Diomedea brachyura.
> (78) P. R. R. Surv., vol. ix, p. 822.

Diomedea chlororhyncha.
> (78) P. R. R. Surv., vol. ix, p. 822.

Diomedea (Diomedea) exulans.
> (78) P. R. R. Surv., vol. ix, p. 821.

Diomedea fuliginosa.
> (78) P. R. R. Surv., vol. ix, p. 823.

Dolichonyx.
> (632½) Birds of N. A., 1874, vol. ii, p. 148.

Doliochonyx oryzivorus.
> (78) P. R. R. Surv., vol. ix, p. 522; (632½) Birds of N. A., 1874; *Ibid.*, App., p. 517.

Donacobius.
> (115) Review of N. A. Birds, June, 1864, Part I, p. 5; *Ibid.*, July, 1864, Part I, p. 56.

Donacobius albo-vittatus.
> (115) Review of N. A. Birds, July, 1864, Part I, p. 57.

Donacobius atricapillus.
> (115) Review of N. A. Birds, July, 1864, Part I, p. 57.

Dulinæ.
> (115) Review of N. A. Birds, May, 1866, Part I, p. 401.

EUSPIZA TOWNSENDII.
(78) P. R. R. Surv., vol. ix, p. 495; (632½) Birds
of N. A., 1874, vol. ii, p. 68, pl. xxviii, fig. 68.

Euthlypis.
(115) Review of N. A. Birds, Apr., 1865, Part I,
p. 237; *Ibid.*, p. 238; *Ibid.*, May, 1865, Part
I, p. 262.

Falcinellus Ordii.
(78) P. R. R. Surv., vol. ix, p. 685.

FALCO.
(632½) Birds of N. A., 1874, vol. iii, p. 106 (cut,
p. 106).

Falco (Falco) anatum.
(78) P. R. R. Surv., vol. ix, p. 7.

Falco aurantius.
(78) P. R. R. Surv., vol. ix, p. 10; (87) Mex.
Bound. Surv., vol. ii, p. 3.

Falco candicans.
(78) P. R. R. Surv., vol. ix, p. 13.

Falco columbarius.
(78) P. R. R. Surv., vol. ix, p. 9; (87) Mex.
Bound. Surv., vol. ii, p. 3.

Falco communis anatum.
(632½) Birds of N. A., 1874, vol. iii, p. 132.

Falco communis Pealei.
(632½) Birds of N. A., 1874, vol. iii, p. 137.

FALCO FEMORALIS.
(78) P. R. R. Surv., vol. ix, p. 11; (87) Mex.
Bound. Surv., vol. ii, p. 3; (632½) Birds of N.
A., 1874, vol. iii, p. 155 (cuts, pp. 154, 155,
157).

FALCO GYRFALCO CANDICANS.
(632½) Birds of N. A., 1874, vol. iii, p. 111 (cut, p.
112).

FALCO GYRFALCO ISLANDICUS.
(632½) Birds of N. A., 1874, vol. iii, p. 113 (cut,
p. 114).

Falco gyrfalco labradora.
(632½) Birds of N. A., 1874, vol. iii, p. 117.

FALCO GYRFALCO SACER.
(632½) Birds of N. A., 1874, vol. iii, p. 115 (cut, p.
110); *Ibid.*, App., p. 522 (cut).

Falco islandicus.
(78) P. R. R. Surv., vol. ix., p. 13.

FALCO LANARIUS POLYAGRUS.
(632½) Birds of N. A., 1874, vol. iii, p. 123 (cuts,
pp. 110, 124); *Ibid.*, App., p. 522 (cut).

FALCO LITHOFALCO COLUMBARIUS.
(632½) Birds of N. A., 1874, vol. iii, p. 144 (cut,
p. 146).

Falco lithofalco Richardsoni.
(632½) Birds of N. A., 1874, vol. iii, p. 148.

Falco lithofalco Suckleyi.
(632½) Birds of N. A., 1874, vol. iii, p. 147.

Falco (Falco) nigriceps.
(78) P. R. R. Surv., vol. ix, p. 8.

FALCO NIGRICEPS.
(104) Birds of N. A., 1860, p. 8, pl. xi.

Falco polyagrus.
(78) P. R. R. Surv., vol. ix, p. 12.

FALCO RICHARDSONI.
(632½) Birds of N. A., 1874, App., p. 522 (cut).

Falco sparverius.
(31-a) Stansbury's Surv. Salt Lake [App. C],
p. 325; (78) P. R. R. Surv., vol. ix, p. 13; (87)
Mex. Bound. Surv., vol. ii, p. 3.

Falco sparverius Isabellinus.
(632½) Birds of N. A., 1874, vol. iii, p. 171.

FALCO SPARVERIUS SPARVERIUS.
(632½) Birds of N. A., 1874, vol. iii, p. 169 (cuts,
pp. 159, 173).

Falconidæ.
(632½) Birds of N. A., 1874, vol. iii, p. 103; *Ibid.*,
App., p. 522.

Falconine.
(632½) Birds of N. A., 1874, vol. iii, p. 106.

Florida cærulea.
(78) P. R. R. Surv., vol. ix, p. 671; (87) Mex.
Bound. Surv., vol. ii, p. 24.

Fratercula arctica.
(78) P. R. R. Surv., vol. ix, p. 903.

Fratercula corniculata.
(78) P. R. R. Surv., vol. ix, p. 902.

Fratercula glacialis.
(78) P. R. R. Surv., vol. ix., p. 903.

Fregetta Lawrencii.
(78) P. R. R. Surv., vol. ix, p. 832.

Fringillidæ.
(632½) Birds of N. A., 1874, vol. i, p. 446.

Fringilla meruloides.
(32) Stansbury's Surv. Salt Lake [App. C],
p. 330.

Fulica americana.
(78) P. R. R. Surv., vol. ix, p. 751; (87) Mex.
Bound. Surv., vol. ii, p. 26; (94) P. R. R.
Surv., vol. x, p. 15; (99) Pr. Acad. Nat. Sci.
Phila., 1859, p. 306.

Fuligula affinis.
(31) Stansbury's Surv. Salt Lake [App. C], p.
324.

Fulix affinis.
(78) P. R. R. Surv., vol. ix, p. 791.

Fulix collaris.
(78) P. R. R. Surv., vol. ix, p. 792; (87) Mex.
Bound. Surv., vol. ii, p. 27.

Fulix marila.
(78) P. R. R. Surv., vol. ix, p. 791.

Fulmarus glacialis.
(78) P. R. R. Surv., vol. ix, p. 825.

Fulmarus Rodgersi.
(129) Chicago Acad. Sci., 1869, p. 323.

Galbulidæ.
(115) Review of N. A. Birds, Aug., 1864, Part
I, p. 165.

Galeoscoptes.
(115) Review of N. A. Birds, June, 1864, Part I,
p. 5; (632½) Birds of N. A., 1874, vol. i, p.
51; (115) Review of N. A. Birds, July,
1864, Part I, p. 54.

GALEOSCOPTES CAROLINENSIS.
(115) Review of N. A. Birds, July, 1864, Part
I, p. 54; (632½) Birds of N. A. 1874, vol. i,
p. 52, pl. iii, fig. 5 (cuts, p. 52).

Gallinago Wilsonii.
(78) P. R. R. Surv., vol. ix, p. 710; (87) Mex.
Bound. Surv., vol. ii, p. 25.

Gallinula (Gallinula) galeata.
(78) P. R. R. Surv., vol. ix, p. 752.

Gallinula galeata.
(87) Mex. Bound. Surv., vol. ii, p. 26.

Gallinula martinica.
(78) P. R. R. Surv., vol. ix, p. 753.

Granatellus renustus.
    (115) Review of N. A. Birds, Apr., 1865, Part I,
        p. 230.
Granatellus francescæ.
    (115) Review of N. A. Birds, Apr., 1865, Part
        I, p. 232.
Granatellus Pelzelnii.
    (115) Review of N. A. Birds, Apr., 1865, Part
        I, p. 230.   .
Granatellus Sallaei.
    (115) Review of N. A. Birds, Apr., 1865, Part
        I, p. 232.
Grus americanus.
    (78) P. R. R. Surv., vol. ix, p. 654.
Grus canadensis.
    (31) Stansbury's Surv. Salt' Lake [App. C],
        p. 319; (78) P. R. R. Surv., vol. ix, p. 655; (87)
        Mex. Bound. Surv., vol. ii, p. 24; (94) P. R.
        R. Surv., vol. x, p. 14.
Grus fraterculus.
    (78) P. R. R. Surv., vol. ix, p. 656; (104) Birds
        of N. A., 1860, p. 656, pl. xxxvii.
Guiraca.
    (632½) Birds of N. A., 1874, vol. ii, p. 76.
GUIRACA CÆRULEA.
    (78) P. R. R. Surv., vol. ix, p. 499; (87) Mex.
        Bound. Surv., vol. ii, p. 16; (632½) Birds of
        N. A., 1874, vol. ii, p. 77, pl. xxix, figg. 4, 5
        (cuts, p. 77); Ibid., App., p. 516.
Guiraca ludoviciana.
    (78) P. R. R. Surv., vol. ix, p. 497.
Guiraca melanocephala.
    (78) P. R. R. Surv., vol. ix, p. 498; (99) Pr. Acad.
        Nat. Sci. Phila., 1859, p. 304.
Gymnokitta.
    (632½) Birds of N. A., 1874, vol. ii, p. 259.
GYMNOKITTA CYANOCEPHALLA.
    (32) Stansbury's Surv. Salt Lake [App. C], p.
        331; (78) P. R. R. Surv., vol. ix, p. 574; (632½)
        Birds of N. A., 1874, vol. ii, p. 260 (cuts, pp.
        259, 260).
Hæmatopus ater.
    (78) P. R. R. Surv., vol. ix, p. 700.
Hæmatopus niger.
    (78) P. R. R. Surv., vol. ix, p. 700.
Hæmatopus palliatus.
    (78) P. R. R. Surv., vol. ix, p. 699.
Haliætus.
    (632½) Birds of N. A., 1874, vol. iii, p. 320.
HALIÆTUS ALBICILLA.
    (78) P. R. R. Surv., vol. ix, p. 43; (632½) Birds of
        N. A., 1874, vol. iii, p. 324 (cut, p. 365).
HALIÆTUS LEUCOCEPHALUS.
    (78) P. R. R. Surv., vol. ix, p. 43; (632½) Birds of
        N. A., 1874, vol. iii, p. 326 (cuts, pp. 312, 321,
        328, 330).
Haliaetus pelagicus.
    (78) P. R. R. Surv., vol. ix, p. 42.
Haliaetus Washingtonii.
    (78) P. R. R. Surv., vol. ix, p. 42.
Harelda glacialis.
    (78) P. R. R. Surv., vol. ix, p. 800.
Harporhynchus.
    (115) Review of N. A. Birds, June, 1864, Part I,
        p. 5; Ibid., p. 43; (632½) Birds of N. A.,
        1874, vol. i, p. 35.

Harporhynchus Bendirei.
    (632½) Birds of N. A., 1874, App., p. 500.
HARPORHYNCHUS CINEREUS.
    (99) Pr. Acad. Nat. Sci. Phila., 1859, pp. 303, 304;
        (113) Review of N. A. Birds, July, 1864.
        Part I, p. 46; (632½) Birds of N. A., 1874,
        vol. i, p. 40, pl. iv, fig. 2.
HARPORHYNCHUS CRISSALIS.
    (78) P. R. R. Surv., vol. ix, p. 351; (104) Birds of
        N. A., 1860, p. 351, pl. lxxxii; (115) Review
        of N. A. Birds, July, 1864, Part I, p. 47;
        (632½) Birds of N. A., 1874, vol. i, p. 47, pl.
        iv, fig. 1; Ibid. App. p. 500.
HARPORHYNCHUS CURVIROSTRIS.
    (78) P. R. R. Surv., vol. ix, p. 351; (87) Mex.
        Bound. Surv., vol. ii, p. 12, pl. xiii; (104)
        Birds of N. A., 1860, p. 351, pl. li; (115) Re-
        view of N. A. Birds, July, 1864, Part I, p.
        45; (632½) Birds of N. A., 1874, vol. i, p. 41,
        pl. iii, fig. 3.
Harporhynchus curvirostris Palmeri.
    (632½) Birds of N. A., 1874, vol. i, p. 43; Ibid.,
        App., p. 500.
HARPORHYNCHUS LECONTII.
    (78) P. R. R. Surv., vol. ix, p. 350; (87) Mex.
        Bound. Surv., vol. ii, p. 12, pl. xii; (104)
        Birds of N. A., 1860, p. 350, pl. 1; (115) Re-
        view of N. A. Birds, July, 1864, Part I, p.
        47.
HARPORHYNCHUS LONGIROSTRIS.
    (78) P. R. R. Surv., vol. ix, p. 352; (87) Mex.
        Bound. Surv., vol. ii, p. 13, pl. xiv; (104)
        Birds of N. A., 1860, p. 352, pl. lii; (115) Re-
        view of N. A. Birds, July, 1864, Part I, p.
        44.
HARPORHYNCHUS REDIVIVUS.
    (78) P. R. R. Surv., vol. ix, p. 349; (115) Review
        of N. A. Birds, July, 1864, Part I, p. 48;
        (632½) Birds of N. A., 1874, vol. i, p. 45,
        pl. iv, fig. 4; Ibid., App., p. 501 (cut, p. 501).
HARPORHYNCHUS REDIVIVUS LECONTEI.
    (632½) Birds of N. A., 1874, vol. i, p. 44, pl. iv,
        fig. 3.
HARPORHYNCHUS RUFUS.
    (78) P. R. R. Surv., vol. ix, p. 353; (115) Review
        of N. A. Birds, July, 1864, Part I, p. 44;
        (632½) Birds of N. A., 1874, vol. i, p. 37, pl.
        iii, fig. 1 (cuts, p. 35); Ibid., App., p.
        500.
HARPORHYNCHUS RUFUS LONGIROSTRIS.
    (632½) Birds of N. A., 1874, vol. i, p. 39, pl. iii,
        fig. 2.
Harporhynchus oscellatus.
    (115) Review of N. A. Birds, July, 1864, Part
        I, p. 59; (632½) Birds of N. A., 1874, App.,
        p. 499.
Harporhynchus vetula.
    (78) P. R. R. Surv., vol. ix, p. 352.
Hedymeles.
    (632½) Birds of N. A., 1874, vol. ii, p, 69.
HEDYMELES LUDOVICIANUS.
    (632½) Birds of N. A., 1874, vol. ii, p. 70, pl. xxx,
        figg. 4, 5.
HEDYMELES MELANOCEPHALUS.
    (632½) Birds of N. A., 1874, vol. ii, p. 73, pl. xxx,
        figg. 1, 2 (cuts, pp. 69, 71).

Himantopus nigricollis.
> (78) P. R. R. Surv., vol. ix, p. 704; (87) Mex.
> Bound. Surv., vol. ii, p. 25.

Hirundinidæ.
> (115) Review of N. A. Birds, Aug., 1864, Part I,
> p. 165; Ibid., May, 1865, Part I, p. 267;
> (632½) Birds of N. A., 1874, vol. i, p. 326.

Hirundo.
> (115) Review of N. A. Birds, May, 1865, Part I,
> p. 270; Ibid., p. 271; Ibid., p. 293; Ibid.,
> p. 294; (632½) Birds of N. A., 1874, vol. i, p.
> 338.

Hirundo albilinea.
> (115) Review of N. A. Birds, May, 1865, Part I,
> p. 300.

Hirundo albiventris.
> (115) Review of N. A. Birds, May, 1865, Part I,
> p. 302.

Hirundo andecola.
> (115) Review of N. A. Birds, May, 1865, Part I,
> p. 320.

HIRUNDO BICOLOR.
> (78) P. R. R. Surv., vol. ix, p. 310; (87) Mex.
> Bound. Surv., vol. ii, p. 11; (115) Review of
> N. A. Birds, May, 1865, Part I, p. 296; (632½)
> Birds of N. A., 1874, vol. i, p. 344, pl. xvi,
> fig. 10 (cut, p. 345).

Hirundo cyaneoviridis.
> (115) Review of N. A. Birds, May, 1865, Part I,
> p. 303.

Hirundo erythrogaster.
> (115) Review of N. A. Birds, May, 1865, Part I,
> p. 295.

Hirundo euchrysea.
> (115) Review of N. A. Birds, May, 1865, Part I,
> p. 304.

HIRUNDO HORREORUM.
> (78) P. R. R. Surv., vol. ix, p. 308; (87) Mex.
> Bound. Surv., vol. ii, p. 11; (115) Review of
> N. A. Birds, May, 1865, Part I, p. 294; (632½)
> Birds of N. A., 1874, vol. i, p. 339, pl. xvi,
> fig. 9 (cuts, pp. 338, 339).

Hirundo leucorrhoa.
> (115) Review of N. A. Birds, May, 1865, Part I,
> p. 301.

Hirundo lunifrons.
> (78) P. R. R. Surv., vol. ix, p. 309.

Hirundo maculosa.
> (115) Review of N. A. Birds, May, 1865, Part I,
> p. 320.

Hirundo Meyeni.
> (115) Review of N. A. Birds, May, 1865, Part I,
> p. 302.

HIRUNDO THALASSINA.
> (78) P. R. R. Surv., vol. ix, p. 311; (87) Mex.
> Bound. Surv., vol. ii, p. 11; (99) Pr. Acad.
> Nat. Sci. Phila., 1859 (1860), p. 303; (115)
> Review of N. A. Birds, May, 1865, Part
> I, p. 299; (632½) Birds of N. A., 1874, vol.
> i, p. 347, pl. xvi, fig. 11 (cut, p. 344).

Hirundo unalaschkensis.
> (115) Review of N. A. Birds, May, 1865, Part I,
> p. 320.

Histrionicus torquatus.
> (78) P. R. R. Surv., vol. ix, p. 799.

Hydrobata mexicana.
> (78) P. R. R. Surv., vol. ix, p. 229.

Hydrochelidon plumbea.
> (78) P. R. R. Surv., vol. ix, p. 864.

Hyemalis.
> (115) Review of N. A. Birds, Aug., 1864, Part I,
> p. 138.

Hyemalis pacificus.
> (115) Review of N. A. Birds, Aug., 1864, Part I,
> p. 138.

Hylemathorus.
> (115) Review of N. A. Birds, Aug., 1864, Part I,
> p. 94.

Hylocichla.
> (115) Review of N. A. Birds, June, 1864, Part I,
> p. 12; Ibid., p. 13.

Hylophilus.
> (115) Review of N. A. Birds, May, 1866, Part I,
> p. 324; Ibid., p. 372.

Hylophilus acuticauda.
> (115) Review of N. A. Birds, May, 1866, Part I,
> p. 378.

Hylophilus aurantiifrons.
> (115) Review of N. A. Birds, May, 1866, Part I,
> p. 377.

Hylophilus cineraceus.
> (115) Review of N. A. Birds, May, 1866, Part I,
> p. 375.

Hylophilus decurtatus.
> (115) Review of N. A. Birds, May, 1866, Part I,
> p. 380.

Hylophilus ferruginifrons.
> (115) Review of N. A. Birds, May, 1866, Part I,
> p. 377.

Hylophilus flaveolus.
> (115) Review of N. A. Birds, May, 1866, Part I
> p. 375.

Hylophilus flavipes.
> (115) Review of N. A. Birds, May. 1866, Part I,
> p. 375.

Hylophilus frontalis.
> (115) Review of N. A. Birds, May, 1866, Part I,
> p. 375.

Hylophilus insularis.
> (115) Review of N. A. Birds, May, 1866, Part I,
> p. 379.

Hylophilus ochraceiceps.
> (115) Review of N. A. Birds, May, 1866, Part I,
> p. 376.

Hylophilus olivaceus.
> (115) Review of N. A. Birds, May, 1866, Part I,
> p. 375.

Hylophilus pœcilotis.
> (115) Review of N. A. Birds, May, 1866, Part I,
> p. 375.

Hylophilus pusillus.
> (115) Review of N. A. Birds, May, 1866, Part I,
> p. 381.

Hylophilus semibrunneus.
> (115) Review of N. A. Birds, May, 1866, Part I,
> p. 372.

Hylophilus thoracicus.
> (115) Review of N. A. Birds, May, 1866, Part I,
> p. 375.

H:.lophilus viridiflavus.
(115) Review of N. A. Birds, May, 1866, Part I, p. 380.
Hylotomus.
(632½) Birds of N. A., 1874, vol. ii, p. 548.
HYLOTOMUS PILEATUS.
(78) P. R. R. Surv., vol. ix, p. 107; (632½) Birds of N. A., 1874, vol. ii, p. 559, pl. lvi, figg. 4, 5 (cuts, pp. 549, 550).
Hypocolius ampelin us.
(32) Stansbury'sSurv.SaltLake[App. C],p.328.
Hypotriorchis aurantius.
(78) P. R. R. Surv., vol. ix, p. 10.
Hypotriorchis columbarius.
(78) P. R. R. Surv., vol. ix, p. 9.
HYPOTRIORCHIS FEMORALIS.
(78) P. R. R. Surv., vol. ix, p. 11; (104) Birds of N. A., 1860, p. 11, pl. i.
.Ibis alba.
(78) P. R. R. Surv., vol. ix, p. 684; (87) Mex. Bound. Surv., vol. ii, p. 24.
IBIS GUARAUNA.
(104) Birds of N. A., 1860, p. 661, pl. lxxxvii.
Ibis Ordii.
(78) P. R. R. Surv., vol. ix, p. 685; (87) Mex. Bound. Surv., vol. ii, p. 24.
Ibis rubra.
(78) P. R. R. Surv., vol. ix, p. 683.
Icteria.
(115) Review of N. A. Birds, Apr., 1865, Part I, p. 228; (632½) Birds of N. A., 1874, vol. i, p. 306.
Icterieæ.
(115) Review of N. A. Birds, Aug. 1854, Part I, p. 166.
Icterinæ.
(632½) Birds of N. A., 1874, vol. ii, p. 179.
Icteria longicauda.
(78) P. R. R. Surv., vol. ix, p. 294; (87) Mex. Bound. Surv., vol. ii, p. 10.
ICTERIA LONGICAUDA.
(104) Birds of N. A., 1860, p. 249, pl. xxxiv, fig. 2; (115) Review of N. A. Birds, Apr., 1865, Part I, p. 230.
Icteria Velasquezii.
(32) Stansbury's Surv. Salt Lake [App. C], p. 328.
ICTERIA VIRENS.
(115) Review of N. A. Birds, Apr., 1865, Part I, p. 228; (632½) Birds of N. A., 1874, vol. i, p. 307, pl. xv, fig. 12 (cuts, pp. 306, 307).
Icteria virens longicauda.
(632½) Birds of N. A., 1874, vol. i, p. 309.
Icteria viridis.
(78) P. R. R. Surv., vol. ix, p. 179.
Icterianæ.
(115) Review of N. A. Birds, Aug., 1864, Part I, p. 166; Ibid., Apr., 1865, Part I, p. 228; (632½) Birds of N. A., 1874, vol. i, p. 306.
Icteridiæ.
(632½) Birds of N. A., 1874, vol. ii, p. 147.
Icterus.
(632½) Birds of N. A., 1874, vol. ii, p. 179.
Icterus Audubonii.
(78) P. R. R. Surv., vol. ix, p. 542; (87) Mex. Bound. Surv., vol. ii, p. 19.

ICTERUS BALTIMORE.
(78) P. R. R. Surv., vol. ix, p. 548; (87) Mex. Bound. Surv., vol. ii, p. 19; (632½) Birds of N. A., 1874, vol. ii, p. 195, pl. xxxv, fig. 5; Ibid., App., p. 518.
ICTERUS BULLOCKII.
(78) P. R. R. Surv., vol. ix, p. 549; (87) Mex. Bound. Surv., vol. ii, p. 20; (632½) Birds of N. A., 1874, vol. ii, p. 199, pl. xxxiv, fig. 3, (cuts. p. 180); Ibid., App., p. 518.
ICTERUS CUCULLATUS.
(32) Stansbury's Surv. Salt Lake [App. C], p. 331; (78) P. R. R. Surv., vol. ix, p. 546; (87) Mex. Bound. Surv., vol. ii, p. 19; (99) Pr. Acad. Nat. Sci. Phila., 1859, p. 305; (632½) Birds of N. A., 1874, vol. ii, p. 193, pl. xxxv, fig. 6; Ibid., App., p. 517.
Icterus frenatus.
(32) Stansbury's Surv. Salt Lake [App.C],p.331.
Icterus melanocephalus.
(32) Stansbury's Surv. Salt Lake [App. C], p. 331; (78) P. R. R. Surv., vol. ix, p. 543.
ICTERUS MELANOCEPHALUS AUDUBONI.
(632½) Birds of N. A., 1874, vol. ii, p. 186, pl. xxxv, fig. 1.
ICTERUS PARISORUM.
(78) P. R. R. Surv., vol. ix, p. 544; (87) Mex. Bound. Surv., vol. ii, p. 19, pl. xix. fig. 1; (99) Pr. Acad. Nat. Sci. Phila., 1859, p. 305; (104) Birds of N. A., 1860, p. 544, pl. lvii, fig. 1; (632½) Birds of N. A., 1874, vol. ii, p. 188, pl. xxxv, fig. 7.
ICTERUS SPURIUS.
(78) P. R. R. Surv., vol. ix, p. 547; (87) Mex. Bound. Surv., vol. ii, p. 19; (632½) Birds of N. A., 1874, vol. ii, p. 190, pl. xxxiv, figg. 4, 5, 6.
Icterus vulgaris.
(32) Stansbury's Surv. Salt Lake [App. C], p. 331; (78) P. R. R. Surv., vol. ix, p. 542; (632½) Birds of N. A., 1874, vol. ii, p. 184.
ICTERUS WAGLERI.
(78) P. R. R. Surv., vol. ix, p. 545; (87) Mex. Bound. Surv., vol. ii, p. 19, pl. xix. fig. 2; (104) Birds of N. A., 1860, p. 545, pl. lvii, fig. 2.
Ictinia.
(632½) Birds of N. A., 1874, vol. iii, p. 202.
ICTINIA MISSISSIPPIENSIS.
(78) P. R. R. Surv., vol. ix, p. 37; (632½) Birds of N. A., 1874, vol. iii, p. 203 (cuts, pp. 202, 205).
Idiotes.
(115) Review of N. A. Birds, Apr., 1865, Part I, p. 238; Ibid., p. 247 (237).
Inquietus.
(115) Review of N. A. Birds, Aug., 1864, Part I, p. 138.
Intermedius.
(115) Review of N. A. Birds, Aug., 1864, Part I, p. 138.
Ixoreus nævius.
(78) P. R. R. Surv., vol. ix, p. 219.
Junco.
(632½) Birds of N. A., 1874, vol. i, p. 578; (632½) Birds of N.., A1874, App., p. 514.

MELANERPES ERYTHROCEPHALUS.
(78) P. R. R. Surv., vol. ix, p. 113; (632½) Birds of N. A. 1874, vol. ii, p. 564, pl. liv, fig. 4 (cut, p. 560).

Melanerpes, formicivorus.
(32) Stansbury's Surv. Salt Lake [App. C], p. 333; (78) P. R. R. Surv., vol. ix, p. 114; (87) Mex. Bound. Surv., vol. ii, p. 6,

MELANERPES FORMICIVORUS ANGUSTIFRONS.
(632½) Birds of N. A., 1874, vol. ii, p. 573, pl. liii, figg. 3, 4.

MELANERPES FORMICIVORUS FORMICIVORUS.
(632½) Birds of N. A., 1874, vol. ii, p. 566, pl. liii, figg. 1, 2 (cut, p. 567).

MELANERPES TORQUATUS.
(78) P. R. R. Surv., vol. ix, p. 115; (632½) Birds of N. A., 1874, vol. ii, p. 561, pl. liv, fig. 5.

Melanetta velvetina.
(78) P. R. R. Surv., vol. ix, p. 805.

Melanoptila.
(115) Review of N. A. Birds, June, 1864, Part I, p. 5; Ibid., July, 1864, Part I, p. 55.

Melanoptila glabrirostris.
(115) Review of N. A. Birds, July, 1864, Part I, p. 55.

Melanotis.
(115) Review of N. A. Birds, June, 1864, Part I. p. 5; Ibid., July, 1864, Part I, p. 56.

Melanotis cærulescens.
(115) Review of N. A. Birds, July, 1864, Part I, p. 56.

Melanotis hypoleucus.
(115) Review of N. A. Birds, July, 1864, Part I, p. 57.

Meleagridæ.
(632½) Birds of N. A., 1874, vol. iii, p. 402.

Meleagris.
(632½) Birds of N. A., 1874, vol. iii, p. 403.

Meleagris gallopavo.
(78) P. R. R. Surv., vol. ix, p. 615.

MELEAGRIS GALLOPAVO GALLOPAVO.
(632½) Birds of N. A., 1874, vol. iii, p. 404 (cut, pp. 403, 404).

Meleagris mexicana.
(78) P. R. R. Surv., vol. ix, p. 618.

Melopelia.
(632½) Birds of N. A., 1874, vol. iii, p. 376.

MELOPELIA LEUCOPTERA.
(78) P. R. R. Surv., vol. ix, p. 603; (87) Mex. Bound. Surv., vol. ii, p. 21; (99) Pr. Acad. Nat. Sci. Phila., 1859, p. 305; (632½) Birds of N. A., 1874, vol. iii, p. 376, pl. lviii, fig. ? (cuts, pp. 376, 377).

Melospiza.
(632½) Birds of N. A., 1874, vol. ii, p. 16.

Melospiza (Melospiza) fallax.
(78) P. R. R. Surv., vol. ix, p. 481.

MELOSPIZA FALLAX.
(104) Birds of N. A., 1860, p. 481, pl. xxvii, fig. 2.

Melospiza (Melospiza) Gouldii.
(78) P. R. R. Surv., vol. ix, p. 479.

MELOSPIZA GOULDII.
(104) Birds of N. A., 1860, p. 479, pl. lxx, fig. 2.

Melospiza (Melospiza) Heermanni.
(78) P. R. R. Surv., vol. ix, p. 478.

MELOSPIZA HEERMANNI.
(104) Birds of N. A., 1860, p. 478, pl. lxx, fig. 1.

MELOSPIZA INSIGNIS.
(129) Chicago Acad. Sci., 1869, p. 319, pl. xxix, fig. 2.

MELOSPIZA LINCOLNII.
(78) P. R. R. Surv., vol. ix, p. 482; (87) Mex. Bound. Surv., vol. ii, p. 16; (632½) Birds of N. A., 1874, vol. ii, p. 31, pl. xxvii, fig. 13; Ibid., App., p. 514.

Melospiza (Melospiza) melodia.
(78) P. R. R. Surv., vol. ix, p. 477.

MELOSPIZA MELODIA.
(87) Mex. Bound. Surv., vol. ii, p. 16; (632½) Birds of N. A., 1874, vol. ii, p. 19, pl. xxvii, fig. 6 (cuts, p. 16).

MELOSPIZA MELODIA FALLAX.
(632½) Birds of N. A., 1874, vol. ii, p. 22, pl. xxvii, fig. 10.

MELOSPIZA MELODIA GUTTATA.
(632½) Birds of N. A., 1874, vol. ii, p. 27, pl. xxvii, fig. 12.

MELOSPIZA MELODIA HEERMANNI.
(632½) Birds of N. A., 1874, vol. ii, p. 24, pl. xxvii, fig. 9.

MELOSPIZA MELODIA INSIGNIS.
(632½) Birds of N. A., 1874, vol. ii, p. 30, pl. xxvii, fig. 8.

MELOSPIZA MELODIA RUFINA.
(632½) Birds of N. A., 1874, vol. ii, p. 29, pl. xxvii, fig. 11.

MELOSPIZA MELODIA SAMUELIS.
(632½) Birds of N. A., 1874, vol. ii, p. 26, pl. xxvii, fig. 7.

MELOSPIZA PALUSTRIS.
(78) P. R. R. Surv., vol. ix., p. 483; (632½) Birds of N.A., 1874, vol. ii, p. 34, pl. xxviii, figg. 1, 2; Ibid., App., p. 515.

Melospiza (Melospiza) rufina.
(78) P. R. R. Surv., vol. ix, p. 480.

Mergellus albellus.
(78) P. R. R. Surv., vol. ix, p. 817.

Mergullus alle.
(78) P. R. R. Surv., vol. ix, p. 918.

Mergulus Cassinii.
(32) Stansbury's Surv. Salt Lake [App. C], p. 335.

Mergulus cirrocephalus.
(32) Stansbury's Surv. Salt Lake [App. C], p. 335.

Mergus americanus.
(78) P. R. R. Surv., vol. ix, p. 813; (87) Bound. Surv., vol. ii, p. 27.

Mergus serrator.
(78) P. R. R. Surv., vol. ix, p. 814.

Merula.
(115) Review of N. A. Birds, June, 1864, Part I, p. 12; Ibid., p. 31.

Merula olivacea.
(32) Stansbury's Surv. Salt Lake [App. C], p. 328.

Micrathene.
(632½) Birds of N. A., 1874, vol. iii, p. 86.

MICRATHENE WHITNEYI.
(632½) Birds of N. A., 1874, vol. iii, p. 87 (cuts, pp. 86, 88).

Metacillina.
(632½) Birds of N. A., 1874, vol. i, p. 165.
Myiadestes.
(115) Review of N. A. Birds, June, 1866, Part I,
p. 417; *Ibid.*, p. 418; (632½) Birds of N. A.,
1874, vol. i, p. 408.
Myiadestes ardesiaceus.
(115) Review of N. A. Birds. June, 1866, Part I,
p. 421.
Myiadestes armillatus.
(115) Review of N. A. Birds, June, 1866, Part I,
p. 422.
Myiadestes Elisabethi.
(115) Review of N. A. Birds, June, 1866, Part I,
p. 425.
Myiadestes genibarbis.
(115) Review of N. A. Birds, June, 1866, Part I,
p. 423.
Myiadestes griseiventer.
(115) Review of N. A. Birds, June, 1866, Part I,
p. 421.
Myiadestes leucotis.
(115) Review of N. A. Birds, June, 1866, Part I,
p. 432.
Myiadestes melanops.
(115) Review of N. A. Birds, June, 1866, Part I,
p. 426.
Myiadestes obscurus.
(115) Review of N. A. Birds, June, 1866, Part I,
p. 430.
Myiadestes solitarius.
(115) Review of N. A. Birds, June, 1866, Part I,
p. 421.
MYIADESTES TOWNSENDII.
(78) P. R. R. Surv., vol. ix, p. 321; (115) Review
of N. A. Birds, June, 1866, Part I, p.
429; (632½) Birds of N. A., 1874, vol. i, p.
409, pl. xviii, figg. 5, 6 (cuts, pp. 408,
410).
Myiadestes unicolor.
(115) Review of N. A. Birds, June, 1866, Part I,
p. 428.
Myiadestes venezuelensis.
(115) Review of N. A. Birds, June, 1866, Part I,
p. 427.
Myiadestinæ.
(115) Review of N. A. Birds, May, 1866, Part I,
p. 409; *Ibid.*, June, 1866, Part I, p. 417.
Myiarchus.
(632½) Birds of N. A., 1874, vol. ii, p. 329; *Ibid.*,
App., p. 519.
Myiarchus Cooperi.
(78) P. R. R. Surv., vol. ix, p. 160.
MYIARCHUS CRINITUS.
(78) P. R. R. Surv., vol. ix, p. 178; (632½) Birds
of N. A., 1874, vol. ii, p. 334, pl. xliii, fig.
3.
MYIARCHUS CRINITUS CINERASCENS.
(632½) Birds of N. A., 1874, vol. ii, p. 337, pl. xliii,
fig. 6 (cut, p. 334).
Myiarchus Lawrencii.
(78) P. R. R. Surv., vol. ix, p. 181; (87) Mex.
Bound. Surv., vol. ii, p. 8, pl. ix, fig. 3; (104)
Birds of N. A., 1860, p. 181, pl. xlvii, fig.
3.

MYIARCHUS MEXICANUS.
(78) P. R. R. Surv., vol. ix, p. 179; (87) Mex.
Bound. Surv., vol. ii, p. 8; (99) Pr. Acad.
Sci. Phila., 1859 (1860), p. 303; (104) Birds
of N. A., 1860, p. 179, pl. v.
Myioborus.
(115) Review of N. A. Birds, Apr., 1865, Part I,
p. 238; *Ibid.*, May, 1865, Part I, p. 257 (237).
Myiodioctes.
(115) Review of N. A. Birds, Apr., 1865, Part
I, p. 236; *Ibid.*, p. 238: (632½) Birds of N.
A., 1874, vol. i, p. 313.
Myiodioctes Bonapartii.
(78) P. R. R. Surv., vol. ix, p. 295.
MYIODIOCTES CANADENSIS.
(78) P. R. R. Surv., vol. ix, p. 294; (115) Review
of N. A. Birds, Apr., 1865, Part I, p. 239;
(632½) Birds of N. A., 1874, vol. i, p. 320, pl.
xvi, fig. 6.
MYIODIOCTES MINUTUS.
(78) P. R. R. Surv., vol. ix, p. 293; (115) Review
of N. A. Birds, May, 1865, Part I, p. 241;
(632½) Birds of N. A., 1874, vol. i, p. 316, pl.
xvi, fig. 2.
MYIODIOCTES MITRATUS.
(78) P. R. R. Surv., vol. ix, p. 292; (115) Review
of N. A. Birds, Apr., 1865, Part I, p. 239;
(632½) Birds of N. A., 1874, vol. i, p. 314, pl.
xv, figg. 10, 11 (cut, p. 313).
MYIODIOCTES PUSILLUS.
(6) Lit. Rec. S. Journ. Linnæan Ass. Penn.
Col., Oct., 1845, p. 252; (78) P. R. R. Surv.,
vol. ix, p. 293; (87) Mex. Bound. Surv., vol.
ii, p. 10; (115) Review of N. A. Birds, Apr.,
1865, Part I, p. 240; (632½) Birds of N. A.,
1874, vol. i, p. 317, pl. xvi, figg. 3. 4
(cut, p. 314).
Myiodioctes pusillus phileolatus.
(632½) Birds of N. A., 1874, App., p. 507.
Myiodioctes pusillus pileolatus.
(632½) Birds of N. A., 1874, vol. i, p. 319.
Myiothlypis.
(115) Review of N. A. Birds., Apr., 1865, Part
I, pp. 237, 238; *Ibid.*, May, 1865, Part I, p. 251.
Nauclerus.
(632½) Birds of N. A., 1874, vol. iii, p. 190.
NAUCLERUS FORFICATUS.
(632½) Birds of N. A., 1874, vol. iii, p. 193 (cuts,
pp. 191, 193).
Nauclerus furcatus.
(78) P. R. R. Surv., vol. ix, p. 36.
Nectris fuliginosus.
(78) P. R. R. Surv., vol. ix, p. 834.
Neochelidon.
(115) Review of N. A. Birds, May, 1865, Part I,
pp. 270, 305, 307.
Neochloe.
(115) Review of N. A. Birds, May, 1866, Part I,
pp. 323, 371.
Neochloe brevipennis.
(115) Review of N. A. Birds, May, 1866, Part
I, p. 372.
Neocorys.
(115) Review of N. A. Birds, Aug., 1864, Part I,
pp. 151, 155; (632½) Birds of N. A., 1874,
vol. i, p. 174.

OREOSCOPTES MONTANUS.
(78) P. R. R. Surv., vol. ix, p. 347; (87) Mex.
Bound. Surv., vol. ii, p.12; (115) Review of
N. A. Birds, July, 1864, Part I, p. 42; (632½)
Birds of N. A., 1874, vol. i, p. 32, pl. iii, fig. 6
(cuts, pp. 31, 32).

Ornismya Costæ.
(32) Stansbury's Surv. Salt Lake [App. C], p.
326.

Ortalida.
(632½) Birds of N. A., 1874, vol. iii, p. 398.

Ortalida McCalli.
(78) P. R. R. Surv., vol. ix, p. 611; (87) Mex.
Bound. Surv., vol. ii, p. 22.

Ortalida vetula.
(32) Stansbury's Surv. Salt Lake [App. C], p.
334.

ORTALIDA VETULA MACCALLI.
(632½) Birds of N. A., 1874, vol. iii, p. 398, pl. lvii,
fig. 1 (cuts, 398, 399).

Ortyginæ.
(632½) Birds of N. A., 1874, vol. iii, p. 466.

Ortyx.
(632½) Birds of N. A., 1874, vol. iii, p. 467.

Ortyx virginianus.
(78) P. R. R. Surv., vol. ix, p. 640: (87) Mex.
Bound. Surv., vol. ii, p, 22.

Ortyx virginianus floridanus.
(632½) Birds of N. A., 1874, App., p. 522.

Ortyx virginianus texanus.
(632½) Birds of N. A., 1874, vol. iii, p. 468.

ORTYX VIRGINIANUS VIRGINIANUS.
(632½) Birds of N. A., 1874, vol. iii, p. 468, pl.
lxiii, figg. 1, 2 (cuts, pp. 467, 469).

ORTYX TEXANUS.
(78) P. R. R. Surv., vol. ix, p. 641; (104) Birds
of N. A., 1860, p. 472, pl. lxii; (87) Mex.
Bound. Surv., vol. ii, p. 22, pl. xxiv.

Ossifraga gigantea.
(78) P. R. R. Surv., vol. ix, p. 825.

Otocoris occidentalis.
(32) Stansbury's Surv. Salt Lake [App. C],
p. 331; (31) Stansbury's Surv. Salt Lake
[App. C], p. 318.

Otus.
(632½) Birds of N. A., 1874, vol. iii, p. 17.

OTUS (BRACHYOTUS) BRACHYOTUS.
(632½) Birds of N. A., 1874, vol. iii, p. 22 (cuts,
pp. 23, 24).

OTUS VULGARIS WILSONIANUS.
(623½) Birds of N. A., 1874, vol. iii, p. 18 (cuts,
pp. 19, 20, 69, 98, 99, 100, 101).

Otus Wilsonianus.
(78) P. R. R. Surv., vol. ix, p. 53; (94) P. R. R.
Surv., vol. x, p. 12.

Oxyechus montanus.
(78) P. R. R. Surv., vol. ix, p. 693.

Oxyechus vociferus.
(78) P. R. R. Surv., vol. ix, p. 692.

PACHYRHAMPHUS AGLAIÆ.
(78) P. R. R. Surv., vol. ix, p. 164; (87)
Mex. Bound Surv., vol. ii, p. 7, pl. ix, fig.
1; (104) Birds of N. A., 1860, p. 164, pl.
xlvii, fig. 1, male.

Pagophila brachytarsi.
(78) P. R. R. Surv., vol. ix, p. 856.

Pagophila eburnea.
(78) P. R. R. Surv., vol. ix, p. 856.

Pandion.
(632½) Birds of N. A., 1874, vol. iii, p. 182.

Pandion carolineusis.
(78) P. R. R. Surv., vol. ix, p. 44; (87) Mex.
Bound. Surv., vol. ii, p. 4.

PANDION HALLÆTUS CAROLINENSIS.
(632½) Birds of N. A., 1874, vol. iii, p. 184 (cuts,
pp. 185, 187).

Panyptila.
(632½) Birds of N. A., 1874, vol. ii, p. 423.

PANYPTILA MELANOLEUCA.
(78) P. R. R. Surv., vol. ix, p. 141; (104) Birds of
N. A., 1860, p. 141, pl. xviii, fig. 1; (632½)
Birds of N. A., 1874, vol. ii, p. 424, pl. xlv,
fig. 5 (cuts, pp. 422, 423, 425); Ibid., App.,
p. 521.

Paridæ.
(115) Review of N. A. Birds, July, 1864, Part
I, p. 77; Ibid., Aug., 1864, Part I, p. 165:
(632½) Birds of N. A., 1874, vol. i, p. 86.

Parinæ.
(115) Review of N. A. Birds, July, 1864, Part I,
p. 77; (632½) Birds of N. A., 1874, vol. i, p.
86.

PAROIDES FLAVICEPS.
(78) P. R. R. Surv., vol. ix, p. 400: (87) Mex.
Bound. Surv., vol. ii, p. 14, pl. xv, fig. 2;
(99) Pr. Acad. Nat. Sci. Phila., 1859, p. 304:
(104) Birds of N. A., 1860, p. 400, pl. liii,
fig. 2.

Parula.
(115) Review of N. A. Birds, Aug., 1864, Part
I, p. 167; Ibid., Nov., 1864, Part I, p. 168:
(632½) Birds of N. A., 1874, vol. i, p. 207.

PARULA AMERICANA.
(78) P. R. R. Surv., vol. ix, p. 238; (115) Review
of N. A. Birds, Aug., 1864, Part I, p. 169:
(632½) Birds of N. A., 1874, vol. i, p. 208, pl.
x, fig. 7 (cuts, pp. 208, 209); Ibid., App.,
p. 504.

Parula gutturalis.
(115) Review of N. A. Birds, Aug., 1864, Part
I, p. 169; Ibid., p. 172.

Parula inornata.
(115) Review of N. A. Birds, Aug., 1864, Part
I, p. 169; Ibid., p. 171.

Parula pitiayumi.
(115) Review of N. A. Birds, Aug., 1864, Part
I, p. 170, 266; Ibid., p. 169.

Parula superciliosa.
(115) Review of N. A. Birds, Aug, 1864, Part
I, p. 169; Ibid., p. 171.

Parus.
(115) Review of N. A. Birds, July, 1864, Part
I, p. 79; (632½) Birds of N. A., 1874, vol. i, p.
93.

PARUS ATRICAPILLUS.
(78) P. R. R. Surv., vol. ix, p. 390; (115) Review
of N. A. Birds, July, 1864, Part I, p. 80;
(632½) Birds of N. A., 1874, vol. i, p. 96, pl.
vii, fig. 1 (cut, p. 95).

PARUS ATRICAPILLUS OCCIDENTALIS.
(632½) Birds of N. A. 1874, vol. i, p. 101, pl. vii,
fig. 3.

PARUS ATRICAPILLUS SEPTENTRIONALIS.
(632½) Birds of N. A., 1874, vol. i, p. 99, pl. vii, fig. 2.

Parus carolinensis.
(78) P. R. R. Surv., vol. ix, p. 392; (115) Review of N. A. Birds, Aug. 1864, Part I, p. 81; (632½) Birds of N. A., 1874, vol. i, p. 102, pl. vii, fig. 4.

Parus Hudsonicus.
(78) P. R. R. Surv., vol. ix, p. 395; (115) Review of N. A. Birds, Aug., 1864, Part I, p. 82; (632½) Birds of N. A., 1874, vol. i, p. 105, pl. vii, fig. 7.

Parus meridionalis.
(78) P. R. R. Surv., vol. ix, p. 392; (115) Review of N.A. Birds, Aug., 1864, Part I, p. 81.

PARUS MONTANUS.
(32) Stansbury's Surv. Salt Lake [App. C], p. 331; (78) P. R. R. Surv., vol. ix, p. 394; (115) Review of N. A. Birds, Aug., 1864, Part I, p. 82; (632½) Birds of N. A., 1874, vol. i, p. 95, pl. vii, fig. 5 (cut, p. 96).

Parus occidentalis.
(78) P. R. R. Surv., vol. ix, p. 391; (115) Review of N. A. Birds, 1864, Part I, p. 81.

PARUS RUFESCENS.
(78) P. R. R. Surv., vol. ix, p. 394; (115) Review of N. A. Birds, Aug., 1864, Part I, p. 83; (632½) Birds of N. A., 1874, vol. i, p. 104, pl. vii, fig. 6; Ibid., App., p. 502.

Parus septentrionalis.
(31) Stansbury's Surv. Salt Lake [App. C], p. 316; (78) P. R. R. Surv., vol. ix, p. 389; (115) Review of N. A. Birds, July, 1864, Part I, p. 79.

Passerculus.
(632½) Birds of N. A., 1874, vol. i, p. 532.

PASSERCULUS ALAUDINUS.
(78) P. R. R. Surv., vol. ix, p. 446; (87) Mex. Bound. Surv., vol. ii, p. 15; (104) Birds of N. A., 1860, p. 446, pl. iv, fig. 1; (632½) Birds of N. A., 1874, App., p. 512; (78) P. R. R. Surv., vol. ix, p. 445.

PASSERCULUS PRINCEPS.
(632½) Birds of N. A., 1874, vol. i, p. 540, pl. xxv, fig. 2; (632½) Birds of N. A., 1874, App., p. 513.

PASSERCULUS ROSTRATUS.
(78) P. R. R. Surv., vol. ix, p. 446; (632½) Birds of N. A., 1874, vol. i, p. 542, pl. xxiv, fig. 12.

PASSERCULUS ROSTRATUS GUTTATUS.
(632½) Birds of N. A., 1874, vol. i, p. 544, pl. xxv, fig. 1.

PASSERCULUS SANDWICHENSIS.
(78) P. R. R. Surv., vol. ix, p. 444; (104) Birds of N. A., 1860, p. 444, pl. xxviii, fig. 2.

PASSERCULUS SAVANNA.
(78) P. R. R. Surv., vol. ix, p. 442; (632½) Birds of N. A., 1874, vol. i, p. 534, pl. xxiv, fig. 8 (cuts, pp. 532, 534).

PASSERCULUS SAVANNA ALAUDINUS.
(632½) Birds of N. A., 1874, vol. i, p. 537, pl. xxiv, fig. 11.

PASSERCULUS SAVANNA ANTHINUS.
(632½) Birds of N. A., 1874, vol. i, p. 539, pl. xxiv, fig. 10.

PASSERCULUS SAVANNA SANDWICHENSIS.
(632½) Birds of N. A., 1874, vol. i, p. 538, pl. xxiv, fig. 9.

Passerella.
(632½) Birds of N. A., 1874, vol. ii, p. 49; Ibid., App., p. 516.

PASSERELLA ILIACA.
(78) P. R. R. Surv., vol. ix, p. 488; (632½) Birds of N. A., 1874, vol. ii, p. 50, pl. xxviii, fig. 7.

Passerella megarhynchus.
(632½) Birds of N. A., 1874, App., p. 516.

Passerella rufina.
(32) Stansbury's Surv. Salt Lake [App. C], p. 331.

PASSERELLA SCHISTACEA.
(78) P. R. R. Surv., vol. ix, p. 490, 925; (104) Birds of N. A., 1860, p. 490, pl. lxix, fig. 3.

PASSERELLA TOWNSENDII.
(78) P. R. R. Surv., vol. ix, p. 489; (632½) Birds of N. A., 1874, vol. ii, p. 53, pl. xxviii, fig. 8 (cuts, pp. 50, 54).

PASSERELLA TOWNSENDI MEGARHYNCHUS.
(632½) Birds of N. A., 1874, vol. ii, p. 57, pl. xxviii, fig. 10 (cut, p. 57).

Passerella Townsendi schistacea.
(632½) Birds of N. A., 1874, vol. ii, p. 56 (cut, p. 56).

Passerella unalaschensis.
(32) Stansbury's Surv. Salt Lake [App. C], p. 331.

Passerellinæ.
(632½) Birds of N. A., 1874, vol. ii, p. 48.

Patagiœnas leucocephala.
(78) P. R. R. Surv., vol. ix, p. 599.

Pediocorys.
(115) Review of N. A. Birds, Aug., 1864, Part I, p. 151; Ibid., p. 157.

Pediœcetes.
(632½) Birds of N. A., 1874, vol. iii, p. 433.

Pediœcetes phasianellus.
(78) P. R. R. Surv., vol. ix, p. 626.

PEDIŒCETES PHASIANELLUS COLUMBIANUS.
(632½) Birds of N. A., 1874, vol. iii, p. 436, pl. lx, fig. 1.

PEDIŒCETES PHASIANELLUS PHASIANELLUS.
(632½) Birds of N. A., 1874, vol. iii, p. 434, pl. lx, fig. 3 (cuts, pp. 433, 444).

PELAGICA VAUXI.
(632½) Birds of N. A., 1874, vol. ii, p. 435, pl. xlv, fig. 8.

Pelecanus erythrorhynchus.
(78) P. R. R. Surv., vol. ix, p. 868; (87) Mex. Bound. Surv., vol. ii, p. 28.

Pelecanus fuscus.
(78) P. R. R. Surv., vol. ix, p. 870.

Pelecanus trachyrrhynchus.
(31) Stansbury's Surv. Salt Lake [App. C], p. 324.

Pelionetta bimaculata.
(78) P. R. R. Surv., vol. ix, p. 808.

Pelionetta perspicillata.
(78) P. R. R. Surv., vol. ix, p. 806.

Pelionetta Trowbridgii.
(78) P. R. R. Surv., vol. ix, p. 806.

Pencœa Lincolnii.
  (31) Stansbury's Surv. Salt Lake [App. C], p. 317.
Pendulinus californianus.
  (32) Stansbury's Surv. Salt Lake [App. C], p. 331.
Penelope policephala.
  (32) Stansbury's Surv. Salt Lake [App. C], p. 334.
Penelopinæ.
  (632½) Birds of N. A., 1874, vol. iii, p. 397.
Perdicidæ.
  (632½) Birds of N. A., 1874, vol. iii, p. 466.
Perisoreus.
  (632½) Birds of N. A., 1874, vol. ii, p. 297.
Perisoreus canadensis.
  (78) P. R. R. Surv., vol. ix, p. 590 ; (94) P. R. R. Surv., vol. x, p. 14.
PERISOREUS CANADENSIS.
  (632½) Birds of N. A., 1874, vol. ii, p. 299, pl. xli, fig. 3; pl. xlii, fig. 4 (cuts, pp. 298, 299).
PERISOREUS CANADENSIS CAPITALIS.
  (632½) Birds of N. A., 1874, vol. ii, p. 302, pl. xli, fig. 4.
Perisoreus canadensis obscurus.
  (632½) Birds of N. A., 1874, vol. ii, p. 302.
Perissoglossa.
  (115) Review of N. A. Birds, Apr., 1865, Part I, p. 180 ; Ibid., p. 181 ; (632½) Birds of N. A., 1874, vol. i, p. 211.
PERISSOGLOSSA CARBONATA.
  (632½) Birds of N. A., 1874, vol. i, p. 214, pl. xii, fig. 3.
PERISSOGLOSSA TIGRINA.
  (115) Review of N. A. Birds, Apr., 1865, Part I, p. 183, fig. 5 ; Ibid., p. 182 ; (632½) Birds of N. A., 1874, vol. i, p. 212, pl. xii, figg. 1, 2 (cut, p. 211).
Petrochelidon.
  (115) Review of N. A. Birds, May, 1865, Part I, pp. 270, 271, 286, 289 ; (632½) Birds of N. A., 1874, vol. i, p. 334.
Petrochelidon fulva.
  (115) Review of N. A. Birds, May, 1865, Part I, p. 291.
PETROCHELIDON LUNIFRONS.
  (115) Review of N. A. Birds, May, 1865, Part I, p. 288 ; (632½) Birds of N. A., 1874, vol. i, p. 334, pl. xvi, fig. 13 (cut, p. 334).
Petrochelidon pœciloma.
  (115) Review of N. A. Birds, May, 1865, Part I, p. 292.
Petrochelidon ruticollaris.
  (115) Review of N. A. Birds, May, 1865, Part I, p. 292.
Petrochelidon Swainsoni.
  (115) Review of N. A. Birds, May, 1865, Part I, p. 290.
Peucæa.
  (632½) Birds of N. A., 1874, vol. ii, p. 37.
PEUCÆA ÆSTIVALIS.
  (78) P. R. R. Surv., vol. ix, p. 484 ; (632½) Birds of N. A., 1874, vol. ii, p. 39, pl. xxviii, fig. 4 (cuts, pp. 37, 39).
Peucæa æstivalis arizonæ.
  (632½) Birds of N. A., 1874, vol. ii, p. 41 ; Ibid., App., p. 515.

Peucæa carpalis.
  (632½) Birds of N. A., 1874, App., p. 515.
PEUCÆA CASSINII.
  (78) P. R. R. Surv., vol. ix, p. 485 ; (87) Mex. Bound. Surv., vol. ii, p. 16 ; (104) Birds of N. A., 1860, p. 485, pl. iv, fig. 2 ; (632½) Birds of N. A., 1874, vol. ii, p. 43, pl. xxviii, fig. 5.
PEUCÆA RUFICEPS.
  (78) P. R. R. Surv., vol. ix, p. 486 ; (632½) Birds of N. A., 1874, vol. ii, p. 42, pl. xxviii, fig. 6.
Phænopepla.
  (632½) Birds of N. A., 1874, vol. i, p. 405 ; (115) Review of N. A. Birds, May, 1866, Part I, p. 415.
PHÆNOPEPLA NITENS.
  (115) Review of N. A. Birds, May, 1866, Part I, p. 416 ; (632½) Birds of N. A., 1874, vol. i, p. 405, pl. xviii, figg. 3, 4 (cuts, p. 406) ; Ibid., App., p. 507.
Phæoprogne.
  (115) Review of N. A. Birds, May, 1865, Part I, pp. 269, 282, 283.
Phæopus borealis.
  (78) P. R. R. Surv., vol. ix, p. 744.
Phæopus hudsonicus.
  (78) P. R. R. Surv., vol. ix, p. 744.
Phtæon flavirostris.
  (78) P. R. R. Surv., vol. ix, p. 855.
PHAINOPEPLA NITENS.
  (78) P. R. R. Surv., vol. ix, pp. 320, 923 ; (87) Mex. Bound. Surv., vol. ii, p. 11 ; (99) Pr. Acad. Nat. Sci. Phila., 1859 (1860), p. 303.
Phalacrocorax carbo.
  (78) P. R. R. Surv., vol. ix, p. 876.
Phalacrocorax cincinnatus.
  (78) P. R. R. Surv., vol. ix, p. 877.
Phalacrocorax dilophus.
  (31) Stansbury's Surv. Salt Lake [App. C], p. 324.
Phalacrocorax penicillatus.
  (32) Stansbury's Surv. Salt Lake [App. C], p. 335.
Phalacrocorax perspicillatus.
  (32) Stansbury's Surv. Salt Lake [App. C], p. 335 ; (78) P. R. R. Surv., vol. ix, p. 877.
Phalaropus fulicarius.
  (78) P. R. R. Surv., vol. ix, p. 707.
Phalaropus hyperboreus.
  (78) P. R. R. Surv., vol. ix, p. 706.
Phalaropus Wilsonii.
  (78) P. R. R. Surv., vol. ix, p. 705.
Phaleris camtschatica.
  (78) P. R. R. Surv., vol. ix, p. 908.
Phaleris cristatellus.
  (78) P. R. R. Surv., vol. ix, p. 906.
Phaleris microceros.
  (78) P. R. R. Surv., vol. ix, p. 908.
Phaleris pusillus.
  (78) P. R. R. Surv., vol. ix, p. 909.
Phaleris tetracula.
  (78) P. R. R. Surv., vol. ix, p. 907.
Pheugopedius.
  (115) Review of N. A. Birds, Aug., 1864, Part I, p. 94 ; Ibid., p. 134.

Picus scalaris lucasanus.
(632½) Birds of N. A., 1874, vol. ii, p. 519.

Picus scapularis.
(32) Stansbury's Surv. Salt Lake [App. C], p. 333.

Picus torquatus.
(31) Stansbury's Surv. Salt Lake [App. C], p. 319.

Picus varius.
(31-a) Stansbury's Surv. Salt Lake [App. C], p. 326.

PICUS VILLOSUS.
(632½) Birds of N. A., 1874, vol. ii, p. 503, pl. xlix, figg. 3, 4, 5 (cut, p. 500).

PICUS VILLOSUS HARRISI.
(632½) Birds of N. A., 1874, vol. ii, p. 507 (cut, p. 502).

Picus villosus major.
(78) P. R. R. Surv., vol. ix, p. 84.

Picus villosus medius.
(78) P. R. R. Surv., vol. ix, p. 84.

Picus villosus minor.
(78) P. R. R. Surv., vol. ix, p. 84.

Pinicola.
(632½) Birds of N. A., 1874, vol. i, p. 452.

Pinicola canadensis.
(78) P. R. R. Surv., vol. ix, p. 410.

PINICOLA ENUCLEATOR.
(632½) Birds of N. A., 1874, vol. i, p. 453, pl. xxi, figg. 1, 2 (cuts, pp. 453, 454); (632½) Birds of N. A., 1874, App., p. 508.

Pipilo.
(632½) Birds of N. A., 1874, vol. ii, p. 104.

PIPILO ABERTII.
(78) P. R. R. Surv., vol. ix, p. 516; (87) Mex. Bound. Surv., vol. ii, p. 18; (104) Birds of N. A., 1860, p. 517, pl. xxx.
(31-a) Stansbury's Surv. Salt Lake [App. C], p. 325; (32) Stansbury's Surv. Salt Lake [App. C], p. 330.
(632½) Birds of N. A., 1874, vol. ii, p. 128, pl. xxxi, fig. 7 (cut, p. 128); (632½) Birds of N. A., 1874, App., p. 517.

Pipilo albigula.
(99) Pr. Acad. Nat. Sci. Phila., 1859, p. 305.

Pipilo arcticus.
(78) P. R. R. Surv., vol. ix, p. 514.

Pipilo chlorura.
(87) Mex. Bound. Surv., vol. ii, p. 18.

PIPILO CHLORURUS.
(632½) Birds of N. A., 1874, vol. ii, p. 131, pl. xxxi, fig. 4 (cuts, p. 132); (78) P. R. R. Surv., vol. ix, p. 519; (632½) Birds of N. A., 1874, App., p. 517.

PIPILO ERYTHROPHTHALMUS.
(78) P. R. R. Surv., vol. ix, p. 512; (632½) Birds of N. A., 1874, vol. ii, p. 109, pl. xxxi, figg. 2, 3 (cuts, pp. 104-9-10-12); Ibid., App., p. 516.

Pipilo erythrophthalmus Alleni.
(632½) Birds of N. A., 1874, vol. ii, p. 112.

Pipilo fusca.
(32) Stansbury's Surv. Salt Lake [App. C], p. 330.

Pipilo fuscus.
(78) P. R. R. Surv., vol. ix, p. 517.

PIPILO FUSCUS ALBIGULA.
(632½) Birds of N. A., 1874, vol. ii, p. 127, pl. xxxi, fig. 11.

PIPILO FUSCUS CRISSALIS.
(632½) Birds of N. A., 1874, vol. ii, p. 122, pl. xxxi, fig. 8 (cut, p. 123).

PIPILO FUSCUS MESOLEUCUS.
(632½) Birds of N. A., 1874, vol. ii, p. 125, pl. xxxi, fig. 10.

PIPILO MACULATUS ARCTICUS.
(632½) Birds of N. A., 1874, vol. ii, p. 119, pl. xxxi, figg. 5, 6.

PIPILO MACULATUS MEGALONYX.
(632½) Birds of N. A., 1874, vol. ii, p. 113, pl. xxxi, fig. 9 (cut, p. 113).

PIPILO MACULATUS OREGONUS.
(632½) Birds of N. A., 1874, vol. ii, p. 116, pl. xxxi, fig. 12 (cut, p. 116).

PIPILO MEGALONYX.
(78) P. R. R. Surv., vol. ix, p. 515; (87) Mex. Bound. Surv., vol. ii, p. 17; (104) Birds of N. A., 1860, p. 515, pl. lxxiii.

PIPILO MESOLEUCUS.
(56) Pr. Acad. Nat. Sci. Phila., 1854, p. 119; (78) P. R. R. Surv., vol. ix, p. 518; (87) Mex. Bound. Surv., vol. ii, p. 18; (104) Birds of N. A., 1860, p. 518, pl. xxix; (632½) Birds of N. A., 1874, App., p. 516.

Pipilo oregona.
(32) Stansbury's Surv. Salt Lake [App. C], p. 330.

Pipilo oregonus
(78) P. R. R. Surv., vol. ix, p. 513.

Planesticus.
(115) Review of N. A. Birds, June, 1864, Part I, p. 12; Ibid., p. 23.

Planesticus migratorius.
(78) P. R. R. Surv., vol. ix, p. 218.

Platalea ajaja.
(78) P. R. R. Surv., vol. ix, p. 686; (87) Mex. Bound. Surv., vol. ii, p. 25.

Platycichla.
(115) Review of N. A. Birds, June, 1864, Part I, p. 4; Ibid., July, 1864, Part I, p. 32; Ibid., June, 1866, Part I, p. 418, 436.

Platycichla brevipes.
(115) Review of N. A. Birds, June, 1864, Part I, p. 32; Ibid., June, 1866, Part I, p. 436.

Plectrophanes.
(632½) Birds of N. A., 1874, vol. i, p. 510.

PLECTROPHANES LAPPONICUS.
(78) P. R. R. Surv., vol. ix, p. 433; (632½) Birds of N. A., 1874, vol. i, p. 515, pl. xxiv, fig. 7 (cut, p. 515).

PLECTROPHANES MACCOWNII.
(32) Stansbury's Surv. Salt Lake [App. C], p. 331; (78) P. R. R. Surv., vol. ix, p. 437; (632½) Birds of N. A., 1874, vol. i, p. 523, pl. xxiv, fig. 1.

PLECTROPHANES MELANOMUS.
(78) P. R. R. Surv., vol. ix, p. 436; (104) Birds of N. A., 1860, p. 432, pl. lxxiv, fig. 2.

Plectrophanes (Plectrophanes) nivalis.
(78) P. R. R. Surv., vol. ix, p. 432.

PLECTROPHANES ORNATUS.
 (78) P. R. R. Surv., vol. ix, p. 435; (632½) Birds of N. A., 1874, vol. i, p. 520, pl. xxiv, fig. 3; Ibid., App., p. 512.
PLECTROPHANES ORNATUS MELANOMUS.
 (632½) Birds of N. A., 1874, vol. i, p. 521, pl. xxiv, fig. 6.
PLECTROPHANES PICTUS.
 (78) P. R. R. Surv., vol. ix, p. 434; (632½) Birds of N. A., 1874, vol. i, p. 518, pl. xxiv, figg. 4, 5.
Plotus anhinga.
 (78) P. R. R. Surv., vol. ix, p. 883.
Pluvialis virginiæus.
 (6) Lit. Rec. and Journ. Linnæan Ass. Penn. Col., Oct., 1845, p. 254.
Podiceps auritus.
 (78) P. R. R. Surv., vol. ix, p. 897.
PODICEPS CALIFORNICUS.
 (78) P. R. R. Surv., vol. ix, p. 896; (104) Birds of N. A., 1860, p. 896, pl. viii, young.
PODICEPS CLARKII.
 (78) P. R. R. Surv., vol. ix, p. 895; (87) Mex. Bound. Surv., vol. ii, p. 28; (104) Birds of N. A., 1860, p. 895, pl. c.
Podiceps cornutus.
 (78) P. R. R. Surv., vol. ix, p. 895.
Podiceps cristatus.
 (78) P. R. R. Surv., vol. ix, p. 893.
PODICEPS DOMINICUS.
 (87) Mex. Bound. Surv., vol. ii, p. 28; (104) Birds of N. A., 1860, p. 897, pl. xcix, fig. 1.
Podiceps griseigena.
 (78) P. R. R. Surv., vol. ix, p. 892.
PODICEPS OCCIDENTALIS.
 (78) P. R. R. Surv., vol. ix, p. 894; (104) Birds of N. A., 1860, p. 894, pl. xxxviii.
PODILYMBUS PODICEPS.
 (104) Birds of N. A., 1860, p. 898, pl. ix, young.
Podylimbus podiceps.
 (78) P. R. R. Surv., vol. ix, p. 898.
Pœcilopternis borealis.
 (78) P. R. R. Surv., vol. ix, p. 25.
Pœcilopternis elegans.
 (78) P. R. R. Surv., vol. ix, p. 28.
Pœcilopternis exypterus.
 (78) P. R. R. Surv., vol. ix, p. 30.
Pœcilopternis lineatus.
 (78) P. R. R. Surv., vol. ix, p. 28.
Pœcilopternis montanus.
 (78) P. R. R. Surv., vol. ix, p. 26.
Pœcilopternis pennsylvanicus.
 (78) P. R. R. Surv., vol. ix, p. 29.
Polioptila.
 (115) Review of N. A. Birds, July, 1864, Part I, p. 67; (632½) Birds of N. A., 1874, vol. i, p. 77.
Polioptila albiloris.
 (115) Review of N. A. Birds, July, 1864, Part I, p. 70.
Polioptila bilineata.
 (115) Review of N. A. Birds, July, 1864, Part I, p. 72.
Polioptila Buffoni.
 (115) Review of N. A. Birds, July, 1864, Part I, p. 70.

POLIOPTILA CŒRULEA.
 (78) P. R. R. Surv., vol. ix, p. 380; (87) Mex. Bound. Surv., vol. ii, p. 13; (115) Review of N. A. Birds, July, 1864, Part I, p. 74; (632½) Birds of N. A., 1874, vol. i, p. 78, pl. vi, fig. 5 (cuts, pp. 77, 79); Ibid., App., p. 501.
Polioptila dumicola.
 (115) Review of N. A. Birds, July, 1864, Part I, p. 73.
Polioptila leucogastra.
 (115) Review of N. A. Birds, July, 1864, Part I, p. 69.
POLIOPTILA MELANURA.
 (78) P. R. R. Surv., vol. ix, p. 382; (99) Pr. Acad. Nat. Sci. Phila., 1859, p. 304; (115) Review of N. A. Birds, July, 1864, Part I, p. 68; (632½) Birds of N. A., 1874, vol. i, p. 81, pl. vi, fig. 7; Ibid., App., p. 502.
Polioptila nigriceps.
 (115) Review of N. A. Birds, July, 1864, Part I, p. 69.
POLIOPTILA PLUMBEA.
 (78) P. R. R. Surv., vol. ix, p. 382; (87) Mex. Bound. Surv., vol. ii, p. 14; (104) Birds of N. A., 1860, p. 382, pl. xxxiii, fig. 1; (115) Review of N. A. Birds, July, 1864, Part I, p. 74; (632½) Birds of N. A., 1874, vol. i, p. 80, pl. vi, fig. 6.
Polioptila superciliaris.
 (115) Review of N. A. Birds, July, 1864, Part I, p. 71.
Polioptilinæ.
 (115) Review of N. A. Birds, July, 1864, Part I, p. 65; (632½) Birds of N. A., 1874, vol. i, p. 77.
Polyborus.
 (632½) Birds of N. A., 1874, vol. iii, p. 176.
Polyborus tharus.
 (78) P. R. R. Surv., vol. ix, p. 45; (87) Mex. Bound. Surv., vol. ii, p. 4.
POLYBORUS THARUS AUDUBONI.
 (632½) Birds of N. A., 1874, vol. iii, p. 178 (cuts, pp. 176, 179).
Polysticta Stelleri.
 (78) P. R. R. Surv., vol. ix, p. 801.
Poœcœtes.
 (632½) Birds of N. A., 1874, vol. i, p. 544.
POŒCETES GRAMINEUS.
 (78) P. R. R. Surv., vol. ix, p. 447; (87) Mex. Bound. Surv., vol. ii, p. 15; (632½) Birds of N. A., 1874, vol. i, p. 545 (cuts, pp. 545, 546), vol. ii, pl. xxix, fig. 1.
Poospiza.
 (632½) Birds of N. A., 1874, vol. i, p. 589.
POOSPIZA BELLI.
 (78) P. R. R. Surv., vol. ix, p. 470; (632½) Birds of N. A., 1874, vol. i, p. 593, pl. xxvi, fig. 9 (cut, p. 595); (632½) Ibid., App., p. 514.
Poospiza Belli nevadensis.
 (632½) Birds of N. A., 1874, vol. i, p. 594.
POOSPIZA BILINEATA.
 (78) P. R. R. Surv., vol. ix, p. 470; (87) Mex. Bound. Surv., vol. ii, p. 15; (632½) Birds of N. A., 1874, vol. i, p. 590, pl. xxvi, fig. 8 (cuts, pp. 589, 590).

22 BD

Porphyrula martinica.
(78) P. R. R. Surv., vol. ix, p. 753.
Porzana carolina.
(6) Lit. Rec. and Journ. Linnæan Assoc. Penn. Col., Oct., 1845, p. 255.
Porzana (Porzana) carolina.
(78) P. R. R. Surv., vol. ix, p. 749.
Porzana carolina.
(87) Mex. Bound. Surv., vol.' ii, p. 26.
Porzana jamaicensis.
(6) Lit. Rec. and Journ. Linnæan Assoc. Penn. Col., Oct., 1845, p. 257; (78) P. R. R. Surv., vol. ix, p. 749.
Porzana noveboracensis.
(6) Lit. Rec. and Journ. Linnæan Assoc. Penn. Col., Oct., 1845, p. 255; (78) P. R. R. Surv., vol. ix, p. 750.
Procellaria gigantea.
(78) P. R. R. Surv., vol. ix, p. 825.
Procellaria glacialis.
(78) P. R. R. Surv., vol. ix, p. 825.
Procellaria meridionalis.
(32) Stansbury's Surv. Salt Lake [App. C], p. 335; (78) P. R. R. Surv., vol. ix, p. 827.
Procellaria pacifica.
(78) P. R. R. Surv., vol. ix, p. 826.
Procellaria pelagica.
(78) P. R..R. Surv., vol. ix, p. 831.
Procellaria tenuirostris.
(78) P. R. R. Surv., vol. ix, p. 826.
Progne.
(115) Review of N. A. Birds, May, 1865, Part I, pp. 269, 271, 272; (632½) Birds of N. A., 1874, vol. i, p. 327.
Progne chalybea.
(115) Review of N. A. Birds, May, 1865, Part I, p. 282.
Progne concolor.
(115) Review of N. A. Birds, May, 1865, Part I, p. 278.
Progne cryptoleuca.
(115) Review of N. A. Birds, May, 1865, Part I, p. 277.
Progne domestica.
(115) Review of N. A. Birds, May, 1865, Part I, p. 282.
Progne dominicensis.
(115) Review of N. A. Birds, May, 1865, Part I, p. 279.
Progne elegans.
(115) Review of N. A. Birds, May, 1865, Part I, p. 275.
Progne furcata.
(115) Review of N. A. Birds, May, 1865, Part I, p. 278.
Progne fusca.
(115) Review of N. A. Birds, May, 1865, Part I, p. 282.
Progne leucogaster.
(115) Review of N. A. Birds, May, 1865, Part I, p. 280.
Progne purpurea.
(78) P. R. R. Surv., vol. ix, p. 314; (87) Mex. Bound. Surv., vol. ii, p. 11; (99) Pr. Acad. Nat. Sci. Phila., 1859 (1860), p. 303.

PROGNE SUBIS.
(115) Review of N. A. Birds, May, 1865, Part I, p. 275; (632½) Birds of N. A., 1874, vol. i, p. 329, pl. xvi, figg. 7, 10 (cuts, pp. 329, 330).
Progne subis cryptoleuca.
(632½) Birds of N. A., 1874, vol. i, p. 332.
Progne tapera.
(115) Review of N. A. Birds, May, 1865, Part I, p. 282.
Protonotaria.
(115) Review of N. A. Birds, Aug., 1864, Part I, p. 173; (632½) Birds of N. A., 1874, vol. i, p. 183.
PROTONOTARIA CITREA.
(78) P. R. R. Surv., vol. ix, p. 239; (115) Review of N. A. Birds, August, 1864, Part I, p. 173; (632½) Birds of N. A., 1874, vol. i, p. 184, pl. x, fig. 8 (cuts, pp. 183, 184).
Psaltria plumbea.
(56) Pr. Acad. Nat. Sci. Phila., 1854, p. 118.
Psaltriparus.
(115) Review of N. A. Birds, Aug., 1864, Part I, p. 84; (632½) Birds of N. A., 1874, vol. i, p. 107.
PSALTRIPARUS MELANOTIS.
(78) P. R. R. Surv., vol. ix, p. 396; (87) Mex. Bound. Surv., vol. ii, p. 14, pl. xv, fig. 3; (104) Birds of N. A., 1860, p. 396, pl. liii, fig. 3; (115) Review of N. A. Birds, Aug., 1864, Part I, p. 84; (632½) Birds of N. A., 1874, vol. i, p. 108, pl. vii, fig. 8 (cut, p. 108).
Psaltriparus minimus.
(78) P. R. R. Surv., vol. ix, p. 397; (115) Review of N. A. Birds, Aug., 1864, Part I, p. 84.
PSALTRIPARUS MINIMUS MINIMUS.
(632½) Birds of N. A., 1874, vol. i, p. 109, pl. vii, fig. 9 (cut, p. 109).
PSALTRIPARUS MINIMUS PLUMBEUS.
(632½) Birds of N. A., 1874, vol. i, p. 110, pl. vii, fig. 10.
PSALTRIPARUS PLUMBEUS.
(78) P. R. R. Surv., vol. ix, p. 398; (104) Birds of N. A., 1860, p. 398, pl. xxxiii, fig. 2; (115) Review of N. A. Birds, Aug., 1864, Part I, p. 84.
Psarocolius auricollis.
(32) Stansbury's Surv. Salt Lake [App. C], p. 331.
Pseudogryphus.
(632½) Birds of N. A., 1874, vol. iii, p. 338.
PSEUDOGRYPHUS CALIFORNIANUS.
632½) Birds of N. A., 1874, vol. iii, p. 338 (cuts, pp. 338-340, 341-355, 356).
Psilorhinus.
(632½) Birds of N. A., 1874, vol. ii, p. 303.
PSILORHINUS MORIO.
(78) P. R. R. Surv., vol. ix, p. 592; (87) Mex. Bound. Surv., vol. ii, p. 21; (632½) Birds of N. A., 1874, vol. ii, p. 304, pl. xlii, fig. 2 (cuts, pp. 303, 304); (104) Birds of N. A., 1860, p. 592, pl. lxviii, figg. 1, 2.
Psittacidæ.
(632½) Birds of N. A., 1874, vol. ii, p. 585.

Pterocyanea Rafflesii.
  (31) Stansbury's Surv. Salt Lake [App. C], p. 322.
Ptilogonatinæ.
  (115) Review of N. A. Birds, May, 1866, Part I, pp. 401, 408; (632½) Birds of N. A., 1874, vol. i, p. 404.
Ptilogonys.
  (115) Review of N. A. Birds, May, 1866, Part I, p. 410.
Ptilogonys caudatus.
  (115) Review of N. A. Birds, May, 1866, Part I, p. 413.
Ptilogonys cinereus.
  (115) Review of N. A. Birds, May, 1866, Part I, p. 410.
Ptychorhamphus aleuticus.
  (32) Stansbury's Surv. Salt Lake [App. C], p. 335; (78) P. R. R. Surv., vol. ix, p. 910.
Puffinus anglorum.
  (78) P. R. R. Surv., vol. ix, p. 834.
Puffinus cinereus.
  (78) P. R. R. Surv., vol. ix, p. 835.
Puffinus fuliginosus.
  (78) P. R. R. Surv., vol. ix, p. 834.
Puffinus major.
  (78) P. R. R. Surv., vol. ix, p. 833.
Puffinus obscurus.
  (78) P. R. R. Surv., vol. ix, p. 835.
PUFFINUS TENUIROSTRIS.
  (129) Chicago Acad. Sci., 1869, p. 322, pl. xxxiv, fig. 2.
Pygochelidon.
  (115) Review of N. A. Birds, May, 1865, Part I, pp. 270, 305, 308.
Pyranga.
  (632½) Birds of N. A., 1874, vol. i, p. 432.
Pyranga æstiva.
  (78) P. R. R. Surv., vol. ix, p. 301; (87) Mex. Bound. Surv., vol. ii, p. 11.
PYRANGA ÆSTIVA ÆSTIVA.
  (632½) Birds of N. A., 1874, vol. i, p. 441, pl. xx, figg. 5, 6 (cut, p. 442).
PYRANGA ÆSTIVA COOPERI.
  (632½) Birds of N. A., 1874, vol. i, p. 444, pl. xx, figg. 1, 2.
PYRANGA HEPATICA.
  (78) P. R. R. Surv., vol. ix, p. 302; (104) Birds of N. A., 1860, p. 302, pl. xxxi; (632½) Birds of N. A., 1874, vol. i, p. 440, pl. xx, figg. 9, 10; Ibid., App., p. 508.
PYRANGA LUDOVICIANA.
  (78) P. R. R. Surv., vol. ix, p. 303; (632½) Birds of N. A., 1874, vol. i, p. 437, pl. xx, figg. 3, 4 (cut, p. 435).
PYRANGA RUBRA.
  (78) P. R. R. Surv., vol. ix, p. 300; (632½) Birds of N. A., 1874, vol. i, p. 435, pl. xx, figg. 7, 8, (cut, p. 432).
Pyrgita.
  (632½) Birds of N. A., 1874, vol. i, p. 525.
PYRGITA DOMESTICA.
  (632½) Birds of N. A., 1874, vol. i, p. 525, pl. xxiii, fig. 12 (cuts, pp. 525, 526).
Pyrgitinæ.
  (632½) Birds of N. A., 1874, vol. i, p. 524.

Pyrocephalus.
  (632½) Birds of N. A., 1874, vol. ii, p. 386.
Pyrocephalus mexicanus.
  (632½) Birds of N. A., 1874, App., p. 520.
Pyrocephalus rubineus.
  (32) Stansbury's Surv. Salt Lake [App. C], p. 329; (78) P. R. R. Surv., vol. ix, p. 201; (87) Mex. Bound. Surv., vol. ii, p. 9.
PYROCEPHALUS RUBINEUS MEXICANUS.
  (632½) Birds of N. A., 1874, vol. ii, p. 387, pl. xliv, fig. 5 (cuts, pp. 386, 388).
Pyrrhula.
  (632½) Birds of N. A., 1874, vol. i, p. 456.
PYRRHULA CASSINI.
  (632½) Birds of N. A., 1874, vol. i, p. 457, pl. xxiii, fig. 11 (cuts, p. 457); Ibid., App., p. 508.
PYRRHULA COCCINEA.
  (129) Chicago Acad. Sci., 1869, p. 316., pl. xxix, fig. 1.
Pyrrhula inornata.
  (32) Stansbury's Surv. Salt Lake [App. C], p. 331.
Pyrrhuloxia.
  (632½) Birds of N. A., 1874, vol. ii, p. 95.
Pyrrhuloxia sinuata.
  (78) P. R. R. Surv., vol. ix, p. 508; (87) Mex. Bound. Surv., vol. ii, p. 17; (99) Pr. Acad. Nat. Sci. Phila., 1859, p. 304; (632½) Birds of N. A., 1874, vol. ii, p. 95, pl. xxx, fig. 3 (cuts, pp. 95, 96).
Querquedula carolinensis.
  (31) Stansbury's Surv. Salt Lake [App. C], p. 322.
Querquedul cyanoptera.
  (78) P. R. R. Surv., vol. ix, p. 780; (87) Mex. Bound. Surv., vol. ii, p. 26.
Querquedula discors.
  (78) P. R. R. Surv., vol. ix, p. 779; (87) Mex. Bound. Surv., vol. ii, p. 26.
Quiscalinæ.
  (632½) Birds of N. A., 1874, vol. ii, p. 202.
Quiscalus.
  (632½) Birds of N. A., 1874, vol. ii, p. 212.
Quiscalus baritus.
  (78) P. R. R. Surv., vol. ix, p. 556; (104) Birds of N. A., 1860, p. 556, pl. xxxii.
Quiscalus macrourus.
  (32) Stansbury's Surv. Salt Lake [App. C], p. 331; (78) P. R. R. Surv., vol. ix, p. 553.
QUISCALUS MACROURA.
  (87) Mex. Bound. Surv., vol. ii, p. 20, pl. xx; (104) Birds of N. A., 1860, p. 553, pl. lviii.
QUISCALUS MAJOR.
  (78) P. R. R. Surv., vol. ix, p. 555; (87) Mex. Bound. Surv., vol. ii, p. 20; (632½) Birds of N. A., 1874, vol. ii, p. 222, pl. xxxvi, figg. 3, 4.
QUISCALUS MAJOR MACRURUS.
  (632½) Birds of N. A., 1874, vol. ii, p. 225, pl. xxxvi, figg. 1, 2.
QUISCALUS PURPUREUS.
  (632½) Birds of N. A., 1874, vol. ii, p. 214, pl. xxxvii, fig. 1 (cuts, pp. 212, 215).
QUISCALUS PURPUREUS ÆNEUS.
  (632½) Birds of N. A., 1874, vol. ii, p. 218 (cut, p. 218).

QUISCALUS PURPUREUS AGLÆUS.
(632½) Birds of N. A., 1874, vol. ii, p. 221, pl. xxxvii, fig. 2 (cut, p. 221).

Quiscalus versicolor.
(78) P. R. R. Surv., vol. ix, p. 555.

Rallus crepitans.
(78) P. R. R. Surv., vol. ix, p. 747.

Rallus elegans.
(78) P. R. R. Surv., vol. ix, p. 746.

Rallus virginianus.
(78) P. R. R. Surv., vol. ix, p. 748; (87) Mex. Bound. Surv., vol. ii, p. 26.

Ramphocinclus.
(115) Review of N. A. Birds, July, 1864, Part I, pp. 39, 41.

Ramphocinclus brachyurus.
(115) Review of N. A. Birds, July, 1864, Part I, p. 41.

Recurvirostra americana.
(31) Stansbury's Surv. Salt Lake [App. C], p. 320; (78) P. R. R. Surv., vol. ix, p. 703; (87) Mex. Bound. Surv., vol. ii, p. 25.

Recurvirostra occidentalis.
(31-a) Stansbury's Surv. Salt Lake [App. C], p. 326; (32) Stansbury's Surv. Salt Lake [App. C], p. 334.

Regulinæ.
(115) Review of N. A. Birds, July, 1864, Part I, p. 65; (632½) Birds of N. A., 1874, vol. i, p. 72.

Regulus.
(115) Review of N. A. Birds, July, 1864, Part I, p. 65; (632½) Birds of N. A., 1874, vol. i, p. 72.

REGULUS CALENDULA.
(78) P. R. R. Surv., vol. ix, p. 226; (87) Mex. Bound. Surv., vol. ii, p. 9; (115) Review of N. A. Birds, July, 1864, Part I, p. 66; (632½) Birds of N. A., 1874, vol. i, p. 75, pl. v, fig. 6; Ibid., App., p. 501 (cut, p. 501).

REGULUS CUVIERI.
(78) P. R. R. Surv., vol. ix, p. 228; (115) Review of N. A. Birds, July, 1864, Part I, p. 66; (632½) Birds of N. A., 1874, vol. i, p. 75, pl. v, fig. 7.

REGULUS SATRAPA.
(78) P. R. R. Surv., vol. ix, p. 227; (115) Review of N. A. Birds, July, 1864, Part I, p. 65; (632½) Birds of N. A., 1874, vol. i, p. 73, pl, v, fig. 8 (cuts, pp. 72, 73).

Rhamphocinclus.
(115) Review of N. A. Birds, June, 1864, Part I, p. 4.

Rhamphopis flammigerus.
(32) Stansbury's Surv. Salt Lake [App. C], p. 330.

Rhinogryphus.
(632½) Birds of N. A.; 174, vol. iii, p. 343.

RHINOGRYPHUS AURA.
(632½) Birds of N. A., 1874, vol. iii, p. 344 (cuts, pp. 343, 346, 355-356.)

Rhodinocincla.
(115) Review of N. A. Birds, Aug., 1864, Part I, p. 91.

Rhodinocichla rosea.
(115) Review of N. A. Birds, Aug., 7, 1864, Part I, p. 91.

Rhodostethia rosea.
(78) P. R. R. Surv., vol. ix, p. 857.

Rhyacophilus solitarius.
(78) P. R. R. Surv., vol. ix, p. 733.

RHYNCHOFALCO FEMORALIS.
(632½) Birds of N. A., 1874, vol. iii, p. 155 (cuts, pp. 154, 155, 157).

Rhynchophanes Maccownii.
(78) P. R. R. Surv., vol. ix, p. 437.

Rhynchops nigra.
(78) P. R. R. Surv., vol. ix, p. 866; (87) Mex. Bound. Surv., vol. ii, p. 28.

Rhynchopsitta pachyrhyncha.
(78) P. R. R. Surv., vol. ix, p. 66; (87) Mex. Bound. Surv., vol. ii, p. 5.

Rissa brevirostris.
(78) P. R. R. Surv., vol. ix, p. 855.

Rissa nivea.
(78) P. R. R. Surv., vol. ix, p. 855.

Rissa septentrionalis.
(78) P. R. R. Surv., vol. ix, p. 854.

Rissa tridactyla.
(78) P. R. R. Surv., vol ix. p. 854.

Rostrhamus.
(632½) Birds of N. A., 1874, vol. iii, p. 207.

ROSTHRAMUS SOCIABILIS.
(32) Stansbury's Surv. Salt Lake [App. C], p. 327; (78) P. R. R. Surv., vol. ix, p. 38; (104) Birds of N. A., 1860, p. 38, pl. lxv, figg. 1, 2, adult and young,

ROSTRHAMUS SOCIABILIS PLUMBEUS.
(632½) Birds of N. A., 1874, vol. iii, p. 209 (cuts, pp. 208, 211).

Sagmatorrhina labradoria.
(78) P. R. R. Surv., vol. ix, p. 904.

Salpinctes.
(115) Review of N. A. Birds, Aug., 1864, Part I, pp. 94, 109; (632½) Birds of N. A., 1874, vol. i, p. 134.

SALPINCTES OBSOLETUS.
(78) P.·R. R. Surv., vol. ix, p. 357; (87) Mex. Bound. Surv., vol. ii, p. 13; (115) Review of N. A. Birds, Aug., 1864, Part I, p. 110; (632½) Birds of N. A., 1874, vol. i, p. 135, pl. viii, fig. 3 (cuts, pp. 135, 136); Ibid., App., p. 503.

Saltator rufiventris.
(32) Stansbury's Surv. Salt Lake [App. C], p. 330.

Saurophagus Bairdii.
(32) Stansbury's Surv. Salt Lake [App. C], p. 329.

Saurophagus sulphuratus.
(32) Stansbury's Surv. Salt Lake [App. C], p. 329.

Saxicola.
(115) Review of N. A. Birds, July, 1864, Part I, p. 61; (632½) Birds of N. A., 1874, vol. i, p. 59.

SAXICOLA ŒNANTHE.
(78) P. R. R. Surv., vol. ix, p. 220; (115) Review of N. A. Birds, July, 1864, Part I, p. 61; (632½) Birds of N. A., 1874, vol. i, p. 60, pl. v, fig. 6 (cuts, pp. 59, 60); (632½) Ibid., App., p. 501.

Saxicola œnanthoides.
(32) Stansbury's Surv. Salt Lake [App.C], p.. 329.

Setophaga aurantiaca.
(115) Review ot N. A. Birds, May, 1865, Part I,
p. 261.
Setophaga Belli.
(32) Stansbury's Surv. Salt Lake [App. C], p.
329.
Setophaga brunneiceps.
(115) Review of N. A. Birds, May, 1865, Part
I, p. 258.
Setophaga castaneo-capilla.
(115) Review of N. A. Birds, May, 1865, Part I,
p. 259.
Septophaga flammea.
(115) Review of N. A. Birds, May, 1865, Part
I, p. 259.
Setophaga lachrymosa.
(115) Review of N. A. Birds, May, 1865, Part
I, p. 263.
Setophaga melanocephala.
(115) Review of N. A. Birds, May, 1865, Part
I, p. 258.
SETOPHAGA MINIATA.
(78) P. R. R. Surv., vol. ix, p. 299; (104) Birds
of N. A., 1860, p. 299, pl.lxxviii, fig. 1; (115)
Review of N. A. Birds, May, 1865, Part I,
p. 259.
Setophaga multicolor.
(115) Review of N. A. Birds, May, 1865, Part
I, p. 256.
Setophaga ornata.
(115) Review of N. A. Birds, May, 1865, Part I,
p. 258.
SETOPHAGA PICTA.
(32) Stansbury's Surv. Salt Lake [App. C],
p. 329; (78) P. R. R. Surv., vol. ix, p. 298;
(87) Mex. Bound. Surv., vol. ii, p. 11; (104)
Birds of N. A., 1860, p. 298, pl. lxxvii, fig.
2; (115) Review of N. A. Birds, May, 1865,
Part I, p. 256; (632½) Birds of N. A., 1874,
App., p. 507.
Setophaga rubra.
(32) Stansbury's Surv. Salt Lake [App. C], p.
329.
Setophaga rubifrons.
(32) Stansbury's Surv.Salt Lake[App.C],p.329.
Setophaga ruficoronata.
(115) Review of N. A. Birds, May, 1865, Part
I, p. 258.
SETOPHAGA RUTICILLA.
(78) P. R. R. Surv., vol. ix, p. 297; (115) Review of N. A. Birds, May, 1865, Part I, p.
256; (632½) Birds of N. A., 1874, vol. i, p.
322, pl, xvi, fig, 1 (cuts, pp. 322, 323).
Setophaga torquata.
(115) Review of N. A. Birds, May, 1865, Part I,
p. 261.
Setophaga verticalis.
(115) Review of N. A. Birds, May, 1865, Part I,
p. 258.
Setophaga vulnerata.
(32) Stansbury's Surv. Salt Lake [App. C|,
p. 329.
Setophaginæ.
(115) Review of N. A. Birds, Aug., 1864, Part
I, p, 167; Ibid., Apr., 1865, Part I, p. 235;
(632½) Birds of N. A., 1874, vol. i, p. 311.

Sialia.
(115) Review of N. A. Birds, July, 1864, Part I,
p. 62; (632½) Birds of N. A , 1874, vol. i, p. 62.
SIALIA ARCTICA.
(78) P. R. R. Surv., vol. ix., p. 224; (87) Mex.
Bound. Surv., vol. ii, p. 9; (94) P. R. R.
Surv., vol. x, p. 13, pl. xxxv; (115) Review
of N. A. Birds, July, 1864, Part I, p. 64;
(632½) Birds of N. A., 1874, vol. i, p. 67, pl.
v, fig. 4.
Sialia azurea.
(115) Review of N. A. Birds, July, 1864, Part I,
p. 62.
Sialia macroptera. •
(31) Stansbury's Surv. Salt Lake [App. C], p.
314; (32) Stansbury's Surv. Salt Lake
[App. C], p. 328.
SIALIA MEXICANA.
(78) P. R. R. Surv., vol. ix, p. 223; (87) Mex.
Bound. Surv., vol. ii, p. 9; (115) Review of
N. A. Birds, July, 1864, Part I, p. 63; (632½)
Birds of N. A., 1874, vol i, p. 65, pl. v, fig. 2;
Ibid., App., p. 501.
SIALIA SIALIS.
(78) P. R. R. Surv., vol. ix, p. 222; (115) Review of N. A. Birds, July, 1864, Part I,p.
62; (632½) Birds of N. A., 1874, vol. i, p. 62,
pl. v, fig. 3 (cuts, pp. 62, 63).
SIMORHYNCHUS CASSINI.
(129) Chicago Acad. Sci., 1869, p. 324, pl. xxxi,
fig. 2.
Simorhynchus cristatellus.
(78) P. R. R. Surv., vol. ix, p. 906.
Sitta.
(115) Review of N. A. Birds, Aug., 1864, Part
I, p. 86; (632½) Birds of N. A., 1874, vol. i,
p. 114.
Sitta aculeata.
(104) Birds of N. A., 1860, p. 375, pl. xxxiii,
fig. 3; (115) Review of N. A. Birds, Aug.,
1864, Part I, p. 86.
SITTA CANADENSIS.
(78) P. R. R. Surv., vol. ix, p. 376; (115) Review
of N. A. Birds, Aug., 1864, Part I, p. 87;
(632½) Birds of N. A., 1874, vol. i, p. 118, pl.
viii, figg. 7, 8.
SITTA CAROLINENSIS.
(78) P. R. R. Surv., vol. ix, p. 374; (104) Birds of
N. A., 1860, p. 374, pl. xxxiii, fig. 4; (115) Review of N. A. Birds, Aug.,1864, Part I, p.86.
SITTA CAROLINENSIS ACULEATA.
(632½) Birds of N. A., 1874, vol. i, p. 117 (cut, p.
115).
SITTA CAROLINENSIS CAROLINENSIS.
(632½) Birds of N. A., 1874, vol. i, p. 114, pl. viii,
figg, 1, 2 (cut, p. 114).
SITTA PUSILLA.
(78) P. R. R. Surv., vol. ix, p. 377; (115) Review
N. A. Birds, Aug., 1864, Part I, p. 88; (632½)
Birds of N. A., 1874, vol. i, p. 122, pl. viii, fig.
9; Ibid., App., p. 502.
SITTA PYGMÆA.
(78) P. R. R. Surv., vol. ix, p. 378; (115) Review
of N. A. Birds, Aug., 1864, Part I, p. 88;
(632½) Birds of N. A., 1874, vol. i, p. 120, pl.
viii, fig. 10; Ibid., App., p. 502.

Sittacinæ.
> (632½) Birds of N. A., 1874, vol. ii, p. 585.

Sittinæ.
> (115) Review of N. A. Birds, July, 1864, Part I, p. 77; *Ibid.*, Aug., 1864, Part I, p. 86; (632½) Birds of N. A., 1874, vol. i, p. 113.

Somateria mollissima.
> (78) P. R. R. Surv., vol. ix, p. 809.

Somateria nigra.
> (78) P. R. R. Surv., vol. ix, p. 810.

Somateria spectabilis.
> (78) P. R. R. Surv., vol. ix, p. 810.

Spatula clypeata.
> (78) P. R. R. Surv., vol. ix, p. 781; (87) Mex. Bound. Surv., vol. ii, p. 27.

Speotyto.
> (632½) Birds of N. A., 1874, vol. iii, p. 88.

SPEOTYTO CUNICULARIA HYPOGÆA.
> (632½) Birds of N. A., 1874, vol. iii, p. 90 (cuts, pp. 89, 93–98, 99, 100, 101).

Spermophila.
> (632½) Birds of N. A., 1874, vol. ii, p. 90.

Spermophila albogularis.
> (32) Stansbury's Surv. Salt Lake [App. C], p. 330.

SPERMOPHILA BADIIVENTRIS.
> (129) Chicago Acad. Sci., 1869, p. 319, pl. xxviii, fig. 3.

SPERMOPHILA MORELETII.
> (78) P. R. R. Surv., vol. ix, p. 506; (87) Mex. Bound. Surv., vol. ii, p. 17, pl. xvi, figg. 2, 3; (104) Birds of N. A., 1860, p. 506, pl. liv, figg. 2, 3; (632½) Birds of N. A., 1874, vol. ii, p. 91, pl. xxix, fig. 17 (cuts, pp. 90, 91).

Sphyrapicus.
> (632½) Birds of N. A., 1874, vol. ii, p. 535.

SPHYRAPICUS NUCHALIS.
> (78) P. R. R. Surv., vol. ix, pp. 103, 921; (104) Birds of N. A., 1860, pp. 103, 921, pl. xxxv, figg. 1, 2.

Sphyrapicus ruber.
> (78) P. R. R. Surv., vol. ix, p. 104.

SPHYRAPICUS THYROIDEUS.
> (78) P. R. R. Surv., vol. ix, p. 106; (632½) Birds of N. A., 1874, vol. ii, p. 547, pl. lvi, fig. 6.

Sphyrapicus varius.
> (78) P. R. R. Surv., vol. ix, p. 103; (632½) Birds of N. A., 1874, App., p. 521.

SPHYRAPICUS VARIUS NUCHALIS.
> (632½) Birds of N. A., 1874, vol. ii, p. 542, pl. li, figg. 3, 4 (cut, p. 535).

SPHYRAPICUS VARIUS RUBER.
> (632½) Birds of N. A., 1874, vol. ii, p. 544, pl. li, fig. 6.

SPHYRAPICUS VARIUS VARIUS.
> (632½) Birds of N. A., 1874, vol. ii, p. 539, pl. li, figg. 1, 2 (cut, 539).

SPHYRAPICUS WILLIAMSONII.
> (78) P. R. R. Surv., vol. ix, p. 105; (104) Birds of N. A., 1860, p. 105, pl. xxxiv, fig. 1; (632½) Birds of N. A., 1874, vol. ii, p. 545, pl. li, fig. 5.

Spizella.
> (632½) Birds of N. A., 1874. vol. ii, p. 1.

SPIZELLA ATRIGULARIS.
> (78). P. R. R. Surv., vol. ix, p. 476; (87) Mex. Bound. Surv., vol. ii, p. 16, pl. xvii, fig. 1; (104) Birds of N. A., 1860, p. 476, pl. lv, fig. 1; (632½) Birds of N. A., 1874, vol. ii, p. 15, vol. i, pl. xxvi, figg. 11, 12.

Spizella Breweri.
> (78) P. R. R. Surv., vol. ix, p. 475; (87) Mex. Bound. Surv., vol. ii, p. 16.

SPIZELLA MONTICOLA.
> (78) P. R. R. Surv., vol. ix, p. 472; (632½) Birds of N. A., 1874, vol. ii, p. 3, pl. xxvii, fig. 5 (cuts, pp. 1, 3); *Ibid.*, App., p. 514.

SPIZELLA PALLIDA.
> (78) P. R. R. Surv., vol. ix, p. 474; (87) Mex. Bound. Surv., vol. ii, p. 16; (632½) Birds of N. A., 1874, vol. ii, p. 11, pl. xxvii, fig. 3.

SPIZELLA PALLIDA BREWERI.
> (632½) Birds of N. A., 1874, vol. ii, p. 13, pl. xxvii, fig. 4; *Ibid.*, App., p. 514.

SPIZELLA PUSILLA.
> (78) P. R. R. Surv., vol. ix, p. 473; (632½) Birds of N. A., 1874, vol. ii, p. 5, pl. xxvii, fig. 2.

SPIZELLA SOCIALIS.
> (78) P. R. R. Surv., vol. ix, p. 473; (632½) Birds of N. A., 1874, vol. ii, p. 7, pl. xxvii, fig. 7; *Ibid.*, App., p. 514.

Spizella socialis arizonæ.
> (632½) Birds of N. A., 1874, vol. ii, p. 11.

Spizellinæ.
> (632½) Birds of N. A., 1874, vol. i, p. 528.

Spizinæ.
> (632½) Birds of N. A., 1874, vol. ii, p. 58.

Squatarola helvetica.
> (78) P. R. R. Surv., vol. ix, p. 697.

Starnœnas.
> (632½) Birds of N. A., 1874, vol. iii, p. 394.

Starnœnas cyanocephala.
> (78) P. R. R. Surv., vol. ix, p. 608.

STARNŒNAS CYANOCEPHALA.
> (632½) Birds of N. A., 1874, vol. iii, p. 395, pl. lviii, fig. 5 (cuts, pp. 395, 396).

Stelgidopteryx.
> (115) Review of N. A. Birds, May, 1865, Part I, pp. 270, 271, 312; (632½) Birds of N. A., 1874, vol. i. p. 350.

Stelgidopteryx fulvigula.
> (115) Review of N. A. Birds, May, 1865, Part I, p. 318.

Stelgidopteryx fulvipennis.
> (115) Review of N. A. Birds, May, 1865, Part I, p. 316.

Stelgidopteryx ruficollis.
> (115) Review of N. A. Birds, May, 1865, Part I, p. 315.

STELGIDOPTERYX SERRIPENNIS.
> (78) P. R. R. Surv., vol. ix, p. 313; (115) Review of N. A. Birds, May, 1865, Part I, p. 315; (632½) Birds of N. A., 1874, vol. i, p. 350, pl. xvi, fig. 12 (cut, p. 350).

Stelgidopteryx uropygialis.
> (115) Review of N. A. Birds, May, 1865, Part I, p. 317.

Stellula.
> (632½) Birds of N. A., 1874, vol. ii, p. 445.

Stellula calliope.
(632½) Birds of N. A., 1874, vol. ii, p. 445, pl. xlvii, fig. 9 (cut, p. 445).
Stercorarius cataractes.
(78) P. R. R. Surv., vol. ix, p. 838.
Stercorarius cepphus.
(78) P. R. R. Surv., vol. ix, p. 840.
Stercorarius parasiticus.
(78) P. R. R. Surv., vol. ix, p. 839.
Stercorarius pomarinus.
(78) P. R. R. Surv., vol. ix, p. 838.
Sterna acuflavida.
(78) P. R. R. Surv., vol. ix, p. 860; (87) Mex. Bound. Surv., vol. ii, p. 27.
STERNA ALEUTICA.
(129) Chicago Acad. Sci., 1869, p. 321, pl. xxxi, fig. 1.
Sterna aranea.
(78) P. R. R. Surv., vol. ix, p. 859.
Sterna caspica.
(32) Stansbury's Surv. Salt Lake [App. C], p. 335; (78) P. R. R. Surv., vol. ix, p. 859.
STERNA ELEGANS.
(32) Stansbury's Surv. Salt Lake [App. C], p. 335; (78) P. R. R. Surv., vol. ix, p. 860; (104) Birds of N. A., 1860, p. 860, pl. xciv.
Sterna Forsteri.
(78) P. R. R. Surv., vol. ix, p. 862.
Sterna frenata.
(78) P. R. R. Surv., vol. ix, p. 864.
Sterna fuliginosa.
(78) P. R. R. Surv., vol. ix, p. 861.
Sterna Havelli.
(78) P. R. R. Surv., vol. ix, p. 861.
Sterna macroura.
(78) P. R. R. Surv., vol. ix, p. 862.
Sterna paradisea.
(78) P. R. R. Surv., vol. ix, p. 863.
STERNA PIKEI.
(78) P. R. R. Surv., vol. ix, p. 863; (104) Birds of N. A., 1860, p. 863, pl. xcv.
Sterna regia.
(78) P. R. R. Surv., vol. ix, p. 859.
Sterna Trudeauii.
(78) P. R. R. Surv., vol. ix, p. 861.
Sterna Wilsoni.
(78) P. R. R. Surv., vol. ix, p. 861.
Strepsilas interpres.
(78) P. R. R. Surv., vol. ix, p. 701; (87) Mex. Bound. Surv., vol. ii, p. 25.
Strepsilas melanocephalus.
(32) Stansbury's Surv. Salt Lake [App. C], p. 334; (78) P. R. R. Surv., vol. ix, p. 702.
STREPSILAS MELANOCEPHALA.
(104) Birds of N. A., 1860, p. 702, pl. vii.
Strigidæ.
(632½) Birds of N. A., 1874, vol. iii, p. 4.
Strix.
(632½) Birds of N. A., 1874, vol. iii, p. 10.
STRIX FLAMMEA PRATINCOLA.
(632½) Birds of N. A., 1874, vol. iii, p. 13 (cuts, pp. 10, 14, 15, 96, 98, 99, 100, 101).
Strix frontalis.
(32) Stansbury's Surv. Salt Lake [App. C], p. 327.

Strix pratincola.
(78) P. R. R. Surv., vol. ix, p. 47; (87) Mex. Bound. Surv., vol. ii, p. 4; (632½) Birds of N. A., 1874, App., p. 522.
Sturnella.
(632½) Birds of N. A., 1874, vol. ii, p. 171.
STURNELLA MAGNA.
(78) P. R. Surv., vol. ix, p. 535; (632½) Birds of N. A., 1874, vol. ii, p. 174, pl. xxxiv, fig. 2 (cuts, p. 171).
STURNELLA MAGNA NEGLECTA.
(632½) Birds of N. A., 1874, vol. ii, p. 176, pl. xxxiv, fig. 1.
Sturnella neglecta.
(31) Stansbury's Surv. Salt Lake [App. C], p. 316; (32) Stansbury's Surv. Salt Lake [App. C], p. 331; (78) P. R. R. Surv., vol. ix, p. 537; (87) Mex. Bound. Surv., vol. ii, p. 19.
Sturnidæ.
(632½) Birds of N. A., 1874, vol. ii, p. 228.
Sturnus.
(632½) Birds of N. A., 1874, vol. ii, p. 228.
STURNUS VULGARIS.
(632½) Birds of N. A., 1874, vol. ii, p. 229, pl. v, fig. 8 (cut, p. 228).
Sula (Sula) bassana.
(78) P. R. R. Surv., vol. ix, p. 871.
Sula fiber.
(78) P. R. R. Surv., vol. ix, p. 872.
Surnia.
(632½) Birds of N. A., 1874, vol. iii, p. 74.
Surnia ulula.
(78) P. R. R. Surv., vol. ix, p. 64.
SURNIA ULULA HUDSONIA.
(632½) Birds of N. A., 1874, vol. iii, p. 75 (cuts, pp. 97, 98, 99, 100, 101, 102).
Sylvia griseicollis.
(115) Review of N. A. Birds, May, 1865, Part I, p. 266.
Sylvia ochroleuca.
(115) Review of N. A. Birds, May, 1865, Part I, p. 266.
Sylvia pumila.
(115) Review of N. A. Birds, May, 1865, Part I, p. 266.
Sylvia russeicauda.
(115) Review of N. A. Birds, May, 1865, Part I, p. 266.
Sylvia semitorquata.
(115) Review of N. A. Birds, May, 1865, Part I, p. 266.
Sylvia virescens.
(115) Review of N. A. Birds, May, 1865, Part I, p. 266.
SYLVICOLA KIRTLANDII.
(40) Ann. Lyc. Nat. Hist. N. Y., v, 1852, p. 217, pl. vi.
Sylvicola olivacea.
(32) Stansbury's Surv. Salt Lake [App. C], p. 328.
Sylvicoleæ.
(115) Review of N. A. Birds, Aug., 1864, Part I, p. 166.
Sylvicolidæ.
(115) Review of N. A. Birds, Aug., 1864, Part I, pp. 160, 164; (632½) Birds of N. A., 1874, vol. i, p. 177.

Sylvicolinæ.
(115) Review of N. A. Birds, Aug., 1864, Part I,
pp. 166, 167; (632½) Birds of N. A., 1874,
vol. i, p. 179.
Sylviidæ.
(115) Review of N. A. Birds, July, 1864, Part I,
p. 64; Ibid., Aug., 1864, Part I, p. 164;
(632½) Birds of N. A., 1874, vol. i, p. 69.
Sylviinæ.
(632½) Birds of N. A., 1874, vol. i, p. 69.
Symphemia semipalmata.
(31) Stansbury's Surv. Salt Lake [App. C], p.
320; (78) P. R. R. Surv., vol. ix, p. 729; (94)
P. R. R. Surv., vol. x, p. 15.
Synthliborhamphus antiquus.
(78) P. R. R. Surv., vol. ix, p. 916.
Syrnium.
(632½) Birds of N. A., 1874, vol. iii, p. 28.
SYRNIUM CINEREUM.
(78) P. R. R. Surv., vol. ix, p. 56; (632½) Birds of
N. A., 1874, vol. iii, p. 30 (cuts, pp. 30, 98, 99,
100, 101, 102).
SYRNIUM NEBULOSUM.
(78) P. R. R. Surv., vol. ix, p. 56; (632½) Birds of
N. A., 1874, vol. iii, p. 34 (cuts, pp. 28, 35).
SYRNIUM OCCIDENTALE.
(104) Birds of N. A., 1860, p. 50, pl. lxvi; (632½)
Birds of N. A., 1874, vol. iii, p. 38 (cut, p. 38).
Tachycineta.
(115) Review of N. A. Birds, May, 1365, Part I,
pp. 294, 296, 270.
Tachypetes aquila.
(78) P. R. R. Surv., vol. ix, p. 873.
Tachytriorchis Cooperi.
(78) P. R. R. Surv., vol. ix, p. 31.
Tanagridæ.
(632½) Birds of N. A., 1874, vol. i, p. 431.
Tantalus loculator.
(78) P. R. R. Surv., vol. ix, p. 682; (87) Mex.
Bound. Surv., vol. ii, p. 24.
Telmatodytes.
(115) Review of N. A. Birds, Aug., 1864, Part I,
p. 94; Ibid., Oct., 1864, Part I, p. 147; Ibid.,
Aug., 1864, Part I, p. 163.
Telmatodytes palustris.
(78) P. R. R. Surv., vol. ix, p. 564.
Teretristeæ.
(115) Review of N. A. Birds, Aug., 1864, Part
I, p. 166.
Teretristis.
(115) Review of N. A. Birds, Apr., 1865, Part
I, p. 233.
Teretristis fernandinæ.
(115) Review of N. A. Birds, Apr., 1865, Part
I, p. 234.
Teretristis Forusii.
(115) Review of N. A. Birds, Apr., 1865, Part
I, p. 235.
TERETRISTIS FORUSII.
(115) Review of N. A. Birds, Aug., 1864, Part
I, p. 163, fig. 8.
Tetrao canadensis.
(78) P. R. R. Surv., vol. ix, p. 622.
Tetrao Franklinii.
(78) P. R. R. Surv., vol. ix, p. 623.

Tetrao obscurus.
(78) P. R. R. Surv., vol. ix, p. 620; (632½) Birds
of N. A., 1874. App., p. 522.
Tetrao urophasianus.
(31) Stansbury's Surv. Salt Lake [App. C],
p. 319.
Tetraonidæ.
(632½) Birds of N. A., 1874, vol. iii, p. 414.
Thalassarche chlororhyncha.
(78) P. R. R. Surv., vol. ix, p. 822.
Thalassidroma fregetta.
(32) Stansbury's Surv. Salt Lake [App. C],
p. 335.
Thalassidroma furcata.
(32) Stansbury's Surv. Salt Lake [App. C],
p. 335; (78) P. R. R. Surv., vol. ix, p. 829.
Thalassidroma Hornbyi.
(78) P. R. R. Surv., vol. ix, p. 829.
Thalassidroma Leachii.
(78) P. R. R. Surv., vol. ix, p. 830.
THALASSIDROMA MELANIA.
(76) P. R. R. Surv., vol. ix, p. 830; (99) Pr.
Acad. Nat. Sci. Phila., 1859, p. 306; (104).
Birds of N. A., 1860, p. 830, pl. xcix, fig. 2.
Thalassidroma pelagica.
(78) P. R. R. Surv., vol. ix, p. 831.
Thalassidroma Wilsoni.
(78) P. R. R. Surv., vol. ix, p. 831.
Thalassoica tenuirostris.
(78) P. R. R. Surv., vol. ix, p. 826.
Thaumatias.
(632½) Birds of N. A., 1874, vol. ii, p. 468.
Thaumatias linnæi.
(632½) Birds of N. A., 1874, vol. ii, p. 468.
Thryomanes.
(115) Review of N. A. Birds, Aug., 1864, Part
I, pp. 94, 126.
Thryophilus.
(115) Review of N. A. Birds. Aug., 1864. Part
I, pp. 94, 127.
Thryophilus albipectus.
(115) Review of N. A. Birds, Aug., 1864, Part
I, p. 132.
Thryophilus castaneus.
(115) Review of N. A. Birds, Aug., 1864, Part
I, p. 133.
Thryophilus Galbraithi.
(115) Review of N. A. Birds, Aug., 1864, Part
I, p. 131.
Thryophilus longirostris.
(115) Review of N. A. Birds, Aug., 1864, Part
I, p. 131.
Thryophylus modestus.
(115) Review of N. A. Birds, Aug., 1864, Part
I, p. 131.
Thryophilus rufalbus poliopleura.
(115) Review of N. A. Birds, Aug., 1864, Part
I, p. 128.
Thryophilus rufalbus rufalbus.
(115) Review of N. A. Birds, Aug., 1864, Part
I, p. 128.
Thryophilus Schotti.
(115) Review of N. A. Birds, Aug., 1864, Part
I, p. 133.

Thryophilus sinaloa.
(115) Review of N. A. Birds, Aug., 1864, Part
I, p. 130.

Thryophilus striolatus.
(115) Review of N. A. Birds, Aug., 1864, Part
I, p. 131.

Thryothorus.
(115) Review of N. A. Birds, Aug., 1864, Part
I, pp. 94, 120, 123; (632½) Birds of N. A.,
1874, vol. i, p. 141.

Thryothorus albinucha.
(115) Review of N. A. Birds, Aug., 1864, Part
I, pp. 125, 149.

Thryothorus albipectus.
(115) Review of N. A. Birds, Aug., 1864, Part
I, p. 122.

THRYOTHORUS BERLANDIERI.
(78) P. R. R. Surv., vol. ix, p. 362; (87) Mex.
Bound. Surv., vol. ii, p. 13; (104) Birds of
N. A., 1860, p. 362, pl. lxxxiii, fig. 1; (115)
Review of N. A. Birds, Aug., 1864, Part I,
pp. 121, 124.

Thryothorus Bewickii.
(78) P. R. R. Surv., vol. ix, p. 363; (87) Mex.
Bound. Surv., vol. ii, p. 13; (115) Review of
N. A. Birds, Aug., 1864, Part I, p. 122.

THRYOTHORUS BEWICKII BEWICKII.
(115) Review of N. A. Birds, Aug., 1864, Part
I, p. 126; (632½) Birds of N. A., 1874, vol. i,
p. 145, pl. ix, figg. 3, 4 (cuts, pp. 142, 145).

Thryothorus Bewickii leucogaster.
(115) Review of N. A. Birds, Aug., 1864, Part I,
p. 127; (632½) Birds of N. A., 1874, vol. i, p.
147; Ibid, App., p. 504.

Thryothorus Bewickii spilurus.
(115) Review of N. A. Birds, Aug., 1864, Part I,
p. 126; (632½) Birds of N. A., 1874, vol. i, p.
147.

Thryothorus castaneus.
(115) Review of N. A. Birds, Aug., 1864, Part
I, p. 123.

Thryothorus fasciato-ventris.
(115) Review of N. A. Birds, Aug., 1864, Part
I, p. 121.

Thryothorus felix.
(115) Review of N. A. Birds, Aug., 1864, Part I,
p. 121.

Thryothorus Galbraithi,
(115) Review of N. A. Birds, Aug., 1864, Part
I, p. 123.

Thryothorus leucogaster.
(115) Review of N. A. Birds, Aug., 1864, Part I,
p. 122.

Thryothorus longirostris.
(115) Review of N. A. Birds, Aug., 1864, Part I,
p. 123.

Thryothorus Ludovicianus.
(78) P. R. R. Surv., vol. ix, p. 361; (115) Review
of N. A. Birds, Aug., 1864, Part I, pp. 121,
123; (632½) Birds of N. A., 1874, App., p. 503.

THRYOTHORUS LUDOVICIANUS BERLANDIERI.
(632½) Birds of N. A., 1874, vol. i, p. 144, pl. ix,
fig. 2.

THRYOTHORUS LUDOVICIANUS LUDOVICIANUS.
(632½) Birds of N. A., 1874, vol. i, p. 142, pl. ix,
fig. 1 (cut, p. 141).

Thryothorus maculipectus.
(115) Review of N. A. Birds, Aug., 1864, Part I,
p. 121.

Thryothorus modestus.
(115) Review of N. A. Birds, Aug., 1864, Part I,
p. 122.

Thryothorus murinus.
(115) Review of N. A. Birds, Aug., 1864, Part I,
p. 123.

Thryothorus nigricapillus.
(115) Review of N. A. Birds, Aug., 1864, Part I,
p. 123.

Thryothorus petenicus.
(115) Review of N. A. Birds, Aug., 1864, Part I,
pp. 122, 125.

Thryothorus pleurostictus.
(115) Review of N. A. Birds, Aug., 1864, Part I,
pp. 121, 123.

Thryothorus poliopleura.
(115) Review of N. A. Birds, Aug., 1864, Part I.
p. 122.

Thryothorus rufalbus.
(115) Review of N. A. Birds, Aug., 1864, Part I,
p. 122.

Thryothorus rutilus.
(115) Review of N. A. Birds, Aug., 1864, Part I,
p. 121.

Thryothorus Schottii.
(115) Review of N. A. Birds, Aug., 1864, Part I,
p. 123.

Thryothorus sinaloa.
(115) Review of N. A. Birds, Aug., 1864, Part I,
p. 122.

Thryothorus spilurus.
(115) Review of N. A. Birds, Aug., 1864, Part I,
p. 122.

Tinnunculus sparverius.
(78) P. R. R. Surv., vol. ix, p. 13; (94) P. R. R.
Surv., vol. x, p. 12; (99) Pr. Acad. Nat. Sci.
Phila., 1859 (1860), p. 302.

Tinnunculus sparverius Isabellinus.
(632½) Birds of N. A., 1874, vol. iii, p. 171.

TINNUNCULUS SPARVERIUS SPARVERIUS.
(632½) Birds of N. A., 1874. vol. iii, p. 169 (cuts,
pp. 159, 173).

Toxostoma curvirostris.
(32) Stansbury's Surv. Salt Lake [App. C], p.
329.

Toxostoma Lecontei.
(32) Stansbury's Surv. Salt Lake [App. C], p.
329.

Toxostoma rediviva.
(32) Stansbury's Surv. Salt Lake [App. C], p.
328.

Trichopicus Gairdneri.
(78) P. R. R. Surv., vol. ix, p. 91.

Trichopicus Harrisii.
(78) P. R. R. Surv., vol. ix, p. 87.

Trichopicus pubescens.
(78) P. R. R. Surv., vol. ix, p. 89.

Trichopicus villosus major.
(78) P. R. R. Surv., vol. ix, p. 84.

Trichopicus villosus medius.
(78) P. R. R. Surv., vol. ix, p. 84.

Trichopicus villosus minor.
(78) P. R. R. Surv., vol. ix, p. 84.

Tringa alpina americana.
(78) P. R. R. Surv., vol. ix, p. 719.
Tringa Bonapartii.
(78) P. R. R. Surv., vol. ix, p. 722.
Tringa (Tringa) canutus.
(78) P. R. R. Surv., vol. ix, p. 715.
Tringa canutus.
(87) Mex. Bound. Surv., vol. ii, p. 25
Tringa (Tringa) Cooperi.
(78) P. R. R. Surv., vol. ix, p. 716.
TRINGA COOPERI.
(104) Birds of N. A., 1860, p. 716, pl. lxxxix,
fig. 1.
Tringa maculata.
(78) P. R. R. Surv., vol. ix, p. 720; (87) Mex.
Bound. Surv., vol. ii, p. 25.
Tringa maritima.
(78) P. R. R. Surv., vol. ix, p. 717.
Tringa subarquata.
(78) P. R. R. Surv., vol. ix, p. 718.
Tringa Wilsonii.
(78) P. R. R. Surv., vol. ix, p. 721.
Tringoides macularius.
(78) P. R. R. Surv., vol. ix, p. 735.
Trochilidæ.
(632½) Birds of N. A., 1874, vol. ii, p. 437.
Trochilus.
(632½) Birds of N. A., 1874, vol. ii, p. 447.
TROCHILUS ALEXANDRI.
(78) P. R. R. Surv., vol. ix, p. 133; (87) Mex.
Bound. Surv., vol. ii, p. 6, pl. v, fig. 3; (104)
Birds of N. A., 1860, p. 133, pl. xliv, fig. 3;
(632½) Birds of N. A., 1874, vol. ii, p. 450, pl.
xlvii, fig. 1 (cuts, p. 451).
TROCHILUS COLUBRIS.
(78) P. R. R. Surv., vol. ix, p. 131; (87) Mex.
Bound. Surv., vol. ii, p. 6; (632½) Birds of
N. A., 1874, vol. ii, p. 448, pl. xlvii, fig. 2
(cuts, p. 447).
Troglodytes.
(115) Review of N. A. Birds, Aug., 1864, Part
I, pp. 94, 137, 138; (632½) Birds of N. A.,
1874, vol. i, p. 148.
TROGLODYTES ÆDON.
(78) P. R. R. Surv., vol. ix, p. 367; (115) Review
of N. A. Birds., Aug. 1864, Part I, p. 138;
(632½) Birds of N. A., 1874, vol. i, p. 149, pl.
ix, fig. 5 (cuts, pp. 149, 150).
Troglodytes ædon aztecus.
(115) Review of N. A. Birds., Aug., 1864, Part
I, p. 139.
Troglodytes ædon Parkmanni.
(632½) Birds of N. A., 1874, vol. i, p. 53.
TROGLODYTES ALASCENSIS.
(129) Chicago Acad. Sci., 1869, p. 315, pl. xxx,
fig. 3.
Troglodytes albifrons.
(32) Stansbury's Surv. Salt Lake [App. C], p.
327.
Troglodytes albinucha.
(115) Review of N. A. Birds. Aug., 1864, Part
I, p. 123.
Troglodytes americanus.
(78) P. R. R. Surv., vol. ix, p. 368; (115) Review
of N. A. Birds, Aug., 1864, Part I, p,
141.

Troglodytes brunneicollis.
(115) Review of N. A. Birds, Aug., 1864, Part
I, p. 144.
Troglodytes hyemalis.
(78) P. R. R. Surv., vol. ix, p. 369; (115) Review
of N. A. Birds, A ug., 1864, Part I, p. 144.
Troglodytes hyemalis pacificus.
(115) Review of N. A. Birds, Aug., 1864, Part
I, p. 145.
Troglodytes inquietus.
(115) Review of N. A. Birds, Aug., 1864, Part
I, p. 143.
Troglodytes intermedius.
(115) Review of N. A. Birds, Aug., 1864, Part
I, p. 142.
Troglodytes Parkmanni.
(78) P. R. R. Surv., vol. ix, p. 367; (87) Mex.
Bound. Surv., vol. ii, p. 13; (115) Review of
N. A. Birds, Aug., 1864, Part I, p. 140.
TROGLODYES PARVULUS ALASCENSIS.
(632½) Birds of N. A., 1874, vol. i, p. 157, pl. ix,
fig. 8.
TROGLODYTES PARVULUS HYEMALIS.
(632½) Birds of N. A., 1874, vol. i, p. 155, pl. ix,
fig. 9; Ibid., App., p. 504 (cut).
Troglodytidæ.
(115) Review of N. A. Birds, Aug., 1864, Part
I, p. 91; (632½) Birds of N. A., 1874, vol. i, p.
130.
TROGON MEXICANUS.
(78) P. R. R. Surv., vol. ix, p. 69; (87) Mex.
Bound. Surv., vol. ii, p. 5, pll. i, ii; (104)
Birds of N. A., 1860, p. 65, pl. xl.
Trupialis militaris.
(78) P. R. R. Surv., vol. ix, p. 533.
Tryngites rufescens.
(78) P. R. R. Surv., vol. ix, p. 739; (104) Birds
of N. A., 1860, p. 739, pl. vi.
Turdidæ.
(115) Review of N. A. Birds, June, 1864, Part
I, p. 1; Ibid., Aug. 1864, Part I, pp. 164,
165; (632½) Birds of N. A., 1874, vol. i, p. 1.
Turdinæ.
(115) Review of N. A. Birds, June, 1864, Part
I, p. 4; (632½) Birds of N. A., 1874, vol. i, p. 3.
Turdus.
(115) Review of N. A. Birds, June, 1864, Part
I, pp. 4, 11, 12; (632½) Birds of N. A., 1874,
vol. i, p. 3.
Turdus (Turdus) Aliciæ.
(78) P. R. R. Surv., vol. ix, p. 217.
TURDUS ALICIÆ.
(104) Birds of N. A., 1860, p. 217, pl. lxxxi, fig.
2; (115) Review of N. A. Birds, June, 1864,
Part I, p. 21; (632½) Birds of N. A., 1874,
vol. i, p. 11, pl. i, fig. 3.
Turdus assimilis.
(115) Review of N. A. Birds, June, 1864, Part
I, p. 24.
Turdus Auduboni, Baird (new specific name)=
Merula silens.
(115) Review of N. A. Birds, June, 1864, Part
I, p. 16.
Turdus confinis.
(115) Review of N. A. Birds, June, 1864, Part
I, p. 29.

Turdus flavirostris.
(115) Review of N. A. Birds, June, 1864, Part I, p. 31.
TURDUS (TURDUS) FUSCESCENS.
(78) P. R. R. Surv., vol. ix, p. 214; (115) Review of N. A. Birds, June, 1864, Part I, p. 17; (632½) Birds of N. A., 1874, vol. 1, p. 9, pl. i, fig. 5.
Turdus Grayi.
(115) Review of N. A. Birds, June, 1864, Part I, p. 26.
Turdus gymnopthalmus.
(115) Review of N. A. Birds, July, 1864, Part I, p. 59.
TURDUS ILIACUS.
(115) Review of N. A. Birds, June, 1864, Part I, p. 23; (632½) Birds of N. A., 1874, vol. i, p. 23, pl. ii, fig. 4 (cut, p. 22).
Turdus infuscatus.
(115) Review of N. A. Birds, June, 1864, Part I, p. 31.
Turdus leucauchen.
(115) Review of N. A. Birds, June, 1864, Part I, p. 24.
Turdus jamaicensis.
(115) Review of N. A. Birds, June, 1864, Part I, p. 23.
Turdus migratorius.
(78) P. R. R. Surv., vol. ix, p. 218; (115) Review of N. A. Birds, June, 1864, Part I, p. 28.
TURDUS MIGRATORIUS CONFINIS.
(632¼) Birds of N. A., 1874, vol. i, p. 27, pl. ii, fig. 1.
TURDUS MIGRATORIUS MIGRATORIUS.
(632½) Birds of N. A., 1774, vol. i, p. 25, pl. ii, fig. 3 (cuts, pp. 24, 25).
Turdus (Turdus) mustelinus.
(78) P. R. R. Surv., vol. ix, p. 212.
TURDUS MUSTELINUS.
(115) Review of N. A. Birds, June, 1864, Part I, p. 13; (632½) Birds of N. A., 1874, vol. i, p. 7, pl. i, fig. 1 (cut, p. 4).
TURDUS NÆVIUS.
(78) P. R. R. Surv., vol. ix, p. 219; (115) Review of N. A. Birds, June, 1864, Part I, p. 32; (632½) Birds of N. A., 1874, vol. i, p. 29, pl. ii, fig. 2 (cuts, pp. 28, 29).
Turdus (Turdus) nanus.
(78) P. R. R. Surv., vol. ix, p, 213.
Turdus nanus.
(87) Mex. Bound. Surv., vol. ii, p. 9; (115) Review of N. A. Birds, June, 1864, Part I, p. 15.
Turdus nigrescens.
(115) Review of N. A. Birds, July, 1864, Part I, p. 58.
Turdus obsoletus.
(115) Review of N. A. Birds, June, 1864, Part I, p. 28.
TURDUS PALLASII.
(115) Review of N. A. Birds, June, 1864, Part I, p. 14; (632½) Birds of N. A., 1874, vol. i, p. 18, pl. i, fig. 6.
TURDUS PALLASI AUDUBONI.
(632½) Birds of N. A., 1874, vol. i, p. 21, pl. i, fig. 21; Ibid., App., p. 499.

TURDUS PALLASI NANUS.
(632½) Birds of N. A., 1874, vol. i, p. 20, pl. i, fig. 7; Ibid., App., p. 499.
Turdus (Turdus) Pallasi silens.
(78) P. R. R. Surv., vol. ix, p. 212.
Turdus phæopygus.
(115) Review of N. A. Birds, July, 1864, Part I, p. 59.
Turdus pinicola.
(115) Review of N. A. Birds, July, 1864, Part I, p. 58.
Turdus plebeius.
(115) Review of N. A. Birds, July, 1864, Part I, p. 58.
Turdus rufitorques.
(115) Review of N. A. Birds, June, 1864, Part I, p. 32.
Turdus rufopalliatus.
(32) Stansbury's Surv. Salt Lake [App. C], p. 328.
Turdus (Turdus) Swainsonii.
(78) P. R. R. Surv., vol. ix, p. 216.
TURDUS SWAINSONII.
(115) Review of N. A. Birds, June, 1864, Part I, p. 19; (632½) Birds of N. A., 1874, vol. i, p. 14, pl. i, fig. 4.
TURDUS SWAINSONI USTULATUS.
(632½) Birds of N. A., 1874, vol. i, p. 16, pl. i, fig. 2 (cut, p. 5).
Turdus (Turdus) ustulatus.
(78) P. R. R. Surv., vol. ix, p. 215.
TURDUS USTULATUS.
(104) Birds of N. A., 1860, p. 215, pl. lxxxi, fig. 1; (115) Review of N. A. Birds, June, 1864, Part I, p. 18.
Turdus xanthoscelis.
(115) Review of N. A. Birds, July, 1864, Part I, p. 59.
Tylorhamphus camtschatica.
(78) P. R. R. Surv., vol. ix, p. 908.
Tylorhamphus tetracula.
(78) P. R. R. Surv., vol. ix, p. 907.
Tyrannidæ.
(115) Review of N. A. Birds, Aug., 1864. Part I, p. 165; (632½) Birds of N. A., 1874, vol. ii, p. 306.
Tyrannula cayanensis.
(32) Stansbury's Surv. Salt Lake [App. C], p. 329.
Tyrannula cinerascens.
(32) Stansbury's Surv. Salt Lake [App. C], p. 329.
Tyrannula flaviventris.
(1) Pr. Acad. Nat. Sci. Phila., 1843, p. 283; (3) Amer. Journ. Sci. and Arts, 1846, p. 274: (32) Stansbury's Surv. Salt Lake [App. C], p. 329.
Tyrannula Lawrenceii.
(32) Stansbury's Surv. Salt Lake [App. C], p. 329.
Tyrannula minima.
(1) Pr. Acad. Nat. Sci. Phila., 1843, p. 284; (3) Amer. Journ. Sci. and Arts, 1846, p. 275; (32) Stansbury's Surv. Salt Lake [App. C], p. 329.
Tyrannus.
(632½) Birds of N. A., 1874, vol. ii, p. 314.

RANNUS CAROLINENSIS.

(78) P. R. R. Surv., vol. ix, p. 171 ; (632½) Birds of N. A., 1874, vol. ii, p. 316, pl. xliii, fig. 4 (cuts pp. 314, 316).

rannus Cassinii.

(32) Stansbury's Surv. Salt Lake [App. C], p. 329.

RANNUS COUCHII.

(78) P. R. R. Surv., vol. ix, p. 175 ; (87) Mex. Bound. Surv., vol. ii, p. 8, pl. xi, fig. 1; (104) Birds of N. A., 1860, p. 1875, pl. xlix, fig. 1.

RANNUS DOMINICENSIS.

(78) P. R. R. Surv., vol. ix, p. 172 ; (632½) Birds of N. A., 1874, vol. ii, p. 319, pl. xliii, fig. 8.

RANNUS MELANCHOLICUS.

(78) P. R. R. Surv., vol. ix, p. 176 ; -(104) Birds of N. A., 1860, p. 176, pl. xlix, fig. 2.

RANNUS MELANCHOLICUS COUCHII.

(632½) Birds of N. A., 1874, vol. ii, p. 329, pl. xliii, fig. 7.

RANNUS VERTICALIS.

(78) P. R. R. Surv., vol. ix, p. 178 ; (632½) Birds of N. A., 1874, vol. ii, p. 324, pl. xliii, fig. 2.

RANNUS VOCIFERANS.

(78) P. R. R. Surv., vol. ix, p. 174 ; (87) Mex. Bound. Surv., vol. ii, p. 8, pl. x ; (104) Birds of N. A., 1860, p. 174, pl. xlviii ; (632½) Birds of N. A., 1874, vol. ii, p. 327, pl. xliii, fig. 5 ; (632½) Birds of N. A., 1874, App., p. 518.

ria arra.

(78) P. R. R. Surv., vol. ix, p. 915.

ria brevirostris.

(32) Stansbury's Surv. Salt Lake [App. C], p. 335.

ria (Uria) carbo.

(78) P. R. R. Surv., vol. ix, p. 913.

RIA CARBO.

(104) Birds of N. A., 1860, p. 913, pl. xcvii.

ria (Uria) columba.

(78) P. R. R. Surv., vol. ix, p. 912.

RIA COLUMBA.

(104) Birds of N. A., 1860, p. 912, pl. xcvi, fig. 4.

ria (Uria) grylle.

(78) P. R. R. Surv., vol. ix, p. 911.

RIA GRYLLE.

(104) Birds of N. A., 1860, p. 911, pl. xcvi, fig. 2.

ria lomvia.

(78) P. R. R. Surv., vol. ix, p. 913.

ria ringvia.

(78) P. R. R. Surv., vol. ix, p. 914.

rile penicillatus.

(78) P. R. R. Surv., vol. ix, p. 880.

rile violaceus.

(78) P. R. R. Surv., vol. ix, p. 881.

Jtamania torda.

(78) P. R. R. Surv., vol. ix, p. 901.

Jermivora brevipennis.

(32) Stansbury's Surv. Salt Lake [App.C], p.328.

Jormivoreæ.

(115) Review of N. A. Birds, Aug , 1864, Part I, p.166.

Jireo.

(115) Review of N. A. Birds, May, 1866, Part I, pp. 323, 250 ; (632½) Birds of N. A.1874, vol. i, pp. 357, 389.

Vireo altiloquus.

(78) P. R. R. Surv., vol. ix, p. 334.

Vireo atricapilla.

(32) Stansbury's Surv. Salt Lake [App.C],p.328.

VIREO ATRICAPILLUS.

(78) P. R. R. Surv., vol. ix, p. 337 ; (87) Mex. Bound. Surv., vol. ii, p. 12 ; (115) Review of N. A. Birds, May, 1866, Part I, p. 350 ; (632½) Birds of N. A., 1874, vol. i, p. 383, pl. xvii, fig. 6 (cut, p. 383).

VIREO BARBATULA.

(115) Review of N. A. Birds, Aug., 1864, Part I, p. 163, fig. 9.

VIREO BELLI.

(32) Stansbury's Surv. Salt Lake [App. C], p. 328 ; (78) P. R. R. Surv., vol. ix, p. 337 ; (87) Mex. Bound. Surv., vol. ii, p. 12 ; (115) Review of N. A. Birds, May, 1866, Part I, p. 358 ; (632½) Birds of N. A., 1874, vol. i, p. 389, pl. xvii, fig. 13 (cut, p. 389).

Vireo Carnioli.

(115) Review of N. A. Birds, May, 1866, Part I, p. 356.

VIREO CASSINII.

(78) P. R. R. Surv., vol. ix, p. 340 ; (104) Birds of N. A., 1860, p. 340, pl. lxxviii, fig. 1.

Vireo crassirostris.

(115) Review of N. A. Birds, May, 1866, Part I, p. 368.

Vireo flavifrons.

(78) P. R. R. Surv., vol. ix, p. 341.

Vireo flavoviridis.

(78) P. R. R. Surv., vol. ix, p. 332 ; (87) Mex. Bound. Surv., vol. ii, p. 12.

Vireo gilvus.

(78) P. R. R. Surv., vol. ix, p. 335.

Vireo Gundlachi.

(115) Review of N. A. Birds, May, 1866, Part I, p. 369.

VIREO HUTTONI.

(32) Stansbury's Surv. Salt Lake [App. C], p. 328 ; (78) P. R. R. Surv., vol. ix., p. 339 ; (87) Mex. Bound. Surv., vol. ii, p. 12 ; (104) Birds of N. A., 1860, p. 339, pl. lxviii, fig. 2 ; (115) Review of N. A. Birds, May, 1866, Part I, p. 357 ; (632½) Birds of N. A., 1874, vol. i, p. 387, pl. xvii, fig. 12 (cut, p. 387).

Vireo hypochryseus.

(115) Review of N. A. Birds, May, 1866, Part I, p. 370.

Vireo Latimeri.

(115) Review of N. A. Birds, May, 1866, Part I, p. 364.

Vireo modestus.

(115) Review of N. A. Birds, May, 1866, Part I, p. 362.

VIREO NOVEBORACENSIS.

(78) P. R. R. Surv., vol. ix, p. 338 ; (87) Mex. Bound. Surv., vol. ii, p. 12 ; (115) Review of N. A. Birds, May, 1866, Part I, p. 354 ; (632½) Birds of N. A., 1874, vol. i, p. 385 ; pl. xvii, fig. ii (cuts, p. 382).

Vireo ochraceus.

(115) Review of N. A. Birds, May, 1866, Part I, p. 366.

Xanthornus mexicanus.
(32) Stansbury's Surv. Salt Lake [App. C], p. 331.
Xanthoura.
(632¼) Birds of N. A., 1874, vol. ii, p. 294.
XANTHOURA INCAS LUXUOSA.
(632½) Birds of N. A., 1874, vol. ii, p. 295, pl. xlii, fig. 1 (cuts, pp. 294, 296).
Xanthoura luxuosa.
(78) P. R. R. Surv., vol. ix, p. 589; (87) Mex. Bound. Surv., vol. ii, p. 21.
Xema Sabinii.
(78) P. R. R. Surv., vol. ix, p. 857.
Xenopicus albolarvatus.
(78) P. R. R. Surv., vol. ix, p. 96.
Zenaida.
(632½) Birds of N. A., 1874, vol. iii, p. 378.
ZENAIDA AMABILIS.
(78) P. R. R. Surv., vol. ix, p. 602; (632½) Birds of N. A., 1874, vol. iii, p. 379, pl. lviii, fig. 3 (cut, p. 379).
Zenaidinæ.
(632½) Birds of N. A., 1874, vol. iii, p. 374.
Zenaidura.
(632½) Birds of N. A., 1874, vol. iii, p. 381.
ZENAIDURA CAROLINENSIS.
(78) P. R. R. Surv., vol. ix, p. 604; (87) Mex. Bound. Surv., vol. ii, p. 21; (632½) Birds of N. A., 1874, vol. iii, p. 383, pl. lviii, fig. 2 (cuts, pp. 382, 383).
Zonotrichia.
(632½) Birds of N. A., 1874, vol. i, p. 565.

ZONOTRICHIA ALBICOLLIS.
(78) P. R. R. Surv., vol. ix, p. 463; (632½) Birds of N. A., 1874, vol. i, p. 574, pl. xxvi, fig. 10.
Zonotrichia Cassinii.
(32) Stansbury's Surv. Salt Lake [App. C], p. 330.
ZONOTRICHIA CORONATA.
(78) P. R. R. Surv., vol. ix, p. 461; (632½) Birds of N. A., 1874, vol. i, p. 573, pl. xxvi, fig. 1.
Zonotrichia fallax.
(56) Pr. Acad. Nat. Sci. Phila., 1854, p. 119.
ZONOTRICHIA GAMBELII.
(32) Stansbury's Surv. Salt Lake [App. C], p. 330; (78) P. R. R. Surv., vol. ix, p. 460; (87) Mex. Bound. Surv., vol. ii, p. 15; (104) Birds of N. A., 1860, p. 460, pl. xix, fig. 1.
ZONOTRICHIA LEUCOPHRYS.
(78) P. R. R. Surv., vol. ix, p. 458; (87) Mex. Bound. Surv., vol. ii, p. 15; (99) Pr. Acad. Nat. Sci. Phila., 1859, p. 304; (104) Birds of N. A., 1860, p. 458, pl. lxix, fig. 2; (632½) Birds of N. A., 1874, vol. i, p. 566, pl. xxv, figg. 9, 10 (cuts, pp. 565, 567).
ZONOTRICHIA LEUCOPHRYS GAMBELI.
(632½) Birds of N. A., 1874, vol. i, p. 569, pl. xxv, figg. 11, 12; Ibid., App., p. 514.
ZONOTRICHIA QUERULA.
(32) Stansbury's Surv. Salt Lake [App. C], p. 330; (78) P. R. R. Surv., vol. ix, p. 462; (632½) Birds of N. A., 1874, vol. i, p. 577, pl. xxvi, fig. 47.

REPTILES.

ABASTOR ERYTHROGRAMMUS.
(39) Cat. N. A. Reptiles, 1853, Part I, p. 125; (92) P. R. R. Surv., vol. x, pl. xxxiii, fig. 93.
Acris acheta.
(55) Pr. Acad. Nat. Sci. Phila., 1854, p. 59.
ACRIS CREPITANS.
(55) Pr. Acad. Nat. Sci. Phila., 1854, p. 59; (88) Mex. Bound. Surv., vol. ii, p. 28, pl. xxxvii, figg. 14–17; (96) P. R. R. Surv., vol. x, p. 44.
Agkistrodon contortrix. (See Ancistrodon.)
(39) Cat. N. A. Reptiles, 1853, Part I, p. 17.
Alligator lucius.
(88) Mex. Bound. Surv., vol. ii, p. 5.
Amblystoma mavortium.
(95) P. R. R. Surv., vol. x, p. 20.
AMBLYSTOMA PROSERPINA.
(36) Pr. Acad. Nat. Sci. Phila., 1852, p. 173. (88) Mex. Bound. Surv., vol. ii, p. 29, pl. xxxv figg. 7–14.
Amblystoma tenebrosum.
(37) Pr. Acad. Nat. Sci. Phila., 1852, p. 174.
AMBLYSTOMA TEXANUM.
(88) Mex. Bound. Surv., vol. ii, p. 29, pl. xxxv, fig. 13.
Ambystoma episcopus.
(10) Journ. Acad. Nat. Sci. Phila., Oct., 1849, pp. 284, 293.
Ambystoma Jeffersoniana.
(10) Journ. Acad. Nat. Sci. Phila., Oct., 1849, p. 283.

Ambystoma lurida.
(10) Journ. Acad. Nat. Sci. Phila., Oct., 1849, p. 284.
Ambystoma macrodactyla.
(10) Journ. Acad. Nat. Sci. Phila., Oct., 1849, pp. 283, 292.
Ambystoma mavortia.
(10) Journ. Acad. Nat. Sci. Phila., Oct., 1849, pp. 284, 292.
Ambystoma opaca.
(10) Journ. Acad. Nat. Sci. Phila., Oct., 1849, p. 283.
Ambystoma punctata.
(10) Journ. Acad. Nat. Sci. Phila., Oct., 1849, p. 283.
Ambystoma tigrina.
(10) Journ. Acad. Nat. Sci. Phila., Oct., 1849, p. 284.
ANCISTRODON CONTORTRIX.
(49½) Serpents of N. Y., 1854, pp. 13, 14; (88) Mex. Bound. Surv., vol. ii, p. 15; (92) P. R. R. Surv., vol. x, pl. xxv, fig. 12.
Aneides lugubris.
(21) Outlines of Gen. Zoology, 1851, p. 256; (38) Pr. Acad. Nat. Sci. Phila., 1853, p. 301.
Anniella pulchra.
(38) Pr. Acad. Nat. Sci. Phila., 1853, p. 301; (88) Mex. Bound. Surv., vol. ii, p. 13.
Anolis carolinensis.
(88) Mex. Bound. Surv., vol. ii, p. 12.

Anolis Cooperi.
>　(74½) Pr. Acad.Nat.Sci. Phila., 1858, p. 254; (100)
>　Pr. Acad. Nat. Sci. Phila., 1858, p. 254.

ARIZONA ELEGANS.
>　(88) Mex. Bound. Surv., vol. ii, p. 18, pl. xiii;
>　(96) P. R. R. Surv., vol. x, p. 42.

Aspidonectes Emoryi.
>　(88) Mex. Bound. Surv., vol. ii, p. 3.

BASCANION CONSTRICTOR.
>　(39) Cat. N. A. Reptiles, 1853, Part I, p. 98; (49½)
>　Serpents of N. Y., 1854, pp. 22, 23; (92)P. R.
>　R. Surv., vol. x, pl. xxxi, fig. 67.

BASCANION FLAVIVENTRIS.
>　(39) Cat. N. A. Reptiles, 1853, Part I, p. 96; (88)
>　Mex. Bound. Surv., vol. ii, p. 20; (92) P. R.
>　R. Surv., vol. x, pl. xxxi, fig. 70.

BASCANION FOXII.
>　(39) Cat. N. A. Reptiles, 1853, Part I, p. 96; (92)
>　P. R. R. Surv., vol. x, pl. xxxi, fig. 69.

BASCANION FREMONTII.
>　(39) Cat. N. A. Reptiles, 1853, Part I, p. 95; (92)
>　P. R. R. Surv., vol. x, pl. xxxi, fig. 68.

BASCANION VESTUSTUS.
>　(38) Pr. Acad. Nat. Sci. Phila., 1853, p. 300; (39)
>　Cat. N. A. Reptiles, 1853, Part I, p. 97; (92)
>　P. R. R. Surv., vol. x, pl. xxxvi, fig. 6.

Batrachoseps quadridigitata.
>　(10) Journ. Acad. Nat. Sci. Phila., Oct., 1849, p.
>　287.

BUFO ALVARIUS.
>　(88) Mex. Bound Surv., vol. ii, p. 26, pl. xli, figg.
>　1-6.

BUFO AMERICANUS.
>　(88) Mex. Bound. Surv., vol. ii, p. 25, pl. xxxix,
>　figg. 1-4; (96) P. R. R. Surv., vol. x, p. 44, pl.
>　xxv, fig. 2.

Bufo boreas.
>　(37) Pr. Acad. Nat. Sci. Phila., 1852, p. 174.

BUFO COGNATUS.
>　(49) Marcy and McClellan's Expl. Red River,
>　La. [App. F.], p.242, pl. xi; (88) Mex. Bound
>　Surv., vol. ii, p. 27; (96) P. R. R. Surv., vol.
>　x, p. 44, pl. xxvi.

Bufo Columbianus.
>　(44) Pr. Acad. Nat. Sci. Phila., 1853, p. 379.

Bufo debilis.
>　(88) Mex. Bound. Surv., vol. ii, p. 27.

Bufo granulosus.
>　(36) Pr. Acad. Nat. Sci. Phila., 1852, p. 173.

BUFO HALOPHILA.
>　(38) Pr. Acad. Nat. Sci. Phila., 1853, p. 301; (88)
>　Mex. Bound. Surv., vol. ii, p. 26, pl. xli, figg.
>　7-12.

BUFO INSIDIOR.
>　(88) Mex. Bound. Surv., vol. ii, p. 26, pl. xli,
>　figg. 13-18.

BUFO NEBULIFER.
>　(88) Mex. Bound. Surv., vol. ii, p. 25, pl. xl, figg.
>　1-4; (96) P. R. R. Surv., vol. x, p. 44.

BUFO PUNCTATUS.
>　(36) Pr. Acad. Nat. Sci. Phila., 1852, p. 173; (88)
>　Mex. Bound. Surv., vol. ii, p. 25, pl. xxxix,
>　figg. 5-7.

BUFO SPECIOSUS.
>　(88) Mex. Bound. Surv., vol. ii, p. 26, pl. xl, figg.
>　5-10.

BUFO WOODHOUSII.
>　(88) Mex. Bound. Surv., vol. ii, p. 27; (95) P. R.
>　R. Surv., vol. x, p. 20; (96) P. R. R. Surv.,
>　vol. x, p. 44, pl. xxv, fig. 1.

Calamaria tenuis.
>　(37) Pr. Acad. Nat. Sci. Phila., 1852, p. 176.

Callisaurus ventralis.
>　(88) Mex. Bound. Surv., vol. ii, p. 8; (95) P. R.
>　R. Surv., vol. x, p. 17.

CELUTA AMŒNA.
>　(39) Cat. N. A. Reptiles, 1853, Part I, p. 129;
>　(49½) Serpents of N. Y., 1854, pp. 25, 26; (92)
>　P. R. R. Surv., vol. x, pl. xxxiii, fig. 95.

CHLOROSOMA VERNALIS.
>　(39) Cat. N. A. Reptiles, 1853, Part I, p. 108;
>　(49½) Serpents of N. Y., 1854, pp. 23, 24; (92)
>　P. R. R. Surv., vol. x, pl. xxxii, fig. 81.

Chorophilus nigritus.
>　(55) Pr. Acad. Nat. Sci. Phila., 1854, p. 60.

Chrysemys oregonensis.
>　(88) Mex. Bound. Surv., vol ii, p. 4.

Churchillia bellona.
>　(27) Pr. Acad. Nat. Sci. Phila., 1852, p. 70; (33)
>　Stansbury's Surv., Salt Lake [App. C], p.
>　350.

CNEMIDOPHORUS GRACILIS.
>　(35) Pr. Acad. Nat. Sci. Phila., 1852, p. 128; (38)
>　Pr. Acad. Nat. Sci. Phila., 1853, p. 301; (88)
>　Mex. Bound. Surv., vol. ii, p. 10, pl. xxxiv,
>　figg. 7-14.

CNEMIDOPHORUS GRAHAMII.
>　(35) Pr. Acad. Nat. Sci. Phila., 1852, p. 128; (88)
>　Mex. Bound. Surv., vol. ii, p. 10, pl. xxxii,
>　figg. 1-6.

CNEMIDOPHORUS GULARIS.
>　(35) Pr. Acad. Nat. Sci. Phila., 1852, p. 128; (49)
>　Marcy and McClellan's Expl. Red River,
>　La. [App. F], 239, pl. x, figg. 1-4; (88) Mex.
>　Bound. Surv., vol. ii, p. 11; (96) P. R. R.
>　Surv., vol. x, p. 38.

Cnemidophorus inornatus.
>　(74½) Pr. Acad. Nat. Sci. Phila., 1858, p. 255; (88)
>　Mex. Bound. Surv., vol. ii, p. 10; (100) Pr.
>　Acad. Nat. Sci. Phila., 1858, p. 255.

Cnemidophorus marmoratus.
>　(35) Pr. Acad. Nat. Sci. Phila., 1852, p. 128.

Cnemidophorus octolineatus.
>　(74½) Pr. Acad. Nat. Sci. Phila., 1858, p. 255; (88)
>　Mex. Bound. Surv., vol. ii, p. 10; (100) Pr.
>　Acad. Nat. Sci. Phila., 1858, p. 255.

Cnemidophorus perplexus.
>　(35) Pr. Acad. Nat. Sci. Phila., 1852, p. 128; (88)
>　Mex. Bound. Surv., vol. ii, p. 10.

Cnemidophorus præsignis.
>　(35) Pr. Acad. Nat. Sci. Phila., 1852, p. 128.

Cnemidophorus sex-lineatus.
>　(96) P. R. R. Surv., vol. x, p. 38.

Cnemidophorus tess'elatus.
>　(95) P. R. R. Surv., vol. x, p. 18.

CNEMIDOPHORUS TIGRIS.
>　(27) Pr. Acad. Nat. Sci. Phila., 1852, p. 69;
>　(33) Stansbury's Surv. Salt Lake [App. C],
>　p. 338, pl. ii; (88) Mex. Bound. Surv., vol.
>　ii, p. 10, pl. xxxiii.

Coluber mormon.
  (27) Pr. Acad. Nat. Sci. Phila., 1852, p. 70; (33) Stansbury's Surv. Salt Lake [App. C], p. 351.
CONTIA MITIS.
  (38) Pr. Acad. Nat. Sci. Phila., 1853, p. 300; (89) Cat. N. A. Reptiles, 1853, Part I, p. 110; (92) P.R.R. Surv..vol x, pl. xxxvi, fig.7.
CROTALUS, SP.
  (92) P. R. R. Surv., vol. x, pl. xxxv, fig. 2; Ibid., fig. 3; Ibid., fig. 5.
CROTALUS ADAMANTEUS.
  (39) Cat. N. A. Reptiles, 1853, Part I, p. 3; (92) P. R. R. Surv., vol. x, pl. xxiv, fig. 2.
CROTALUS ATROX.
  (39) Cat. N. A. Reptiles, 1853, Part I, p. 5; (88) Mex. Bound. Surv., vol. ii, p. 14, pl. i; (92) P. R. R. Surv., vol. x, pl. xxiv, fig. 3; (96) P. R. R. Surv., vol. x, p. 39.
CROTALUS CERASTES.
  (88) Mex. Bound. Surv., vol. ii, p. 14, pl. iii; (92) P. R. R. Surv., vol. x, pl. xxxv, fig. 4.
CROTALUS CONFLUENTUS.
  (89) Cat. N. A. Reptiles, 1853, Part I, p. 8; (49) Marcy and McClellan's Expl. Red River, La. [App. F], p. 217, pl. 1, 3d vol.; (88) Mex. Bound. Surv., vol. ii, p. 14; (92) P. R. R. Surv., vol. x, pl. xxiv, fig. 4; (96) P. R. R. Surv., vol. x, p. 40.
CROTALUS DURISSUS.
  (39) Cat. N. A. Reptiles, 1853, Part I, p. 1; (49½) Serpents of N. Y., 1854, pp. 9, 10, 11; (92) P. R. R. Surv, vol. x, pl. xxiv, fig. 1; (96) P. R. R. Surv., vol. x, p, 39.
CROTALUS LUCIFER.
  (37) Pr. Acad. Nat. Sci. Phila., 1852, p. 176; (38) Pr. Acad. Nat. Sci. Phila., 1853, p. 300; (39) Cat. N. A. Reptiles, 1853, Part I, p. 6; (92) P. R. R. Surv., vol. x, pl. xxxvi, fig. 1.
CROTALUS MOLOSSUS.
  (39) Cat. N. A. Reptiles, 1853, Part I, p. 10; (88) Mex. Bound. Surv., vol. ii, p. 14, pl. ii; (92) P. R. R. Surv., vol. x, pl. xxiv, fig. 5.
CROTALUS OREGONUS.
  (39) Cat. N. A. Reptiles, 1853, Part I, p. 145; (92) P. R. R. Surv., vol. x, pl. xxiv, fig. 6.
CROTALUS TIGRIS.
  (88) Mex. Bound. Surv., vol. ii, p. 14, pl. iv; (92) P. R. R. Surv., vol. x, pl. xxxv, fig. 1.
Crotalophorus consors.
  (39) Cat. N. A. Reptiles, 1853, Part I, p. 12; (88) Mex. Bound. Surv., vol. xii, p. 15; (92) P.R. R. Surv.,vol. x, pl. xxvi, fig. 8.
CROTALOPHORUS EDWARDSII
  (39) Cat. N. A. Reptiles, 1853, Part I, p. 15; (88) Mex. Bound. Surv., vol. ii, p. 15, pl. v, fig. 1; (92) P. R. R. Surv., vol. x, pl. xxv, fig. 10.
CROTALOPHORUS KIRTLANDII.
  (39) Cat. N. A. Reptiles, 1853, Part I, p. 16; (92) P. R. R. Surv., vol. x, pl. xxv, fig. 11 (adult).
  (92) P. R. R. Surv., vol. x, pl.xxv, fig.11 (young).
CROTALOPHORUS MILIARIUS.
  (39) Cat. N. A. Reptiles, 1853, Part I, p. 11; (92) P. R. R. Surv., vol. x, pl. xxiv, fig. 7; (96) P. R. R. Surv., vol. x, p. 40.

CROTALOPHORUS TERGEMINUS.
  (39) Cat. N. A. Reptiles, 1853, Part I, p. 14; (49½) Serpents of N. Y., 1854, pp. 11, 12, 13; (92) P. R. R. Surv., vol. x, pl. xxv, fig. 9.
CROTAPHYTUS COLLARIS.
  (49) Marcy and McClellan's Expl. Red River La. [App. F.], p 235; (88) Mex. Bound. Surv., vol. ii, p. 6; (95) P. R. R. Surv., vol. x, p. 17, pl. xxiv, fig. 1; (96) P. R. R. Surv., vol. x, p. 37.
Crotaphytus dorsalis.
  (35) Pr. Acad. Nat. Sci. Phila., 1852, p. 126; (38) Pr. Acad. Nat. Sci. Phila., 1853, p. 301.
Crotaphytus Gambelii.
  (35) Pr. Acad. Nat. Sci. Phila., 1852, p. 126.
Crotaphytus reticulatus.
  (74½) Pr. Acad. Nat. Sci. Phila., 1858, p. 253; (88) Mex. Bound. Surv., vol. ii, p. 6; (100) Pr. Acad. Nat. Sci. Phila., 1858, p. 253.
CROTAPHYTUS WIZLIZENII.
  (27) Pr. Acad. Nat. Sci. Phila., 1852, p. 69; (33) Stansbury's Surv. Salt Lake [App. C], p. 340, pl. iii; (88) Mex. Bound. Surv., vol. ii, p.7, pl. xxxi; (95) P. R. R. Surv., vol. x, p. 17; (96) P. R. R. Surv., vol. x, p. 37.
Desmognathus.
  (10) Journ. Acad. Nat. Sci. Phila., Oct., 1849, 2d ser., p. 282.
Desmognathus auriculatus.
  (10) Journ. Acad. Nat. Sci. Phila., Oct., 1849, p. 285.
Desmognathus fuscus.
  (10) Journ. Acad. Nat. Sci. Phila., Oct. 1849. p. 285.
Desmognathus niger.
  (10) Journ. Acad. Nat. Sci. Phila., 1849, p. 285.
DIADOPHIS, SP.
  (92) P. R. R. Surv., vol. x, pl. xxxii, fig. 2.
DIADOPHIS AMABILIS.
  (38) Pr. Acad. Nat. Sci. Phila., 1853, p. 300; (39) Cat. N. A. Reptiles, 1853, Part I, p. 113; (92) P. R. R. Surv., vol. x, pl. xxxiii, fig. 83.
DIADOPHIS DOCILIS.
  (39) Cat. N. A. Reptiles, 1853, Part I, p. 114; (88) Mex. Bound. Surv., vol. ii, p. 22, pl. xxi, fig. 3; (92) P. R. R. Surv., vol. x, pl. xxxii, fig. 1; Ibid., pl. xxxiii, fig. 84; (96) P. R. R. Surv., vol. x, p. 43.
Diadophis pulchellus.
  (39) Cat. N. A. Reptiles, 1853, Part I, p. 115; (92) P. R. R. Surv., vol. x, pl. xxxiii, fig. 85.
Diadophis punctatus.
  (92) Cat. N. A. Reptiles, 1853, Part I, .p. 112; (49½) Serpents of N. Y., 1854, pp. 24, 25; (92) P. R. R. Surv., vol. x, pl. xxxiii, fig. 82.
Diadophis regalis.
  (39) Cat. N. A. Reptiles, 1853, Part I, p. 115; (88) Mex. Bound. Surv., vol. ii, p. 22; (92 P. R. R. Surv., vol. x, pl. xxxiii, fig. 86.
DIPSAS SEPTENTRIONALIS.
  (88) Mex. Bound. Surv., vol. ii, p. 16, pl. viii, fig. 1; (92) P. R. R. Surv., vol. x, pl. xxv, fig. 18.

DIPSOSAURUS DORSALIS.
(88) Mex. Bound. Surv., vol. ii, p. 8, pl. xxxii, figg. 7-18.

DOLIOSAURUS McCALLI.
(88) Mex. Bound. Surv., vol. ii, p. 9, pl. xxvii, figg. 4-6.

Doliosaurus modestus.
(88) Mex. Bound. Surv., vol. ii, p. 10; (96) P. R. R. Surv., vol. x, p. 38.

Doliosaurus platyrhinos.
(95) P. R. R. Surv., vol. x, p. 18.

Elaps fulvius.
(39) Cat. N. A. Reptiles, 1853, Part I, p. 21; (92) P. R. R. Surv., vol. x, pl. xxv, fig. 15.

ELAPS TENER.
(39) Cat. N. A. Reptiles, 1853, Part I, p. 22; (88) Mex. Bound. Surv., vol. ii, p. 15, pl. vii, fig. 1; (92) P. R. R. Surv., vol. x, pl. xxv, fig. 16.

Elaps tristis.
(39) Cat. N. A. Reptiles, 1853, Part I, p. 23; (92) P. R. R. Surv., vol. x, pl. xxv, fig. 17.

Elgaria formosa.
(37) Pr. Acad. Nat. Sci. Phila., 1852, p. 175.

Elgaria grandis.
(37) Pr. Acad. Nat. Sci. Phila., 1852, p. 176.

Elgaria nobilis.
(35) Pr. Acad. Nat. Sci. Phila., 1852, p. 128.

Elgaria principis.
(37) Pr. Acad. Nat. Sci. Phila., 1852, p. 175.

Elgaria scincicauda.
(27) Pr. Acad. Nat. Sci. Phila., 1852, p. 69; (33) Stansbury's Surv. Salt Lake [App. C], p. 348, pl. iv, figg. 1-3; (38) Pr. Acad. Nat. Sci. Phila., 1853, p. 301.

Emys marmorata.
(37) Pr. Acad. Nat. Sci. Phila., 1852, p. 176.

Euphryne.
(74½) Pr. Acad. Nat. Sci. Phila., 1858, p. 253; (100) Pr. Acad. Nat. Sci. Phila., 1858, p. 253.

EUPHRYNE OBESA.
(88) Mex. Bound. Surv., vol. ii, p. 6, pl. xxvii.

Euphryne obesus.
(74½) Pr. Acad. Nat. Sci. Phila., 1858, p. 253; (100) Pr. Acad. Nat. Sci. Phila., 1858, p. 253.

EUTÆNIA DORSALIS.
(92) P. R. R. Surv., vol. x, pl. xxvi, fig. 2; (96) P. R. R. Surv., vol. x, p. 40.

EUTÆNIA FAIREYI.
(92) P. R. R. Surv., vol. x, pl. xxvi, fig. 20.

EUTÆNIA LEPTOCEPHALA.
(92) P. R. R. Surv., vol. x, pl. xxxvi, fig. 2.

EUTÆNIA MARCIANA.
(49) Marcy and McClellan's Expl. Red River La. [App. F], p. 221, pl. iii; (88) Mex. Bound. Surv., vol. ii, p. 17; (96) P. R. R. Surv., vol. x, p. 41; (92) P. R. R. Surv., vol. x, pl. xxvi, fig. 26.

EUTÆNIA ORDINATA.
(92) P. R. R. Surv., vol. x, pl. xxvi, fig. 24.

EUTÆNIA ORDINOIDES.
(92) P. R. R. Surv., vol. x, pl. xxvi, fig. 3; (95) P. R. R. Surv., vol. x, p. 19.

EUTÆNIA ORNATA.
(88) Mex. Bound. Surv., vol. ii, p. 16, pl. ix; (92) P. R. R. Surv., vol. x, pl. xxvi, fig. 22.

EUTÆNIA PICKERINGII.
(92) P. R. R. Surv., vol. x, pl. xxxvi, fig. 3.

EUTÆNIA PROXIMA.
(49) Marcy and McClellan's Expl. Red River La. [App. F], p. 220, pl. ii; (88) Mex. Bound. Surv., vol. ii, p. 16; (96) P. R. R. Surv., vol. x, p. 40; (92) P. R. R. Surv., vol. x, pl. xxvi, fig. 21.

EUTÆNIA RADIX.
(92) P. R. R. Surv., vol. x, pl. xxxiv, fig. 5; Ibid., pl. xxvi, fig. 25.

EUTÆNIA SAURITA.
(92) P. R. R. Surv., vol. x, pl. xxvi, fig. 19.

Eutænia saurita.
(49½) Serpents of N. Y., 1854, p. 14.

EUTÆNIA SIRTALIS.
(92) P. R. R. Surv., vol. x, pl. xxvi, fig. 23.

Eutænia sirtalis.
(49½) Serpents of N. Y., 1854, pp. 15, 16.

EUTÆNIA VAGRANS.
(95) P. R. R. Surv., vol. x, p. 19, pl. xvii; (96) P. R. R. Surv., vol. x, p. 41.

Eutainia concinna.
(39) Cat. N. A. Reptiles, 1853, Part I, p. 146.

Eutainia dorsalis.
(39) Cat. N. A. Reptiles, 1853, Part I, p. 31.

Eutainia elegans.
(39) Cat. N. A. Reptiles, 1853, Part I, p. 34.

Eutainia Faireyi.
(39) Cat. N. A. Reptiles, 1853, Part I, p. 25.

Eutainia infernalis.
(39) Cat. N. A. Reptiles, 1853, Part I, p. 26.

Eutainia leptocephala.
(39) Cat. N. A. Reptiles, 1853, Part I, p. 29.

Eutainia Marciana.
(39) Cat. N. A. Reptiles, 1853, Part I, p. 36.

Eutainia ordinata.
(39) Cat. N. A. Reptiles, 1853, Part I, p. 32.

Eutainia ordinoides.
(38) Pr. Acad. Nat. Sci. Phila., 1853, p. 300; (39) Cat. N. A. Reptiles, 1853, Part I, p. 33.

Eutainia parietalis.
(39) Cat. N. A. Reptiles, 1853, Part I, p. 28.

Eutainia Pickeringii.
(39) Cat. N. A. Reptiles, 1853, Part I, p. 27.

Eutainia proxima.
(39) Cat. N. A. Reptiles, 1853, Part I, p. 25.

Eutainia radix.
(39) Cat. N. A. Reptiles, 1853, Part I, p. 34.

Eutainia saurita.
(39) Cat. N. A. Reptiles, 1853, Part I, p. 24.

Eutainia sirtalis.
(39) Cat. N. A. Reptiles, 1853, Part I, p. 30.

Eutainia vagrans.
(39) Cat. N. A. Reptiles, 1853, Part I, p. 35.

FARANCIA ABACURA.
(39) Cat. N. A. Reptiles, 1853, Part I, p. 123; (92) P. R. R. Surv., vol. x, pl. xxxiii, fig. 92.

GEORGIA COLPERI.
(39) Cat. N. A. Reptiles, 1853, Part I, p. 92; (92) P. R. R. Surv., vol. x, pl. xxxiv, fig. 1.

GEORGIA OBSOLETA.
(39) Cat. N. A. Reptiles, 1853, Part I, p. 158;
(88) Mex. Bound. Surv., vol. ii, p. 20, pl. xv
(92) P. R. R. Surv., vol. x, pl. xxxi, fig. 66.
GERRHONOTUS INFERNALIS.
(74½) Pr. Acad. Nat. Sci. Phila., 1858. p. 255;
(88) Mex. Bound. Surv., vol. ii, p. 11; (100)
Pr. Acad. Nat. Sci. Phila., 1858, p. 255.
GERRHONOTUS NOBILIS.
(88) Mex. Bound. Surv., vol. ii, p. 11, pl. xxv,
figg. 1-8.
Gerrhonotus olivaceus.
(74½) Pr. Acad. Nat. Sci. Phila., 1858, p. 255;
(88) Mex. Bound. Surv., vol. ii, p. 11; (100)
Pr. Acad. Nat. Sci. Phila., 1858, p. 225.
GERRHONOTUS WEBBII.
(74½) Pr. Acad. Nat. Sci. Phila., 1858, p. 255;
(88) Mex. Bound. Surv., vol. ii, p. 11, pl.
xxiv, figg. 1-10; (100) Pr. Acad. Nat. Sci.
Phila., 1858, p. 255.
Gypochelys lacertina.
(88) Mex. Bound. Surv., vol. ii, p. 3.
HALDEA STRIATULA.
(39) Cat. N. A. Reptiles, 1853, Part I, p. 120;
(92) P. R. R. Surv., vol. x, pl. xxxiii, fig. 91.
HELOCŒTES CLARKII.
(88) Mex. Bound. Surv., vol. ii, p. 28, pl. xxxvii,
figg. 4-9.
Helocœtes feriarum.
(55) Pr. Acad. Nat. Sci. Phila., 1854, p. 60.
Helocœtes triseriatus.
(55) Pr. Acad. Nat. Sci. Phila., 1854, p. 60.
(55) Pr. Acad. Nat. Sci. Phila., 1854, p. 60.
HELODERMA HORRIDUM.
(88) Mex. Bound. Surv., vol. ii, p. 11, pl. xxvi,
(96) P. R. R. Surv., vol. x, p. 38.
HETERODON ATMODES.
(39) Cat. N. A. Reptiles, 1853, Part I, p. 57; (92)
P. R. R. Surv., vol. x, pl. xxviii, fig. 41.
HETERODON COGNATUS.
(39) Cat. N. A. Reptiles, 1853, Part I, p. 54; (88)
Mex. Bound. Surv., vol. ii, p. 17; (92) P. R.
R. Surv., vol. x, pl. xxviii, fig. 39.
HETERODON NASICUS.
(27) Pr. Acad. Nat. Sci. Phila, 1852, p. 70 ; (53)
Stansbury's Surv. Salt Lake [App. C], p.
352 ; (39) Cat. N. A. Reptiles, 1853, Part I, p.
61 ; (49) Marcy and McClellan's Expl. Red
River La. ·[App. F], p. 222, pl. iv ; (88)
Mex. Bound. Surv., vol. ii, p. 18, pl. xi, fig. 1;
(92) P. R. R. Surv., vol. x, pl. xxviii, fig. 43;
(95) P. R. R. Surv., vol. x, p. 19; (96) P. R.
R. Surv., vol. x, p. 41. .
HETERODON NIGER.
(39) Cat. N. A. Reptiles, 1853, Part I, p. 55; (92)
P. R. R. Surv., vol. x, pl. xxviii, fig. 40.
HETERODON PLATYRHINOS.
(39) Cat. N. A. Reptiles, 1853, Part I, p. 51 ; (49½)
Serpents of N. Y., 1854, pp. 18, 19 ; (92) P. R.
R. Surv., vol. x, pl. xxviii, fig. 38.
HETERODON SIMUS.
(39) Cat. N. A. Reptiles, 1853, Part I, p. 59 ; (92)
P. R. R. Surv., vol. x, pl. xxviii, fig. 42.
Holbrookia affinis.
(35) Pr. Acad. Nat. Sci. Phila., 1852, p. 125; (88)
Mex. Bound. Surv. vol. ii, p. 8.

Holbrookia approximans.
(74½) Pr. Acad. Nat. Sci. Phila., 1858, p. 253;
(88) Mex. Bound. Surv., vol. ii, p. 8; (100)
Pr. Acad. Nat. Sci. Phila., 1858, p. 253.
HOLBROOKIA MACULATA.
(33) Stansbury's Surv. Salt Lake [App. C], p.
342, pl. vi, figg. 1-3 ; (49) Marcy and McClellan's Expl. Red River La. [App. F], p. 236;
(88) Mex. Bound. Surv., vol. ii, p. 8; (95) P.
R. R. Surv., vol. x, p. 18; (96) P. R. R. Surv.,
vol. x, p. 38.
Holbrookia propinqua.
(35) Pr. Acad. Nat. Sci. Phila., 1852, p. 126; (88)
Mex. Bound. Surv., vol. ii, p. 8.
HOLBROOKIA TEXANA.
(35) Pr. Acad. Nat. Sci. Phila., 1852, p. 125; (88)
Mex. Bound. Surv., vol. ii, p. 8, pl. xxx ; (96)
P. R. R. Surv., vol. x, p. 38.
HYLA AFFINIS.
(55) Pr. Acad. Nat. Sci. Phila., 1854, p. 61 ; (88)
Mex. Bound. Surv., vol. ii, p. 29, pl. xxxviii,
figg. 4-7.
Hyla Andersonii.
(55) Pr. Acad. Nat. Sci. Phila., 1854, p. 60.
HYLA EXIMIA.
(55) Pr. Acad. Nat. Sci. Phila., 1854, p. 61 ; (88)
Mex. Bound. Surv., vol. ii, p. 29, pl. xxxviii,
figg. 8-10.
Hyla regilla.
(37) Pr. Acad. Nat. Sci. Phila., 1852, p. 174; (38)
Pr. Acad. Nat. Sci. Phila., 1853, p. 301.
Hyla Richardii.
(55) Pr. Acad. Nat. Sci. Phila., 1854, p. 60.
Hyla semifasciata.
(88) Mex. Bound. Surv., vol. ii, p. 28.
HYLA VANVLIETTI.
(55) Pr. Acad. Nat. Sci. Phila., 1854, p. 61 ; (88)
Mex. Bound. Surv., vol. ii, p. 29, pl. xxxviii,
figg. 1-3.
LAMPROSOMA EPISCOPUM.
(88) Mex. Bound. Surv., vol. ii, p. 22, pl. viii, fig.
2.
LAMPROSOMA OCCIPITALE.
(88) Mex. Bound. Surv., vol. ii, p. 21, pl. xxi, fig.
1; (92) P. R. R. Surv., vol. x, pl. xxxv, fig.
6; (92) P. R. R. Surv., vol. x, pl. xxxv,
fig. 7.
Lepidosternon floridanum.
(74½) Pr. Acad. Nat. Sci. Phila., 1858, p. 255;
(100) Pr. Acad. Nat. Sci. Phila., 1858, p. 255.
LEPTOPHIS ŒSTIVUS.
(39) Cat. N. A. Reptiles, 1853, Part I, p. 106;
(92) P. R. R. Surv., vol. x, pl. xxxii, fig.
79.
LEPTOPHIS MAJALIS.
(39) Cat. N. A. Reptiles, 1853, Part I, p. 107; (49)
Marcy and McClellan's Expl. Red River
La. [App. F], p. 232, pl. ix; (88) Mex. Bound.
Surv., vol. ii, p. 21 ; (92) P. R. R. Surv., vol.
x, pl. xxxii, fig. 80 ; (96) P. R. R. Surv., vol.
x, p. 43.
Litoria occidentalis.
(38) Pr. Acad. Nat. Sci. Phila., 1853, p. 301.
LODIA TENUIS.
(39) Cat. N. A. Reptiles, 1853, Part I, p. 116; (92)
P. R. R. Surv., vol. x, pl. xxxvi, fig. 8.

Lygosoma laterale.
- (88) Mex. Bound. Surv., vol. ii, p. 13 ; (96) P. R. R. Surv., vol. x, p. 39 ; (49) Marcy and Mc-Clellan's Expl. Red River La. [App. F], p.241.

MASTICOPHIS FLAGELIFORMIS.
- (92) P. R. R. Surv., vol. x, pl. xxxii, pl. xxxi, fig. 71 (old), fig. 72 (young) ; (39) Cat. N. A. Reptiles, 1853, Part I, pp. 98, 149.

Masticophis flavigularis.
- (39) Cat. N. A. Reptiles, 1853, Part I, p. 99 ; (49) Marcy and McClellan's Expl. Red River, La. [App. F], p. 230.
- (92) P. R. R. Surv., vol. x, pl. xxxii, fig. 73 (young).

MASTICOPHIS MORMON.
- (39) Cat. N. A. Reptiles, 1853, Part I, p. 101; (92) P. R. R. Surv., vol. x, pl. xxxii, fig. 74.

MASTICOPHIS ORNATUS.
- (39) Cat. N. A. Reptiles, 1853, Part I, p. 102 ; (88) Mex. Bound. Surv., vol. ii, p. 20, pl. xvii; (93) P. R. R. Surv., vol. x, pl. xxxii, fig. 75.

MASTICOPHIS SCHOTTII.
- (88) Mex. Bound. Surv., vol. ii, p. 20, pl. xviii ; (39) Cat. N. A. Reptiles, 1853, Part I, p. 160 ; (92) P. R. R. Surv., vol. x, pl. xxxii, fig. 77.

MASTICOPHIS TÆNIATUS.
- (39) Cat. N. A. Reptiles, 1853, Part I, p. 103 ; (92) P. R. R. Surv., vol. x, pl. xxxii, fig. 76 ; (95) P. R. R. Surv., vol. x, p. 20, pl. xxiii.

MASTICOPHIS TESTACEUS.
- (88) Mex. Bound. Surv., vol. ii, p. 20, pl. xvi ; (96) P. R. R. Surv., vol. x, p. 43.

MICROPS LINEATUS.
- (92) P. R. R. Surv., vol. x, pl. xxxiv, fig. 6.

Necturus lateralis.
- (10) Journ. Acad. Nat. Sci. Phila., Oct., 1849, p. 290; (96) P. R. R. Surv., vol. x, p. 45.

Necturus maculatus.
- (10) Journ. Acad. Nat. Sci. Phila., Oct., 1849, p. 290.

Nerodia Agassizii.
- (39) Cat. N. A. Reptiles, 1853, Part I, p. 41.

NERODIA ERYTHROGASTER.
- (39) Cat. N. A. Reptiles, 1853, Part I, p. 40 ; (92) P. R. R. Surv., vol. x, pl. xxvii, fig. 28 ; (95) P. R. R. Surv., vol. x, p. 19, pl. xviii ; (96) P. R. R. Surv., vol. x, p. 41.

NERODIA FASCIATA.
- (39) Cat. N. A. Reptiles, 1853, Part I, p. 39 ; (92) P. R. R. Surv., vol. x, pl. xxxiv, fig. 4.

NERODIA HOLBROOKII.
- (39) Cat. N. A. Reptiles, 1853, Part I, p. 43 ; (92) P. R. R. Surv., vol. x, pl. xxvii, fig. 30.

NERODIA NIGER.
- (39) Cat. N. A. Reptiles, 1853, Part I, p. 147 ; (92) P. R. R. Surv., vol. x, pl. xxvii, fig. 31.

NERODIA RHOMBIFER.
- (39) Cat. N. A. Reptiles, 1853, Part I, p. 147 ; (92) P. R. R. Surv., vol. x, pl. xxxiv, fig. 2.

NERODIA SIPEDON.
- (33) Cat. N. A. Reptiles, 1853, Part I, p. 38 ; (49½) Serpents of N. Y., 1854, pp. 16, 17 ; (92) P. R. R. Surv., vol. x, pl. xxvii, fig. 27.

NERODIA TAXISPILOTA.
- (89) Cat. N. A. Reptiles, 1853, Part I, p. 43 ; (92) P. R. R. Surv., vol. x, pl. xxvii, fig. 29.

NERODIA TRANSVERSA.
- (39) Cat. N. A. Reptiles, 1853, Part I, p. 148 ; (92) P. R. R. Surv., vol. x, pl. xxvi, fig. 1.

NERODIA WOODHOUSII.
- (39) Cat. N. A. Reptiles, 1853, Part I, p. 42 ; (88) Mex. Bound. Surv., vol. ii, p. 17 ; (92) P. R. R. Surv., vol. x, pl. xxxiv, fig. 3 ; (96) P. R. R. Surv., vol. x, p. 41.

NINIA DIADEMATA.
- (39) Cat. N. A. Reptiles, 1853, Part I, p. 49 ; (92) P. R. R. Surv., vol. x, pl. xxvii, fig. 37.

Notophthalmus torosus.
- (10) Journ. Acad. Nat. Sci. Phila., Oct., 1849, p. 284.

Notophthalmus viridescens.
- (10) Journ. Acad. Nat. Sci. Phila., Oct., 1849, p. 284.

OPHIBOLUS BOYLII.
- (39) Cat. N. A. Reptiles, 1853, Part I, p. 82 ; (88) Mex. Bound. Surv., vol. ii, p. 20 ; (92) P. R. R. Surv., vol. x, pl. xxx, fig. 57.

OPHIBOLUS CLERICUS.
- (39) Cat. N. A. Reptiles, 1853, Part I, p. 88 ; (92) P. R. R. Surv., vol. x, pl. xxx, fig. 62.

OPHIBOLUS DOLIATUS.
- (39) Cat. N. A. Reptiles, 1853, Part I, p. 89 ; (92) P. R. R. Surv., vol. x, pl. xxx, fig. 63.

Ophibolus Evansii.
- (96) P. R. R. Surv., vol. x, p. 43.

OPHIBOLUS EXIMIUS.
- (39) Cat. N. A. Reptiles, 1853, Part I, p. 87 ; (49½) Serpents of N. Y., 1854, pp. 21, 22 ; (92) P. R. R. Surv., vol. x, pl. xxx, fig. 61.

OPHIBOLUS GENTILIS.
- (39) Cat. N. A. Reptiles, 1853, Part I, p. 90 ; (49) Marcy and McClellan's Expl. Red River La. [App. F], p. 229, pl. viii ; (92) P. R. R. Surv., vol. x, pl. xxx, fig. 64.

OPHIBOLUS GETULUS.
- (39) Cat. N. A. Reptiles, 1853, Part I, p. 85 ; (49½) Serpents of N. Y., 1854, pp. 20, 21 ; (92) P. R. R. Surv., vol. x, pl. xxxi, fig. 65.

OPHIBOLUS RHOMBOMACULATUS.
- (39) Cat. N. A. Reptiles, 1853, Part I, p. 86 ; (92) P. R. R. Surv., vol. x, pl. xxx, fig. 60.

OPHIBOLUS SAYI.
- (39) Cat. N. A. Reptiles, 1853, Part I, p. 84 ; (49) Marcy and McClellan's Expl. Red River La. [App. F], p. 228, pl. vii ; (88) Mex. Bound. Surv., vol. ii, p. 20 ; (92) P. R. R. Surv., vol. x, pl. xxx, fig. 59.

OPHIBOLUS SPLENDIDUS.
- (39) Cat. N. A. Reptiles, 1853, Part I, p. 83 ; (88) Mex. Bound. Surv., vol. ii, p. 20, pl. xiv ; (92) P. R. R. Surv., vol. x, pl. xxx, fig. 58 ; (96) P. R. R. Surv., vol. x, p. 43.

OSCEOLA ELAPSOIDEA.
- (39) Cat. N. A. Reptiles, 1853, Part I, p. 133 ; (92) P. R. R. Surv., vol. x, pl. xxxiii, fig. 97.

Ozotheca tristycha.
- (88) Mex. Bound. Surv., vol. ii, p. 3.

Phrynosoma cornutum.
- (49) Marcy and McClellan's Expl. Red River La. [App. F], p. 233 ; (88) Mex. Bound. Surv., vol. ii, p. 9 ; (96) P. R. R. Surv., vol. x, p. 38 ; (38) Pr. Acad. Nat. Sci. Phila., 1852, p. 301.

Phrynosoma modestum.
- (27) Pr. Acad. Nat. Sci. Phila., 1852, p. 69.
Phrynosoma platyrhinos.
(27) Pr. Acad. Nat. Sci. Phila., 1852, p. 69.
PHRYNOSOMA REGALE.
(88) Mex. Bound. Surv., vol. ii, p. 9, pl. xxviii, figg. 1-3.
PHYLLODACTYLUS TUBERCULOSUS.
(88) Mex. Bound. Surv., vol. ii, p. 12, pl. xxiii, figg. 1-8.
Pituophis annectens.
(38) Pr. Acad. Nat. Sci. Phila., 1853, p. 300; (39) Cat. N. A. Reptiles, 1853, Part I, p. 72.
Pituophis bellona.
(39) Cat. N. A. Reptiles, 1853, Part I, p. 66.
Pituophis catenifer.
(39) Cat. N. A. Reptiles, 1853, Part I, p. 69.
PITUQPHIS McCLELLANII.
(39) Cat. N. A. Reptiles, 1853, Part I, p. 68 ; (49) Marcy and McClellan's Expl. Red River La. [App. F], p. 225, pl. v.
Pituophis melanoleucus.
(39) Cat. N. A. Reptiles, 1853, Part I, p. 65.
Pituophis Wilkesii.
(39) Cat. N. A. Reptiles, 1853, Part I, p. 71.
PITYOPHIS ANNECTENS.
(92) P. R. R. Surv., vol. x, pl. xxix, fig. 48.
PITYOPHIS BELLONA.
(88) Mex. Bound. Surv., vol. ii, p. 18 ; (92) P. R. R. Surv., vol. x, pl. xxix, fig. 46 ; (95) P. R. R. Surv., vol. x, p. 19 ; (96) P. R. R. Surv., vol. x, p. 42.
PITYOPHIS CATENIFER.
(92) P. R. R. Surv., vol. x, pl. xxxvi, fig. 4.
PITYOPHIS McCLELLANII.
(92) P. R. R. Surv., vol. x, pl. xxix, fig. 47.
PITYOPHIS MELANOLEUCUS.
(92) P. R. R. Surv., vol. x, pl. xxix, fig. 44.
PITYOPHIS SAYI.
(92) P. R. R. Surv., vol. x, pl. xxix, fig. 45.
PITYOPHIS WILKESII.
(92) P. R. R. Surv., vol. x, pl. xxxvi, fig. 5.
Platythyra flavescens.
(88) Mex. Bound. Surv., vol. ii, p. 3.
Plestiodon anthracinus.
(11) Journ. Acad. Nat. Sci. Phila., Oct., 1849, p. 294.
Plestiodon egregius.
(74½) Pr. Acad. Nat. Sci. Phila., 1858, p. 256; (100) Pr. Acad. Nat. Sci. Phila., 1858, p. 256.
Plestiodon fasciatus.
(96) P. R. R. Surv., vol. x, p. 39.
Plestiodon guttulatus.
(88) Mex. Bound. Surv., vol. ii, p. 12 ; (95) P. R. R. Surv., vol. x, p. 18.
Plestiodon inornatus.
(74½) Pr. Acad. Nat. Sci. Phila., 1858, p. 256; (100) Pr. Acad. Nat. Sci. Phila., 1858, p. 256.
Plestiodon leptogrammus.
(74½) Pr. Acad. Nat. Sci. Phila., 1858, p, 256; (100) Pr. Acad. Nat. Sci. Phila., 1858, p. 256.
Plestiodon obsoletum.
(35) Pr. Acad. Nat. Sci. Phila., 1852, p. 128.

PLESTIODON OBSOLETUS.
(88) Mex. Bound. Surv., vol. ii, p. 12, pl. xxv, figg. 9-16; (96) P. R. R. Surv., vol. x, p. 39.
PLESTIODON SEPTENTRIONALIS.
(74⅔) Pr. Acad. Nat. Sci. Phila., 1858, p. 256 ; (95) P. R. R. Surv., vol. x, p. 18, pl. xxiv, fig. 2 ; (100) Pr. Acad. Nat. Sci. Phila., 1858, p. 256.
PLOSTIODON SKILTONIANUM.
(27) Pr. Acad. Nat. Sci. Phila., 1852, p. 69; (33) Stansbury's Surv. Salt Lake [App. C], p. 349, pl. iv, figg. 4-6; (38) Pr.Acad. Nat. Sci. Phila., 1853, p. 301; (95) P. R. R. Surv., vol. x, p. 18.
Plestiodon tetragrammus.
(74½) Pr. Acad. Nat. Sci. Phila., 1858, p. 256; (88) Mex. Bound. Surv., vol. ii, p. 12; (100) Pr. Acad. Nat. Sci. Phila., 1858, p. 256.
Plethodus erythronota.
(10) Journ. Acad. Nat. Sci. Phila., Oct., 1849, p. 285.
Pseudotriton montanus.
(10) Journ. Acad. Nat. Sci. Phila., Oct., 1849, p. 287 ; (10) Journ. Acad. Nat. Sci. Phila., Oct., 1849, p. 293.
Pseudotriton salmoneus.
(19) Journ. Acad. Nat. Sci. Phila., Oct., 1849, p. 287.
Ptychemys mobilensis.
(88) Mex. Bound. Surv., vol. ii, p. 3.
RANA AREOLATA.
(36) Pr. Acad Nat. Sci. Phila., 1852, p. 173; (88) Mex. Bound. Surv., vol. ii, p. 28, pl. xxxvi, figg. 11, 12.
Rana aurora.
(37) Pr. Acad. Nat. Sci. Phila., 1852, p. 174.
RANA BERLANDIERI.
(88) Mex. Bound. Surv., vol. ii, p. 27, pl. xxxvi, figg. 7-10; (96) P. R. R. Surv., vol. x, p. 45.
Rana Boylii.
(55) Pr. Acad. Nat. Sci. Phila., 1854, p. 62.
Rana cantabrigensis.
(55) Pr. Acad. Nat. Sci. Phila., 1854, p. 62.
Rana Catesbiana.
(88) Mex. Bound. Surv., vol. ii, p. 27; (96) P. R. R. Surv., vol. x, p. 45.
Rana clamitans.
(96) P. R. R. Surv., vol. x, p. 45.
Rana Draytonii.
(37) Pr. Acad. Nat. Sci. Phila., 1852, p. 174.
Rana halecina.
(96) P. R. R. Surv., vol. x, p. 45.
Rana Lecontei.
(38) Pr. Acad. Nat. Sci. Phila., 1853, p. 301.
RANA MONTEZUMÆ.
(55) Pr. Acad. Nat. Sci. Phila., 1854, p. 61; (88) Mex. Bound. Surv., vol. ii, p. 27, pl. xxxvi, figg. 1-6.
Rana pipiens.
(49) Marcy and McClellan's Expl. Red River La. [App. F], p. 243.
Rana pretiosa.
(44) Pr. Acad. Nat. Sci. Phila., 1853, p. 378; (55) Pr. Acad. Nat. Sci. Phila., 1844, p. 62.
Rana septentrionalis.
(55) Pr. Acad. Nat. Sci. Phila., 1854, p. 61.

Rana sinuata.
(55) Pr. Acad. Nat. Sci. Phila., 1854, p. 61.
REGINA CLARKII.
(39) Cat. N. A. Reptiles, 1853, Part I, p. 48; (88) Mex. Bound. Surv., vol. ii, p. 17, pl. x, adult; pl. xi, fig. 2, young; (92) P. R. R. Surv., vol. x, pl. xxvii, fig. 35.
REGINA GRAHAMII.
(39) Cat. N. A. Reptiles, 1853, Part I, p. 47; (88) Mex. Bound. Surv., vol. ii, p. 17, pl. vii, fig. 2; (92) P. R. R. Surv.,vol. x, pl. xxvii, fig. 34.
REGINA KIRTLANDII.
(92) P. R. R. Surv., vol. x, pl. xxvii, fig. 36.
REGINA LEBERIS.
(39) Cat. N. A. Reptiles, 1853, Part I, p. 45; (49½) Serpents of N. Y., 1854, pp. 17, 18; (92) P. R. R. Surv., vol. x, pl. xxvii, fig. 32.
REGINA RIGIDA.
(39) Cat. N. A. Reptiles, 1853, Part I, p. 46; (92) P. R. R. Surv., vol. x, pl. xxvii, fig. 33.
RENA DULCIS.
(39) Cat. N. A. Reptiles, 1853, Part I, p. 142; (88) Mex. Bound. Surv., vol. ii, p. 24; (92) P. R. R. Surv., vol. x, pl. xxxiii, fig. 100.
Rena humilis.
(38) Pr. Acad. Nat. Sci. Phila., 1853, p. 300; (39) Cat. N. A. Reptiles, 1853, Part I, p. 143.
RHINOCHEILUS LECONTEI.
(38) Pr. Acad. Nat. Sci. Phila., 1853, p. 300; (92) P. R. R. Surv., vol. x, pl. xxxiii, fig. 90; (39) Cat. N. A. Reptiles, 1853, Part I, p. 120; (88) Mex. Bound. Surv., vol. ii, p. 21, pl. xx.
RHINOSTOMA COCCINEA.
(39) Cat. N. A. Reptiles, 1853, Part I, p. 118; (92) P. R. R. Surv., vol. x, pl. xxxiii, fig. 89.
SALVADORA GRAHAMIÆ.
(39) Cat. N. A. Reptiles, 1853, Part I, p. 104; (88) Mex. Bound. Surv., vol. ii, p. 21, pl. v, fig. 2; (92) P. R. R. Surv., vol. x, pl. xxxii, fig. 78.
SCAPHIOPUS COUCHII.
(55) Pr. Acad. Nat. Sci. Phila., 1854, p. 62; (88) Mex. Bound. Surv., vol. ii, p. 28, pl. xxxv, figg. 1, 7.
Sceloporus Clarkii.
(35) Pr. Acad. Nat. Sci. Phila., 1852, p. 127; (88) Mex. Bound. Surv., vol. ii, p. 5.
SCELOPORUS CONSOBRINUS.
(49) Marcy and McClellan's Expl. Red River La. [App. F], p. 237, pl. x, figg. 5–12; (88) Mex. Bound. Surv., vol. ii, p. 5; (96) P. R. R. Surv., vol. x, p. 37.
Sceloporus Couchii.
(74½) Pr. Acad. Nat. Sci. Phila., 1858, p. 254; (88) Mex. Bound. Surv., vol. ii, p. 6; (100) Pr. Acad. Nat. Sci. Phila., 1858, p. 254.
Sceloporus dispar.
(35) Pr. Acad. Nat. Sci. Phila., 1852, p. 127.
Sceloporus floridanus.
(74½) Pr. Acad. Nat. Sci. Phila., 1858, p. 254; (100) Pr. Acad. Nat. Sci. Phila., 1858, p. 254.
Sceloporus frontalis.
(37) Pr. Acad. Nat. Sci. Phila., 1852, p. 175.
Sceloporus gracilis.
(37) Pr. Acad. Nat. Sci. Phila., 1852, p. 175.

SCELOPORUS GRACIOSUS.
(27) Pr. Acad. Nat. Sci. Phila., 1852, p. 69; (33) Stansbury's Surv. Salt Lake [App. C], p. 346, pl. v, figg. 1–3; (95) P. R. R. Surv., vol. x, p. 17.
Sceloporus longipes.
(95) P. R. R. Surv., vol. x, p. 17; (100) Pr. Acad. Nat. Sci. Phila., 1858, p. 257; (74½) Pr. Acad. Nat. Sci. Phila., 1858, p. —.
Sceloporus marmoratus.
(88) Mex. Bound. Surv., vol. ii, p. 6.
Sceloporus accidentalis.
(37) Pr. Acad. Nat. Sci. Phila., 1852, p. 175; (38) Pr. Acad. Nat. Sci. Phila., 1853, p. 301; (95) P. R. R. Surv., vol. x, p. 17.
Sceloporus ornatus.
(74½) Pr. Acad. Nat. Sci. Phila., 1858, p. 254; (88) Mex. Bound. Surv., vol. ii, p. 5; (100) Pr. Acad. Nat. Sci. Phila., 1858, p. 254.
SCELOPORUS POINSETTI.
(35) Pr. Acad. Nat. Sci. Phila., 1852, p. 127; (88) Mex. Bound. Surv., vol. ii, p. 5, pl. xxix, figg. 1–3.
Sceloporus scalaris.
(88) Mex. Bound. Surv., vol. ii, p. 6.
SCELOPORUS SPINOSUS.
(88) Mex. Bound. Surv., vol. ii, p. 5, pl. xxix, figg. 4–6; (96) P. R. R. Surv., vol. x, p. 37.
Sceloporus Thayerii.
(35) Pr. Acad. Nat. Sci. Phila., 1852, p. 127; (88) Mex. Bound. Surv., vol. ii, p. 6; (96) P. R. R. Surv., vol. x, p, 37.
Sceloporus torquatus.
(88) Mex. Bound. Surv., vol. ii, p. 5.
Sceloporus undulatus.
(96) P. R. R. Surv., vol. x, p. 37.
SCOTOPHIS ALLEGHANIENSIS.
• (39) Cat. N. A. Reptiles, 1853, Part I, p. 73; (49½) Serpents of N. Y., 1854, pp. 19, 20; (92) P. R. R. Surv., vol. x, pl. xxix, fig. 49; (96) P. R. R. Surv., vol. x, p. 42.
SCOTOPHIS CONFINIS.
(39) Cat. N. A. Reptiles, 1853, Part I, p. 76; (92) P. R. R. Surv., vol. x, pl. xxx, fig. 52.
SCOTOPHIS EMORYI.
(39) Cat. N. A. Reptiles, 1853, Part I, p. 157; (88) Mex. Bound. Surv., vol. ii, p. 19, pl. xii; (92) P. R. R. Surv., vol. x, pl. xxx, fig. 56; (96) P. R. R. Surv., vol. x, p. 43.
SCOTOPHIS GUTTATUS.
(39) Cat. N. A. Reptiles, 1853, Part I, p. 78; (92) P. R. R. Surv., vol. x, pl. xxx, fig. 54.
SCOTOPHIS LÆTUS.
(39) Cat. N. A. Reptiles, 1853, Part I, p. 77; (49) Marcy and McClellan's Expl. Red River La. [App. F], p. 227, pl. vi; (92) P. R. R. Surv., vol. x, pl. xxx, fig 53.
SCOTOPHIS LINDHEIMERI.
(39) Cat. N. A. Reptiles, 1853, Part I, p. 74; (88) Mex. Bound. Surv., vol. ii, p.19; (92) P. R. R. Surv., vol. x, pl. xix, fig. 50.
SCOTOPHIS QUADRIVITTATUS.
(39) Cat. N. A. Reptiles, 1853, Part I, p. 80; (92) P. R. R. Surv., vol. x. pl. xxxx, fig. 55.

SCOTOPHIS VULPINUS.
   (39) Cat. N. A. Reptiles, 1853, Part I, p. 75;
      (92) P. R. R. Surv., vol. x, pl. xxix, fig. 51.
Siren lacertina.
   (88) Mex. Bound. Surv., vol. ii, p. 29.
SIREDON LICHENOIDES.
   (27) Pr. Acad. Nat. Sci. Phila., 1852, p. 68; (33)
      Stansbury's Surv. Salt Lake [App. C], p.
      336, pl. i; (95) P. R. R. Surv., vol. x, p. 20.
Siredon maculatus.
   (10) Journ. Acad. Nat. Sci. Phila., Oct., 1849 p.
      292.
SONORA SEMI-ANNULATA.
   (39) Cat. N. A. Reptiles, 1853, Part I, p. 117;
      (88) Mex. Bound. Surv., vol. ii, p. 21, pl.
      xix, fig. 3. (92) P. R. R. Surv., vol. x, pl.
      xxxiii, fig. 88.
Spelerpes bilineata.
   (10) Journ. Acad. Nat. Sci. Phila., Oct., 1849, p.
      287.
Spelerpes cirrigera.
   (10) Journ. Acad. Nat. Sic. Phila., Oct., 1849, p.
      287.
Spelerpes guttolineata.
   (10) Journ. Acad. Nat. Sci. Phila., Oct., 1849, p.
      287.
Spelerpes longicauda.
   (10) Journ. Acad. Nat. Sci. Phila., Oct., 1849, p.
      287.
SPHÆRIODACTYLUS NOTATUS.
   (74½) Pr. Acad. Nat. Sci. Phila., 1858, p. —; (88)
      Mex. Bound. Surv., vol. ii, p. 12, pl. xxiv,
      figg. 29-37; (100) Pr. Acad. Nat. Sci. Phila.,
      1858, p. 254.
STENODACTYLUS VARIEGATUS.
   (74½) Pr. Acad Nat. Sci. Phila., 1858, p. —; (88)
      Mex. Bound. Surv., vol. ii, p. 12, pl. xxiii,
      figg. 9-27, pl. xxiv, figg. 11-19; (100) Pr.
      Acad. Nat. Sci. Phila., 1858, p. 254.
STORERIA DEKAYI.
   (39) Cat. N. A. Reptiles, 1853, Part I, p. 135;
      (49½) Serpents of N. Y., 1854, p. 26; (92) P.
      R. R. Surv., vol. x, pl. xxxiii, fig. 98.
STORERIA OCCIPITO-MACULATA.
   (39) Cat. N. A. Reptiles, 1853, Part I, p. 137;
      (49½) Serpents of N. Y., 1854, pp. 26, 27,
      28; (92) P. R. R. Surv., vol. x, pl. xxxiii,
      fig. 99.
TÆNIOPHIS IMPERIALIS.
   (88) Mex. Bound. Surv., vol. ii, p. 23, pl. xix, fig.
      1; (92) P. R. R. Surv., vol. xxx, pl. xiii, fig. 87.
TANTILLA CORONATA.
   (39) Cat. N. A. Reptiles, 1853, Part I, p. 131;
      (92) P. R. R. Surv., vol. x, pl. xxxiii, fig. 96.
Tantilla gracilis.
   (39) Cat. N. A. Reptiles, 1853, Part I, p. 132;
      (88) Mex. Bound. Surv., vol. ii, p. 23.
Tapaya breviostris.
   (95) P. R. R. Surv., vol. x, p. 18.
Tapaya Douglassii.
   (95) P. R. R. Surv., vol. x, p. 18.
Tapaya Hernandezii.
   (88) Mex. Bound. Surv., vol. ii, p. 8; (96) P. R.
      R. Surv., vol. x, p. 38.
Tapaya ornatissima.
   (88) Mex. Bound. Surv., vol. ii, p. 9; (96) P. R.
      R. Surv., vol. x, p. 38.

Taricha laevis.
   (38) Pr. Acad. Nat. Sci. Phila., 1853, p. 301.
Thyrosternum sonoriense.
   (88) Mex. Bound. Surv., vol. ii, p. 3.
TOLUCA LINEATA.
   (88) Mex. Bound. Surv., vol. ii, p. 23, pl. xxi, fig.
      2; (92) P. R. R. Surv., vol. x, pl. xxxv,
      fig. 8.
TOXICOPHIS PISCIVORUS.
   (39) Cat. N. A. Reptiles, 1853, Part I, p. 19; (92)
      P. R. R. Surv., vol. x, pl. xxv, fig. 13; (96)
      P. R. R. Surv., vol. x, p. 40.
TOXICOPHIS PUGNAX.
   (39) Cat. N. A. Reptiles, 1853, Part I, p. 20; (88)
      Mex. Bound. Surv., vol. ii, p. 15, pl. vi.; (92)
      P. R. R. Surv., vol. x, pl. xxv, fig. 14.
Trachemys elegans.
   (88) Mex. Bound. Surv., vol. ii, p. 3.
Tropidonotus ordinoides.
   (37) Pr. Acad. Nat. Sci. Phila., 1852, p. 176.
Uma.
   (74½) Pr. Acad. Nat. Sci. Phila., 1858, p. 253; (100)
      Pr. Acad. Nat. Sci. Phila., 1858, p. 253.
Uma notata.
   (74½) Pr. Acad. Nat. Sci. Phila., 1858, p. 253; (100)
      Pr. Acad. Nat. Sci. Phila., 1858, p. 253.
Uta graciosa.
   (88) Mex. Bound. Surv., vol. ii, p. 7.
Uta ornata.
   (35) Pr. Acad. Nat. Sci. Phila., 1852, p. 126; (88)
      Pr. Acad. Nat. Sci. Phila., 1853, p. 301; (88)
      Mex. Bound. Surv., vol. ii, p. 7.
Uta ornata linearis.
   (88) Mex. Bound. Surv., vol. ii, p 7.
Uta Schottii.
   (74½) Pr. Acad. Nat. Sci. Phila., 1858, p. 253; (88)
      Mex. Bound. Surv., vol. ii, p. 7; (100) Pr.
      Acad. Nat. Sci. Phila., 1858, p. 253.
UTA STANSBURIANA.
   (27) Pr. Acad. Nat. Sci. 1852, p. 69; (33) Stans-
      bury's Surv. Salt Lake [App. C], p. 345, pl.
      v, figg. 4-6; (38) Pr. Acad. Nat. Sci. Phila.,
      1853, p. 301; (88) Mex. Bound. Surv., vol.
      ii, p. 7; (96) P. R. R. Surv., vol. x, p. 37.
Uta symmetrica.
   (74½) Pr. Acad. Nat. Sci. Phila., 1858, p. 253; (88)
      Mex. Bound. Surv., vol. ii, p. 7; (100) Pr.
      Acad. Nat. Sci. Phila., 1858, p. 253.
VIRGINIA VALERIA.
   (92) P. R. R. Surv., vol. x, pl. xxxiii, fig. 94;
      (39) Cat. N. A. Reptiles, 1853, Part I, p.
      127.
WENONA.
   (92) P. R. R. Surv., vol. x, pl. xxxii, fig. 3.
Wenona Isabella.
   (37) Pr. Acad. Nat. Sci. Phila, 1852, p. 176; (39)
      Cat. N. A. Reptiles, 1853, Part I, p. 140.
Wenona plumbea.
   (39) Cat. N. A. Reptiles, 1853, Part I, p. 139.
Xantusia.
   (74) P. Acad. Nat. Sci. Phila., 1858, p. 255; (100)
      Pr. Acad. Nat. Sci Phila., 1858, p. 255.
Xantusia vigilis.
   (74½) Pr. Acad. Nat. Sci. Phila., 1858, p. 255; (100)
      Pr. Acad. Nat. Sci. Phila., 1858, p. 255.
Xerobates Berlandieri.
   (88) Mex. Bound. Surv., vol. ii, p. 4.

FISHES.*

Acanthocottus virginianus.
   (63) Ann. Rep. Smith. Inst., 1854, p. (14) 328.
Achirus mollis.
   (63) Ann. Rep. Smith. Inst., 1854, p. (C6) 350.
Acipenser ruthenus.
   (628) Rep. U. S. F. C. II, 1874, p. lxxviii.
Ailurichthys marinus.
   (51) Pr. Acad. Nat. Sci. Phila., 1854, p. 26. (63)
     Ann. Rep. Smith. Inst., 1854, p. (27)341.
Alosa mattowaca.
   (63) Ann. Rep. Smith. Inst., 1854, p. (35) 349.
Alosa menhaden.
   (63) Ann. Rep. Smith. Inst., 1854, p. (33) 347.
Alosa sapidissima.
   (628) Rep. U. S. F. C., Part II, 1874, pp. xii, xvii,
     xxvi, xxxix, xlviii; (890) Ibid., III, 1876,
     p. xviii; (1003) Ibid., IV, 1878, p. 20;* (1031)
     Ibid., V, 1879, p. *29; (1052) Ibid., VI, 1880;
     pp. xxxv, liii, liv.
Alosa teres.
   (63) Ann. Rep. Smith. Inst., 1854, p. (35) 349.
Anguilla tenuirostris.
   (33) Ann. Rep. Smith. Inst., 1854, p. (36) 350.
Anguilla vulgaris.
   (628) Rep. U. S. F. C., II, 1874, p. xxxix.
Aphredoderus Sayanus.
   (63) Ann. Rep. Smith. Inst., 1854, p. (12) 326.
Argyreiosus capillaris.
   (63) Ann. Rep. Smith. Inst., 1854, p. (23) 337.
Arius equestris.
   (51) Pr. Acad. Nat. Sci. 'Phila., 1854, p. 26.
Astyanax argentatus.
   (51) Pr. Acad. Nat. Sci.Phila., 1854, p. 27.
Atherinopsis notatus.
   (63) Ann. Rep. Smith. Inst., 1854, p. (24) 338.
Batrachus variegatus.
   (63) Ann. Rep. Smith. Inst., 1854, p. (26) 340.
Belone truncata.
   (63) Ann. Rep. Smith. Inst., 1854, p. (32) 346.
Boleosoma fusiformis.
   (63) Ann. Rep. Smith. Inst., 1854, p. (14) 328.
Boleosoma lepida.
   (45) Pr. Acad. Nat. Sci.Phila., 1853, p. 338.
Bryttus longulus.
   (51) Pr. Acad. Nat. Sci.Phila., 1854, p. 25.
Caranx chrysos.
   (63) Ann. Rep. Smith. Inst., 1854, p. (22) 336.
Carcharias cœruleus.
   (63) Ann. Rep. Smith. Inst., 1854, p. (38) 352.
Carpiodes tumidus.
   (51) Pr. Acad. Nat. Sci. Phila., 1854, p. 28.
Catostomus Clarkii.
   (51) Pr. Acad. Nat. Sci. Phila., 1854, p. 27.
Catostomus congestus.
   (51) Pr. Acad. Nat. Sci. Phila., 1854, p. 27.
Catostomus gibbosus.
   (63) Ann. Rep. Smith. Inst., 1854, p. (27) 341.
Catostomus insignis.
   (51) Pr. Acad. Nat. Sci. Phila., 1854, p. 28.
Catostomus latipinnis.
   (45) Pr. Acad. Nat. Sci. Phila., 1853, p. 388.

Catostomus plebeius.
   (51) Pr. Acad. Nat. Sci. Phila., 1854, p. 28.
Centropristes nigricans.
   (63) Ann. Rep. Smith. Inst., 1854, p. (9) 323.
Centrarchus pomotis.
   (63) Ann. Rep. Smith. Inst., 1854, p. (11) 325.
Ceratichthys vigilax.
   (46) Pr. Acad. Nat. Sci. Phila., 1853, p. 391.
Chatoessus signifer.
   (63) Ann. Rep. Smith. Inst., 1854, p. (35) 349.
Clupea harengus.
   (628) Rep. U. S. F. C., II, 1874, p. xiii; (1031)
     Ibid., V, 1879, p. *44; (1052) Ibid., VI, 1880,
     p. xxxix.
Cochlognathus ornatus.
   (52) Pr. Acad. Nat. Sci. Phila. 1848, p. 156; (58)
     Pr. Acad. Nat. Sci. Phila., 1854, p. 156.
Conger occidentalis.
   (63) Ann. Rep. Smith. Inst., 1854, p. (37) 351.
Coregonus albus.
   (628) Rep. U. S. F. C., II, 1874, p. xxv; (890)
     Ibid., III, 1876, p. xxxii; (1003) IV, 1878, p.
     xxxvi; (1031) Ibid., V, 1879, p. *39; (1052)
     Ibid., V, 1880, p. xxxv.
Corvina argyroleuca.
   (63) Ann. Rep. Smith. Inst., 1854, p. (17) 331.
Cybium maculatum.
   (63) Ann. Rep. Smith. Inst., 1854, p. (21) 335.
Cyprinodon bovinus.
   (45) Proc. Acad. Nat. Sci. 1853, p. 389.
Cyprinodon elegans.
   (45) Pr. Acad. Nat. Sci. Phila., 1853, p. 389.
Cyprinodon gibbosus.
   (45) Pr. Acad. Nat. Sci. Phila., 1853, p. 390.
Cyprinodon macularius.
   (45) Pr. Acad. Nat. Sci. Phila., 1853, p. 389.
Cyprinodon ovinus.
   (63) Ann. Rep. Smith. Inst., 1854, p. (31) 345.
Cyprinodon parvus.
   (63) Ann. Rep. Smith. Inst., 1854, p. (31) 345.
Cyprinus carpio.
   (628) Rep. U. S. F. C., II, 1874, p. lxxvi; (890)
     Ibid., III, 1876, p. xxxii; (1003) Ibid., IV,
     1878, p. 27; (1031) Ibid. V, 1879, p. 40; (1052)
     Ibid., VI, 1880, p. xl, lvi.
Diodon fuliginosus.
   (63) Ann. Rep. Smith. Inst., 1854, p. (37) 351.
Diodon maculato-striatus.
   (63) Ann. Rep. Smith. Inst., 1854, p. (37) 351.
Engraulis vittata.
   (63) Ann. Rep. Smith. Inst., 1854, p. (33) 347.
Esox fasciatus.
   (63) Ann. Rep. Smith. Inst., 1854, p. (31) 345.
Esox reticulatus.
   (63) Ann. Rep. Smith. Inst., 1854, p. (32) 346.
Eucinostomus.
   (63) Ann. Rep. Smith. Inst., 1854, p. (20) 334.
Eucinostomus argenteus.
   (63) Ann. Rep. Smith. Inst., 1854, p. (21) 335.
Fundulus diaphanus.
   (63) Ann. Rep. Smith. Inst., 1854, p. (29) 343.

*In addition to the citations here given, many discussions of individual species of fishes may be found by examining the list of papers on fishes and fisheries published in the ANNUAL RECORD OF SCIENCE AND INDUSTRY, vols. i–viii, and enumerated above (pp. 255–269).

Fundulus grandis.
  (45) Pr. Acad. Nat. Sci. Phila., 1853, p. 389.
Fundulus multifasciatus.
  (63) Ann. Rep. Smith. Inst., 1854, p. (30) 344.
Fundulus tenellus.
  (45) Pr. Acad. Nat. Sci. Phila., 1853, p. 389.
Fundulus zebra.
  (63) Ann. Rep. Smith. Inst., 1854, p. (28) 342.
Gadus morrhua.
  (1052) Rep. U. S. F. C., VI, 1880, pp. xlii, lvi.
Gasterosteus quadracus.
  (63) Ann. Rep. Smith. Inst., 1854, p. (14) 328.
Gila conocephala.
  (57) Pr. Acad. Nat. Sci. Phila., 1854, p. 135.
Gila elegans.
  (43) Pr. Acad. Nat. Sci. Phila., 1853, p. 369; (47)
    Sitgreaves' Rep. Exp. down Zuñi and Col.
    Rivers, p. 150, pl. ii.
Gila Emoryi.
  (45) Pr. Acad. Nat. Sci. Phila., 1853, p. 388.
Gila gibbosa.
  (51) Pr. Acad. Nat. Sci. Phila., 1854, p. 28.
Gila gracilis.
  (43) Pr. Acad. Nat. Sci. Phila., 1853, p. 369; (47)
    Sitgreaves' Rep. Exp. down Zuñi and Col.
    Rivers, p. 151, pl. jii.
Gila Grabamii.
  (45) Pr. Acad. Nat. Sci. Phila., 1853, p. 389.
Gila pulchella.
  (51) Pr. Acad. Nat. Sci. Phila., 1854, p. 29.
Gila robusta.
  (43) Pr. Acad. Nat. Sci. Phila., 1853, p. 369; (47)
    Sitgreaves' Rep. Exp. down Zuñi and Col.
    Rivers, p. 148, pl. i.
Glyptocephalus cynoglossus.
  (1031) Rep. U. S. F. C., V, 1879, p, *4.
Gobius alepidotus.
  (63) Ann. Rep. Smith. Inst., 1854, p. (25) 339.
Grystes nucensis.
  (51) Pr. Acad. Nat. Sci. Phila., 1854, p. 25.
Herichthys cyanoguttatus.
  (51) Pr. Acad. Nat. Sci. Phila., 1854, p. 25.
Heterandria affinis.
  (45) Pr. Acad. Nat. Sci. Phila., 1853, p. 390.
Heterandria nobilis.
  (45) Pr. Acad. Nat. Sci. Phila., 1853, p. 390.
Heterandria occidentalis.
  (45) Pr. Acad. Nat. Sci. Phila., 1853, p. 390.
Heterandria patruelis.
  (45) Pr. Acad. Nat. Sci. Phila., 1853, p. 390.
Hydrargyra flavula.
  (63) Ann. Rep. Smith. Inst., 1854, p. (30) 344.
Hydrargyra Luciæ.
  (63) Ann. Rep. Smith. Inst., 1854, p. (30) 344.
Hydrargyra similis.
  (45) Pr. Acad. Nat. Sci. Phila., 1853, p. 389.
Idus melanotus.
  (1031) Rep. U. S. F. C., V, 1879, p. *44.
Labrax lineatus.
  (63) Ann. Rep. Smith. Inst., 1854, p. (7) 321.
Labrax mucronatus.
  (63) Ann. Rep. Smith. Inst., 1854, p. (8) 322.
Lavinia crassicauda.
  (57) Pr. Acad. Nat. Sci. Phila., 1854, p. 137.
Lavinia exilicauda.
  (57) Pr. Acad. Nat. Sci. Phila., 1854, p. 137.

Leiostomus obliquus.
  (63) Ann. Rep. Smith. Inst., 1854, p. (15) 329.
Leuciscus bubulinus.
  (46) Pr. Acad. Nat. Sci. Phila., 1853, p. 391; (50)
    Marcy and McClellan's Expl. Red River
    La. [App. F], p. 249, pl. xiv, figg. 5–8.
Leuciscus lutrensis.
  (46) Pr. Acad. Nat. Sci. Phila., 1853, p. 391; (50)
    Marcy and McClellan's Expl. Red River
    La. [App. F], p. 251, pl. xiv, figg. 9–12.
LEUCISCUS VIGILAX.
  (50) Marcy and McClellan's Expl. Red River
    La. [App. F], p, 248, pl. xiv, figg. 1–4.
Leucosomus americanus.
  (63) Ann. Rep. Smith. Inst., 1854, p. (27) 341.
Leucosomus occidentalis.
  (57) Pr. Acad. Nat. Sci. Phila., 1854, p. 137.
Lichia carolina.
  (63) Ann. Rep. Smith. Inst., 1854. p. (21) 335.
Lichia spinosa.
  (63) Ann. Rep. Smith. Inst., 1854, p. (22) 336.
Lobotes emarginatus.
  (63) Ann. Rep. Smith. Inst., 1854, p. (18) 332.
Melanura pygmæa.
  (63) Ann. Rep. Smith. Inst., 1854, p. (28) 342.
Micropterus nigricans, &c.
  (628) Rep. U. S. F. C., IV, 1874, p. xxxvi.
Mugil albula.
  (63) Ann. Rep. Smith. Inst , 1854, p. (25) 339.
Mustelus canis.
  (63) Ann. Rep. Smith. Inst., 1854, p. (39) 353.
Ophidium marginatum.
  (63) Ann. Rep. Smith. Inst., 1854, p. (37) 351.
Osphromenus gourami.
  (628) Rep. U. S. F. C., II, p. lxxvii.
Otolithus regalis.
  (63) Ann. Rep. Smith. Inst.,1855, p. (15) 329.
Pagrus argyrops.
  (63) Ann. Rep. Smith. Inst., 1854, p. (19) 333.
Pastinaca hastata.
  (63) Ann. Rep. Smith. Inst., 1854, p. (39) 353.
Peprilus triacanthus.
  (63) Ann. Rep. Smith. Inst., 1854, p. (24) 338.
Pileoma carbonaria.
  (45) Pr. Acad. Nat. Sci. Phila., 1853, p. 387.
Pimelodus affinis.
  (51) Pr. Acad. Nat. Sci. Phila., 1854, p. 26.
Platessa ocellaris.
  (63) Ann. Rep. Smith. Inst., 1854, p. (35) 349.
Platessa plana.
  (63) Ann. Rep. Smith. Inst., 1854, p. (35) 349.
Pogonias fasciatus.
  (63) Ann. Rep. Smith. Inst., 1854, p. (18) 332.
Pogonichthys inæquilobus.
  (57) Pr. Acad. Nat. Sci. Phila., 1854, p. 136.
Pogonichthys symmetricus.
  (57) Pr. Acad. Nat. Sci. Phila., 1854, p. 136.
Pomolobus pseudoharengus.
  (628) Rep. U. S. F. C., Part II, 1874, pp. xii,
    lix.
Pomotis aquilensis.
  (45) Pr. Acad. Nat. Sci. Phila., 1853, p. 388.
Pomotis breviceps.
  (46) Pr. Acad. Nat. Sci. Phila., 1853, p. 390; (50)
    Marcy and McClellan's Expl. Red River
    La. [App. F], p. 245, pl. xiii.

Pomotis chætodon.

(63) Ann. Rep. Smith. Inst., 1854, p. (10) 324.

Pomotis convexifrons.

(51) Pr. Acad. Nat. Sci. Phila., 1854, p. 24.

Pomotis fallax.

(51) Pr. Acad. Nat. Sci. Phila., 1854, p. 24.

Pomotis heros.

(51) Pr. Acad. Nat. Sci. Phila., 1854, p. 25.

Pomotis longulus.

(46) Pr. Acad. Nat. Sci. Phila., 1853, p. 391; (50) Marcy and McClellan's Expl. Red River La. [App. F], p. 245, pl. xii.

Pomotis nefastus.

(51) Pr. Acad. Nat. Sci. Phila, 1854, p. 24.

Pomotis obesus.

(63) Ann. Rep. Smith. Inst.,1854, p. (10) 324.

Pomotis speciosus.

(51) Pr. Acad. Nat. Sci. Phila., 1854, p. 24.

Prionotus pilatus.

(63)Ann. Rep. Smith. Inst., 1854, p. (13) 327.

Rhombus lævis.

(1031) Rep. U. S. F. C., V, 1879, p. *46·

Rhombus maculatus.

(63) Ann. Rep. Smith. Inst., 1854, p. (36) 350.

Roccus lineatus.

Rep. U. S. F. C., II, 1874, pp. xxvii, xlvii.

Salmo hucho.

(628) Rep. U. S. F. C., II, 1874, pp. xix.

Salmo quinnat.

(628) Rep. U. S. F. C., II, 1874, xxiii, lxix; (890) Ibid., III, p. xxii; (1003) Ibid., IV, 1878, p. 21; (1031) Ibid., V, 1879, p. *31; (1052) Ibid., VI, 1880, pp. xxv, lv.

Salmo salar.

(628) Rep. U. S. F. C., II, 1874, pp. xii, xviii, xxxix, lxi, lxxi; (890) Ibid., III, p. xxx; (1003) Ibid., IV, 1878, p. 25; (1031) Ibid., V. 1879, p. 36; (1052) Ibid., VI, 1880, pp. xxix, liv.

Salvelinus fontinalis.

(628) Rep. U. S. F. C., II, 1877, p. lxxiii; (1052) Ibid., VI, 1880, p. lii.

Salvelinus oquassa.

(628) Rep. U. S. F. C., II, 1874, p. lxxiii.

Salvelinus salvelinus.

(628) Rep. U. S. F. C., II, 1874, p. lxxiv.

Saurus mexicanus.

(63) Ann. Rep. Smith. Inst., 1854, p. (32) 346.

Solea vulgaris.

(1031) Rep. U. S. F. C , V, 1879, p. *46·

Sphyræna borealis.

(63) Ann. Rep. Smith. Inst., 1854, p. (12) 326.

Syngnathus viridescens.

(63) Ann. Rep. Smith. Inst., 1854, p. (37) 351.

Tautoga americana.

(63) Ann. Rep. Smith. Inst., 1854, p. (26) 340.

Temnodon saltator.

(63) Ann. Rep. Smith. Inst., 1854, p. (23) 337.

Tetraodon turgidus.

(63) Ann. Rep. Smith. Inst., 1854, p. (38) 352.

Thymallus tricolor.

(628) Rep. U. S. F. C., II, 1874, p. lxxiv.

Tinca vulgaris.

(1031) Rep. U. S. F. C., V, 1878, p. *44; (1052) Ibid., VI, 1880, p. xliv.

Umbrina alburnus.

(63) Ann. Rep. Smith. Inst., 1854, p. (17) 331.

Zygæna tiburo.

(63) Ann. Rep. Smith. Inst., 1854, p. (39) 53.